Nanobiohybrids for Advanced Wastewater Treatment and Energy Recovery

Nanobiohybrids for Advanced Wastewater Treatment and Energy Recovery

Edited by
Piet N. L. Lens and Priyanka Uddandarao

IWA PUBLISHING

Published by **IWA Publishing**
Unit 104–105, Export Building
1 Clove Crescent
London E14 2BA, UK
Telephone: +44 (0)20 7654 5500
Fax: +44 (0)20 7654 5555
Email: publications@iwap.co.uk
Web: www.iwapublishing.com

First published 2023

British Library Cataloguing in Publication Data
A CIP catalogue record for this book is available from the British Library

ISBN: 9781789063585 (paperback)
ISBN: 9781789063592 (eBook)
ISBN: 9781789063608 (ePub)

Doi: 10.2166/9781789063592

This eBook was made Open Access in December 2023.

Contents

List of Contributors xiii

Preface xvii

Part 1: Concepts of Microbial Synthesis, Water Purification and Energy Storage 1

Chapter 1
Introduction to wastewater treatment and energy recovery 3
Priyanka Uddandarao and Piet N. L. Lens

1.1 Introduction 3
1.2 Process Fundamentals 3
1.3 Building Blocks of NBs 4
1.4 Environmental Remediation 4
1.5 Wastewater Treatment 5
References 7

Chapter 2
Addressing the global water crisis: a comprehensive review of nanobiohybrid applications for water purification 9
I. Kalyan, A. K. Nayak and M. U. Khobragade

2.1 Introduction 9
2.2 Root Cause Behind Continuous Freshwater Shrinking 11
2.3 Methodical Handling of Water Pollution 11
  2.3.1 Treatment technologies 11
  2.3.2 Major drawbacks of current water purification techniques 11
2.4 Nanobiohybrid (NBIOH) Catalyst in Water Purification 13
  2.4.1 Use of nanoparticles in water purification and their problems 13
  2.4.2 Enzymes in water purification and their problems 13

2.4.3 Use of NBIOH catalyst for water purification . . . 14
2.4.3.1 Capacity of NBIOH to treat water . . . 14
2.4.3.2 Problems associated with nanobiohybrid . . . 15
2.5 Conclusion . . . 15
References . . . 17

Chapter 3

Biological production of nanoparticles and their application in photocatalysis . . . *21*
Priyanka Uddandarao and Piet N. L. Lens

3.1 Introduction . . . 21
3.2 Green Synthesis of Nanoparticles . . . 22
3.3 Biological Nanoparticles . . . 23
3.3.1 Plants . . . 23
3.3.2 Bacteria . . . 28
3.4 Fungi . . . 28
3.5 Algae . . . 33
3.6 Photocatalysis . . . 33
3.6.1 Batch degradation of organic pollutants using NPs . . . 39
3.6.2 Photobioreactors . . . 44
3.6.3 Nanobiohybrids . . . 44
3.7 Challenges . . . 46
3.7.1 Toxicity . . . 46
3.7.2 Nanoparticles detection . . . 47
3.7.3 Light accessibility . . . 47
3.8 Conclusion . . . 47
References . . . 47

Chapter 4

Energy storage devices: batteries and supercapacitors . . . *61*
Taniya Rose Abraham, Kevin Sunil, Muhammed Shah, Neelesh Ashok and Sabu Thomas

4.1 Introduction . . . 61
4.2 Batteries: Principles and Operation . . . 62
4.2.1 Battery basics . . . 62
4.2.1.1 Structure and components . . . 62
4.2.1.2 Electrochemical reactions in batteries . . . 63
4.2.2 Battery performance metrics . . . 64
4.2.2.1 Cell, module, and pack level . . . 64
4.2.2.2 Energy density . . . 64
4.2.2.3 Power density . . . 64
4.2.2.4 Specific energy (or gravimetric energy density) . . . 65
4.2.2.5 Specific power (or gravimetric power density) . . . 65
4.2.2.6 Cycle life . . . 65
4.2.2.7 Charge–discharge efficiency . . . 65
4.2.2.8 Self-discharge rate . . . 65
4.2.2.9 Operating temperature . . . 65

4.2.2.10 Impedance . . . 65
4.2.2.11 Round-trip efficiency . . . 65
4.3 Types of Batteries . . . 66
4.3.1 Nickel–cadmium batteries . . . 66
4.3.2 Lead–acid batteries . . . 67
4.3.2.1 Lead–acid battery composition . . . 67
4.3.2.2 Working principle of lead acid battery . . . 67
4.3.2.3 Market perspective . . . 69
4.3.3 Lithium-ion batteries . . . 69
4.3.3.1 Lithium-ion battery composition . . . 69
4.3.3.2 Working principle of lithium-ion battery . . . 69
4.3.3.3 Market perspective . . . 70
4.3.4 Sodium-ion batteries . . . 71
4.3.5 Zinc–air batteries . . . 72
4.4 Supercapacitors . . . 73
4.4.1 Principles and operations . . . 73
4.4.1.1 Electric double-layer capacitance . . . 73
4.4.1.2 Faradaic capacitance . . . 73
4.4.2 Supercapacitor electrode materials . . . 74
4.4.2.1 Electrode materials for EDLC . . . 74
4.4.2.2 Electrode materials for pseudocapacitor . . . 74
4.4.2.3 Electrode materials for hybrid supercapacitor . . . 75
4.5 Types of Supercapacitors . . . 75
4.5.1 Electrochemical double-layer capacitors . . . 76
4.5.2 Pseudocapacitors . . . 76
4.5.3 Hybrid capacitor . . . 76
4.6 Applications of Batteries and Supercapacitors . . . 77
4.6.1 Portable electronics and consumer applications . . . 77
4.6.2 Mobility of the future . . . 78
4.6.2.1 Electric vehicles and hybrid vehicles . . . 78
4.6.2.2 Aerospace applications . . . 79
4.6.3 New energy technologies . . . 80
4.6.3.1 Renewable energy integration . . . 80
4.6.3.2 Grid-scale energy storage . . . 80
4.6.4 Defence application . . . 81
4.7 Conclusion . . . 81
References . . . 81

Part 2: Utility of Organic, Inorganic and Magnetic Nanoparticles . . . 85

Chapter 5
Nanobiohybrids using organic nanoparticles for applications in water and wastewater treatment . . . 87
Upasana Priyadarshini and Remya Neelancherry

5.1 Introduction . . . 87
5.2 Production of Nanobiohybrids . . . 88
5.2.1 Nanohybrids based on cellulose . . . 89
5.2.2 Nanohybrids based on gelatin . . . 90
5.2.3 Nanohybrids based on chitosan . . . 90

5.2.4 Nanohybrids based on pectin ... 90
5.2.5 Nanohybrid based on silk protein ... 90
5.3 Nanobiohybrid Applications in Water and Wastewater Treatment ... 91
5.3.1 Nanobiohybrids as adsorbent ... 91
5.3.2 Nanobiohybrids as catalyst (nanobiocatalysis) ... 92
5.3.2.1 Polymeric nanobiocatalyst ... 95
5.3.2.2 Silica-based nanobiocatalysts ... 95
5.3.2.3 Carbon-based nanobiocatalysts ... 95
5.3.2.4 Metal-based nanobiocatalysts ... 96
5.4 Conclusion ... 96
References ... 96

Chapter 6
Assessing the feasibility of inorganic nanomaterials for nanohybrids formation ... 99
Ranjana Pande and Anirudh Gururaj Patil

6.1 Introduction ... 99
6.1.1 Production of nanoparticles ... 99
6.1.2 Microbial nanohybrids ... 101
6.1.3 Nanohybrid materials for wastewater treatment with respect to microbes ... 102
6.2 Biosynthesis of Metal NPs with Different Microbes ... 102
6.2.1 Bacteria ... 102
6.2.2 Algae ... 103
6.2.3 Fungi ... 103
6.3 Feasibility of Microbe-Based Biogenic NPs for Wastewater Treatment ... 104
6.3.1 Use of biogenic NPs to treat wastewater ... 104
6.3.2 Biogenic inorganic NPs ... 104
6.3.2.1 Bio-Fe and Bio-Mn NPs ... 104
6.3.2.2 Bio-Pd NPs ... 105
6.3.2.3 Bio-Au and Bio-Ag NPs ... 105
6.3.2.4 Bio-bimetal NPs ... 105
6.3.2.5 Composite Bio-Me NPs ... 106
6.4 Conclusions ... 106
Acknowledgement ... 107
References ... 107

Chapter 7
Sustainable wastewater treatment using magnetic nanohybrids ... 113
Anirudh Gururaj Patil, Vivekanandan Alangadu Kothandan, A. B. Hemavathi, Shweta Suri, Ranjana Pande, Sneha Kagale, Snehal Yadav and Gouri Patil

7.1 Introduction ... 114
7.2 Source of Pollutants ... 115
7.2.1 Ore extraction ... 115
7.2.2 Electroplating ... 115
7.2.3 Water pollution ... 116
7.2.3.1 Pharmaceutical waste ... 116
7.2.3.2 Dyes ... 116
7.2.4 Radionuclides ... 116

7.3 Sustainable Wastewater Treatment with Nanohybrids . . . 116
7.4 Magnetic Nanohybrids Materials for Water Contaminant Removal . . . 117
7.4.1 Preparation of magnetic nanohybrid materials . . . 117
7.4.2 Magnetic nanohybrid development . . . 118
7.4.3 Mechanism of adsorptive removal of pollutants using magnetic nanohybrid materials . . . 119
7.5 Factors Influencing Adsorption by Magnetic Nanohybrid Adsorbent . . . 120
7.6 Removal of Water Pollutants Based on Magnetic Nanohybrid Catalyst . . . 121
7.6.1 Carbon-based magnetic nanohybrid adsorbents . . . 121
7.6.1.1 Activated charcoal/biochar-based materials . . . 121
7.6.1.2 Carbon nanotubes . . . 121
7.6.1.3 Graphene-based nanoadsorbents . . . 121
7.6.1.4 Chitosan-based magnetic nanohybrid catalyst . . . 122
7.6.2 Metal-based magnetic nanohybrid catalyst . . . 122
7.6.2.1 Zeolites . . . 122
7.6.2.2 Multi-metals-based magnetic nanohybrid catalyst . . . 122
7.7 Future Prospectives with Challenges . . . 122
Acknowledgements . . . 123
References . . . 124

Chapter 8
Feasibility of nanomaterials to support electroactive microbes in nanobiohybrids . . . 129
Zheng Chen, Yanqiong Zeng, Feng Wang, Peng Huang, Jian Li and Yibin Chen

8.1 Introduction . . . 129
8.2 Inherent Bottlenecks for Electron Transfer in Natural EAB Cells . . . 130
8.3 Nanomaterial Selection for Constructing Efficient Nanobiohybrids . . . 131
8.3.1 Favorable electrical conductivity of NMs . . . 131
8.3.1.1 Metal/metal oxide-based NPs and conductive carbon-based NMs . . . 131
8.3.1.2 Conductive organic nanopolymers . . . 132
8.3.2 Large specific surface area of NMs . . . 132
8.3.3 Photocatalysis capability of NMs . . . 132
8.3.3.1 Metal-based semiconductor NPs . . . 132
8.3.3.2 Carbon-based semiconductor NPs . . . 134
8.3.4 NMs stimulate production of cellular components related to electron transfer . . . 134
8.3.4.1 Increased production of c-Cyts in the presence of NMs . . . 134
8.3.4.2 Increased EPS production in the presence of NMs . . . 134
8.3.5 Special functionalized NMs used for cytoprotection in engineered nanobiohybrids . . . 135
8.3.5.1 Biomimetic inorganic NPs . . . 135
8.3.5.2 Nano-hydrogels . . . 135
8.3.5.3 Hybrid coordination NMs . . . 135
8.3.5.4 Artificial nanoenzymes . . . 135
8.4 Assembly Protocols and Synthetic Strategies Employed for Different Functional Nanobiohybrid Systems . . . 136
8.4.1 Internal bioaugmentation on an individual cell scale . . . 136
8.4.2 External bioaugmentation on an individual cell scale . . . 138
8.4.3 External bioaugmentation on the biofilm scale . . . 139
8.5 Future Directions . . . 139
8.5.1 Present challenges for nanobiohybrid development . . . 139

8.5.2 Outlook for nanobiohybrid development ... 140
Acknowledgments ... 141
References ... 141

Part 3: Environmental Remediation Using NBs ... 145

Chapter 9
Nanobiohybrids: a promising approach for sensing diverse environmental water pollutants ... 147
Angelene Hannah Jebarani D. and Sruthi Ann Alex

9.1 Introduction ... 147
9.2 Importance of Nanomaterials in the Nanobiohybrids ... 148
9.3 Choice of Nanomaterial ... 149
9.3.1 Metallic and metal oxide nanostructures ... 150
9.3.2 Carbonaceous nanomaterials ... 151
9.3.3 Quantum dots ... 151
9.3.4 Polymers ... 152
9.4 Nanobiohybrid Types: Based on Recognition Elements ... 152
9.4.1 Proteins and peptides ... 152
9.4.2 Nucleic acids ... 153
9.4.3 Carbohydrates ... 153
9.4.4 Whole cells ... 154
9.5 Nanobiohybrid Sensor Types Based on Transduction Pathways ... 155
9.5.1 Electrochemical nanobiohybrid sensors ... 155
9.5.2 Optical nanobiohybrid sensors ... 157
9.5.3 Magnetic nanobiohybrid sensors ... 160
9.5.4 Gravimetric nanobiohybrid sensors ... 162
9.5.5 Calorimetric nanobiohybrid sensors ... 165
9.6 Conclusion ... 165
References ... 166

Chapter 10
Unlocking the potential of nanobiohybrids to combat environmental pollution ... 169
Amol D. Gholap, Navnath T. Hatvate, Sampada D. Sawant, Palak N. Khemka and Lalitkumar K. Vora

10.1 Introduction ... 169
10.1.1 Need for environmental bioremediation ... 169
10.1.2 Nanotechnology for environmental remediation ... 170
10.1.3 Nanohybrids ... 171
10.2 Metal NPs ... 173
10.2.1 Properties of metal NPs ... 173
10.2.2 MNP composite applications in membrane technology ... 175
10.2.3 MNP and degradation of environmental pollutants ... 175
10.3 Metal Oxide NPs ... 176
10.3.1 Properties of metal oxide NPs ... 176
10.3.2 Metal oxide NP composite and hybrid applications ... 177
10.3.2.1 Titanium-based composites ... 177

10.3.2.2 Composites based on carbon materials ... 178
10.3.2.3 Magnetic nanocomposites ... 178
10.4 Carbon-Based Nanomaterials ... 179
10.4.1 Properties of carbon-based nanomaterials ... 179
10.4.2 Carbon-based nanomaterial composite and hybrid applications ... 180
10.4.2.1 Composite materials ... 180
10.4.2.2 Hybrid composite materials ... 182
10.5 Nanobiohybrids Assisting Novel Biological Functions ... 182
10.5.1 Extending biofunctions by nanobiohybrids ... 182
10.5.2 Application of advanced nanobionics ... 183
10.5.2.1 Microbial cells ... 183
10.5.2.2 Silk production ... 184
10.5.2.3 Plants ... 184
10.6 Conclusion ... 184
References ... 184

Part 4: Wastewater Treatment Using NBs ... 191

Chapter 11
Nanobiohybrids for pollution control and resource recovery ... ***193***
P. Kumari, R. Kumari, R. Prabhakar and S. R. Samadder

11.1 Introduction ... 193
11.2 Synthesis Methods for NBHs ... 195
11.2.1 In-situ synthesis ... 196
11.2.2 Ex-situ synthesis ... 197
11.3 Role of NBHs in Pollution Control ... 197
11.3.1 Water and wastewater purification ... 197
11.3.2 Air pollution control ... 199
11.3.3 Soil remediation ... 201
11.4 Role of NBHs in Resource Recovery ... 202
11.5 Limitations and Challenges ... 202
Acknowledgments ... 203
References ... 204

Chapter 12
Photoelectrocatalysis and nanohybrids in wastewater treatment ... ***209***
Sweta Sharma and Pooja Devi

12.1 Introduction ... 209
12.2 PEC in Wastewater Treatment ... 210
12.2.1 Principle of photoelectrocatalysis ... 210
12.2.2 PEC process ... 211
12.2.3 PEC systems ... 211
12.2.3.1 p-Type semiconductors ... 212
12.2.3.2 n-Type semiconductor ... 212
12.3 Nanohybrids in PEC Wastewater Treatment ... 213
12.3.1 Synthesis strategies ... 213
12.3.2 Parameter optimization ... 214
12.4 PEC–HER/OER for Waste Water Treatment with Nanohybrids ... 215

12.4.1 Nanohybrid categorization ... 215
12.4.2 Types of wastewater used ... 215
12.5 Conclusions ... 216
Acknowledgements ... 217
References ... 217

Index ... 221

List of Contributors

Taniya Rose Abraham
School of Energy Materials, Mahatma Gandhi University, Kottayam, Kerala, India

Neelesh Ashok
Amrita School of Artificial Intelligence, Amrita Vishwa Vidyapeetham, Coimbatore, Tamil Nadu, India

Sruthi Ann Alex
Centre for Nano Science and Technology, Anna University, Chennai, Tamil Nadu, India

Zheng Chen
School of Public Health & Management, Wenzhou Medical University, Wenzhou, People's Republic of China
Fujian Provincial Key Lab of Coastal Basin Environment, Fujian Polytechnic Normal University, Fuqing, People's Republic of China

Yibin Chen
Fujian Provincial Key Lab of Coastal Basin Environment, Fujian Polytechnic Normal University, Fuqing, People's Republic of China

Pooja Devi
Academy of Scientific and Industrial Research, Ghaziabad, Uttar Pradesh, India
Materials Science and Sensor Application, Central Scientific Instruments Organisation, Chandigarh, India

Amol D. Gholap
Department of Pharmaceutics, St. John Institute of Pharmacy and Research, Palghar 401404, Maharashtra, India

Navnath T. Hatvate
Institute of Chemical Technology Mumbai, Marathwada Campus, Jalna 431213, Maharashtra, India

A. B. Hemavathi
Sri Jayachamarajendra College of Engineering (SJCE), Jagadguru Sri Shivarathreeshwara Science and Technology University, Department of Polymer Science and Technology, Mysore, Karnataka, India

Peng Huang
School of Public Health & Management, Wenzhou Medical University, Wenzhou, People's Republic of China

D. Angelene Hannah Jebarani
Department of Biotechnology, National Institute of Technology (NIT) Andhra Pradesh, Tadepalligudem, India

Sneha Kagale
Kolhapur Institute of Technology's College of Engineering (Autonomous), Department of Biotechnology Engineering, Kolhapur, Maharashtra, India

I. Kalyan
Indian Institute of Technology Kharagpur, School of Nano Science and Technology, Kharagpur, West Bengal 721302, India

Palak N. Khemka
Department of Pharmaceutics, St. John Institute of Pharmacy and Research, Palghar 401404, Maharashtra, India

M. U. Khobragade
COEP Technological University (COEPTECH) Pune, Civil Engineering Department, Pune 411005, Maharashtra, India

P. Kumari
Indian Institute of Technology (Indian School of Mines), Department of Environmental Science and Engineering, Dhanbad 826004, Jharkhand, India

R. Kumari
Indian Institute of Technology (Indian School of Mines), Department of Environmental Science and Engineering, Dhanbad 826004, Jharkhand, India

Vivekanandan Kothandan
Annasaheb Dange College of Engineering & Technology, Department of Aeronautical Engineering, Ashta, Maharashtra, India

Piet N. L. Lens
Department of Microbiology, National University of Ireland, University Road, Galway, Ireland

Jian Li
Wenzhou Medical University, School of Public Health & Management, Wenzhou, People's Republic of China

Ashish. K. Nayak
Prasad V. Potluri Siddhartha Institute of Technology, Vijayawada, Andhra Pradesh, 520007, India

Remya Neelancherry
School of Infrastructure, Indian Institute of Technology Bhubaneswar, Argul 752050, Odisha, India

Ranjana Pande
School of Basic Sciences and Research, Life Sciences Department, Sharda University, Greater Noida, Uttar Pradesh, India

R. Prabhakar
Indian Institute of Technology (Indian School of Mines), Department of Environmental Science and Engineering, Dhanbad 826004, Jharkhand, India

Anirudh Gururaj Patil
Annasaheb Dange College of Engineering & Technology, Department of Food Technology, Ashta, Maharashtra, India

Gouri Patil
Sri Jayachamarajendra College of Engineering (SJCE), Jagadguru Sri Shivarathreeshwara Science and Technology University, Department of Polymer Science and Technology, Mysore, Karnataka, India

Upasana Priyadarshini
Indian Institute of Technology Bhubaneswar, Argul 752050, Odisha, India

S. R. Samadder
Indian Institute of Technology (Indian School of Mines), Department of Environmental Science and Engineering, Dhanbad 826004, Jharkhand, India

Sampada D. Sawant
Department of Pharmaceutics, St. John Institute of Pharmacy and Research, Palghar 401404, Maharashtra, India

Muhammed Shah
School of Energy Materials, Mahatma Gandhi University, Kottayam, Kerala, India

Sweta Sharma
Academy of Scientific and Industrial Research, Ghaziabad, Uttar Pradesh, India
Materials Science and Sensor Application, Central Scientific Instruments Organisation, Chandigarh, India

Kevin Sunil
School of Energy Materials, Mahatma Gandhi University, Kottayam, Kerala, India

Shweta Suri
Amity University, Amity Institute of Food Technology, Noida, Uttar Pradesh, India.

Sabu Thomas
School of Energy Materials, Mahatma Gandhi University, Kottayam, Kerala, India

Priyanka Uddandarao
Department of Microbiology, National University of Ireland, University Road, Galway, Ireland
Department of Biotechnology, National Institute of Technology, Andhra Pradesh, India

Lalitkumar K. Vora
School of Pharmacy, Queen's University Belfast, 97 Lisburn Road, Belfast BT9 7BL, UK

Feng Wang
Wenzhou Medical University, School of Public Health & Management, Wenzhou, People's Republic of China

Snehal Yadav
Annasaheb Dange College of Engineering & Technology, Department of Food Technology, Ashta, Maharashtra, India

Yanqiong Zeng
Wenzhou Medical University, School of Public Health & Management, Wenzhou, People's Republic of China

doi: 10.2166/9781789063592_xvii

Preface

In the face of growing global challenges such as water scarcity and environmental pollution, the quest for innovative and sustainable solutions has become more critical than ever before. The convergence of nanotechnology and biotechnology has opened up remarkable possibilities in addressing these complex issues. This book *Nanobiohybrids for Advanced Wastewater Treatment and Energy Recovery* delves into the fascinating realm where nanoscience and bioprocess engineering intersect to offer novel approaches for tackling two pressing concerns simultaneously: wastewater treatment and energy recovery.

The pollution of water bodies, a consequence of industrialization and urbanization, poses a severe threat to ecosystems and human health. Concurrently, the escalating demand for energy in an era of depleting fossil fuel reserves necessitates exploring alternative avenues for energy generation. This book explores the synergy of cutting-edge nanomaterials and biological entities to craft transformative solutions that not only purify wastewater, but also harness their latent energy potential.

In recent years, there has been a notable surge in the investigation of biohybrid and biomimetic systems inspired by photosynthesis. This approach aims to explore novel photon-to-fuel strategies resembling the process of photosynthesis, thereby advancing our understanding while also facilitating the practical utilization of solar energy. This heightened interest is fueled by the remarkable potential offered by photosynthesis-inspired biohybrids. These systems offer the unique prospect of combining the advantageous attributes of both natural and artificial photosynthetic systems. This is achieved by leveraging biologically evolved protein architectures to tackle synthetic challenges, alongside the integration of synthetic molecules to creatively explore and harness the mechanisms found in nature.

The journey through the chapters of this book will unveil the intricate world of nanobiohybrids, where nanoparticles and biomolecules collaborate harmoniously. These hybrid systems leverage the unique properties of nanomaterials to enhance the efficiency of biological processes, offering unprecedented opportunities for pollutant removal and resource recovery. From fundamental principles to practical applications, readers will traverse the theoretical underpinnings, experimental methodologies, and real-world implementations of nanobiohybrids.

Authored by experts and researchers at the forefront of nanotechnology, bioprocess engineering, and environmental science, this comprehensive volume aims to provide an authoritative resource for academics, researchers, industry professionals, and policymakers. Its pages will empower readers with knowledge about the latest advancements, challenges, and breakthroughs in the realm of nanobiohybrids for wastewater treatment and energy recovery. Each contributed chapter is presented on a stand-alone basis, so that the reader will find it helpful to consider only the theme of each chapter. There are nevertheless many connections between what may at first seem to be quite different topics. As in all the books of the *Integrated Environmental Technology* series, it was one of our purposes to draw out and emphasize these interdisciplinary linkages. For this reason, a comprehensive index is included to facilitate cross-referencing. We hope that the work described in this book will inspire those working in the field and will encourage those who are beginning to investigate this field.

We wish to thank all contributors to this book for their valuable contributions by sharing their expertise in the form of the various chapters included in the book. We also thank all past and present co-workers as well as all collaborators who joined in unravelling different areas of nanobiohybrids and their application in environmental technology as described in this book, especially those at National University Ireland Galway and UNESCO-IHE. We would also like to thank all the reviewers who put a lot of effort in improving the quality of this book. In addition, the national and international granting agencies who supported our work on various aspects of metal removal and recovery over the years are gratefully acknowledged, in particular the Science Foundation Ireland (SFI), who financially supported the open access publication of this book volume through the SFI Research Professorship Programme *Innovative Energy Technologies for Biofuels, Bioenergy and a Sustainable Irish Bioeconomy* (IETSBIO^3; grant number 15/RP/2763). We are also grateful to the editorial team of IWA Publishing, in particular Mark Hammond, Andrew Peart and Katharine Allenby for their help and editorial support in realizing this book.

Priyanka Uddandarao
Piet N.L. Lens
National University of Ireland Galway, Ireland

Part 1

Concepts of Microbial Synthesis, Water Purification and Energy Storage

doi: 10.2166/9781789063592_0003

Chapter 1

Introduction to wastewater treatment and energy recovery

Priyanka Uddandarao[1,2]* *and Piet N. L. Lens*[1]

[1]*Department of Microbiology, National University of Ireland, University Road, Galway, Ireland*
[2]*Department of Biotechnology, National Institute of Technology, Andhra Pradesh, India*
**Corresponding author: uddandaraopriyanka@gmail.com*

1.1 INTRODUCTION

In an era where environmental sustainability and resource scarcity are becoming increasingly critical concerns, innovative approaches to wastewater treatment and energy recovery have gained significant attention. The convergence of nanotechnology and biotechnology has paved the way for revolutionary solutions that address these challenges in unprecedented ways. This book *Nanobiohybrids for Advanced Wastewater Treatment and Energy Recovery* delves into the fascinating realm of nanobiohybrids (NBs), offering a comprehensive exploration of their potential to reshape the landscape of wastewater management and energy harnessing. The book is divided into four parts, where Part 1 covers key concepts of microbial synthesis, water purification and energy storage. Part 2 encompasses utility of organic, inorganic and magnetic nanoparticles (NPs) as well as electroactive bacteria that constitute NBs. Further parts 3 and 4 deal with, respectively, environmental remediation and wastewater treatment using NBs.

1.2 PROCESS FUNDAMENTALS

Nanobiohybrid (NB) system design amalgamates the finest characteristics of biological systems with synthetic materials. This integration encompasses the spectrum of biosystems, ranging from individual biomacromolecules (such as enzymes) to intricate biological entities (such as living cells and plants) (Liang & Liang, 2023). The synergy of biosystems and synthetic materials significantly impacts the ultimate functionality of NBs. Until now, these NBs have been created and utilized across diverse applications, including biocatalysis (Liu *et al.*, 2019; Shen *et al.*, 2022).

The journey into the world of NBs begins with a journey through the fundamental principles of nanotechnology and biotechnology. As we embark on this exploration, we will uncover the intricacies of nanomaterials and their novel applications in wastewater treatment. From NPs that act as efficient adsorbents to nanocomposites that catalyze degradation processes, these innovations have the potential to revolutionize pollutant removal (Manikandan *et al.*, 2022). But the realm of NBs does not stop at wastewater treatment. It also dives into the realm of energy recovery, a crucial element of sustainability (Nasrollahzadeh *et al.*, 2021). The principles of biophotovoltaics, microbial fuel cells, and enzyme-assisted systems will come to life as we explore how these innovations bridge the gap between waste and energy. The integration of NBs into real-world applications presents both challenges and opportunities (Andriukonis *et al.*, 2021; Teodor *et al.*, 2020).

This book is the result of collaborative efforts by experts from diverse fields, including nanotechnology, biotechnology, environmental science, and engineering. Their collective expertise shines through as they present the latest research findings, case studies, and practical insights that highlight the promise of NBs for advanced wastewater treatment and energy recovery. As you turn the pages of '*Nanobiohybrids for Advanced Wastewater Treatment and Energy Recovery*,' prepare to embark on a journey that explores the fusion of tiny particles and living organisms – a journey that has the potential to redefine the boundaries of what is achievable in environmental stewardship. This book serves as a guide to understanding the transformative impact of NBs inviting readers to join the endeavor to build a cleaner, more sustainable future.

1.3 BUILDING BLOCKS OF NBS

To address a wide spectrum of pollutants, diverse water treatment technologies have been developed and implemented, both in experimental and real-world contexts. These technologies are typically categorized into primary processes (such as screening, filtration, centrifugation, separation, sedimentation, coagulation, and flocculation), secondary treatments (including aerobic and anaerobic methods), and tertiary methods (like distillation, crystallization, evaporation, solvent extraction, oxidation, precipitation, ion exchange, microfiltration, ultrafiltration, nanofiltration, reverse osmosis, adsorption, electrolysis, electrodialysis, and others) (Chaudhary *et al.*, 2021; Yadav *et al.*, 2022). Among these, only a handful have demonstrated the capability to effectively address water treatment challenges. Several techniques have proven to be energy-intensive and operationally complex, making them economically unfeasible on a commercial scale (Wicker *et al.*, 2023). Although adsorption methods are straightforward and simple, their effectiveness in water purification remains inadequate.

To begin with, NBs are a viable option for purifying water in cases where alternative chemical methods for transforming water pollutants are inapplicable. Additionally, this technology demonstrates exceptional efficacy in breaking down aromatic water pollutants containing benzene rings, a capability that surpasses even photocatalysis. Moreover, the NBs can be reclaimed and reused multiple times (Chen *et al.*, 2020; Daliran *et al.*, 2022), resulting in an economically feasible approach. Notably, the technique finds utility in industrial settings, generating commercially significant compounds by mitigating organic pollutants in wastewater. Part 2 majorly addresses utility of NBs for wastewater treatment and energy recovery for organic (Chapter 5), inorganic (Chapter 6) and magnetic (Chapter 7) NPs as well as electroactive microbes (Chapter 8).

1.4 ENVIRONMENTAL REMEDIATION

Bioremediation has garnered significant interest as a technique that employs biological agents to eliminate hazardous waste from polluted soil. This method is user-friendly, cost-effective, and environmentally conscious, yielding highly satisfactory outcomes (Chaudhary *et al.*, 2021). Bioremediation involves biodegradation wherein organic pollutants undergo full mineralization into inorganic compounds, carbon dioxide, and water. The common pollutants include toxic organic compounds like aliphatic and aromatic compounds (Yadav & Chauhan, 2022).

Consequently, there is a growing interest among individuals to explore sustainable methods of microbial bioremediation, aiming to rehabilitate polluted environments. Benzene, toluene, and xylene (BTX) represent volatile aromatic compounds widely used in industries, yet their environmental release poses serious hazards. To address this concern, the degradation of BTX using a combination of *Aspergillus niger* cells and ZnS NPs (Figure 1.1) was investigated (Uddandarao & Lens, 2022b). Remarkably, complete degradation of BTX was accomplished within 75 min and 60 min, by NBs composed of chemical and biological ZnS NPs, respectively (Uddandarao & Lens, 2022a).

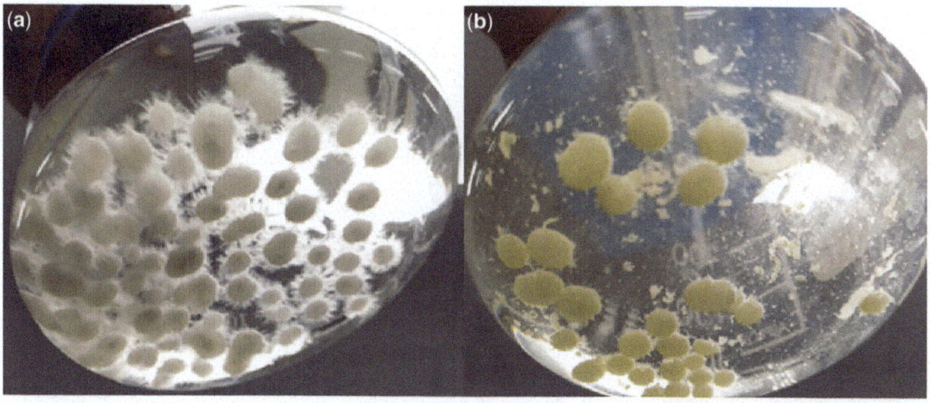

Figure 1.1 Images depicting (a) *Aspergillus niger* pellets and (b) after 4 h, a change in the color of the *Aspergillus niger* hyphae from white to slight yellow was observed due to the formation of *Aspergillus niger*-ZnS NBs. (*Source*: Uddandarao & Lens 2022a).

The future of groundwater remediation seems to lie in the fusion of nanoremediation with other remediation technologies, promoting sustainable and environmentally friendly practices. Leveraging nitrogen-fixing hydrocarbon-degrading bacteria presents another avenue to enhance bioremediation efficiency. This approach involves combining these bacteria with magnetic NPs, creating a synergistic assembly that amplifies the bioremediation process (Safonov *et al.*, 2018). This approach can be extended to address various organic pollutants, including chlorinated pesticides, polycyclic aromatic hydrocarbons, chlorinated hydrocarbons, and polychlorinated biphenyls, all of which commonly accumulate in groundwater (Cao *et al.*, 2020). Lastly, the amalgamation of NPs with oil-degrading bacteria showcases potential to elevate the efficiency of oil spill cleanup (Alabresm *et al.*, 2018). Part 3 deals with detection of a variety of pollutants in the environment's water sources (Chapter 9) and capability of NBs to address environmental pollution (Chapter 10).

1.5 WASTEWATER TREATMENT

One notable instance involves the creation of nano bacteria-hybrids using bacteria that engage in photosynthesis. These hybrids are capable of harnessing solar energy to generate acetic acid from CO_2. More recently, NBs have found application across a wide array of fields, spanning catalysis, energy storage, and electronics (Tao *et al.*, 2022). Efforts have been undertaken to create bacterial cells through the assembly of NBs. The integration of semiconductor ZnS NPs with microbial *Aspergillus niger* cells (Figure 1.2) significantly enhanced the removal of the dye methyl orange with 97.5% and 98% removal achieved within 90 and 60 min, respectively, for chemically and biologically synthesized ZnS NBs (Uddandarao & Lens, 2022b).

This hybrid system effectively overcomes the constraints seen in biological systems, such as their inefficiency in converting light and the expensive requirements for carbon and electron sources to support cell growth and maintenance. It also addresses the limitations of inorganic photocatalysis, which include poor selectivity and the tendency for photocatalytic materials to corrode under the influence of light during operation (Sakimoto *et al.*, 2016). These hybrid systems carry out specific multi-enzyme cascade biocatalysis. They have the ability to adapt to a broad spectrum of light absorption, and a high degree of efficiency in utilizing light (Zuo *et al.*, 2021). The design and assessment of an airlift reactor featuring a polymer-based photocatalytic film, specifically tailored for the photocatalytic degradation of synthetic dye-containing wastewater has been developed (Das & Mahalingam, 2020;

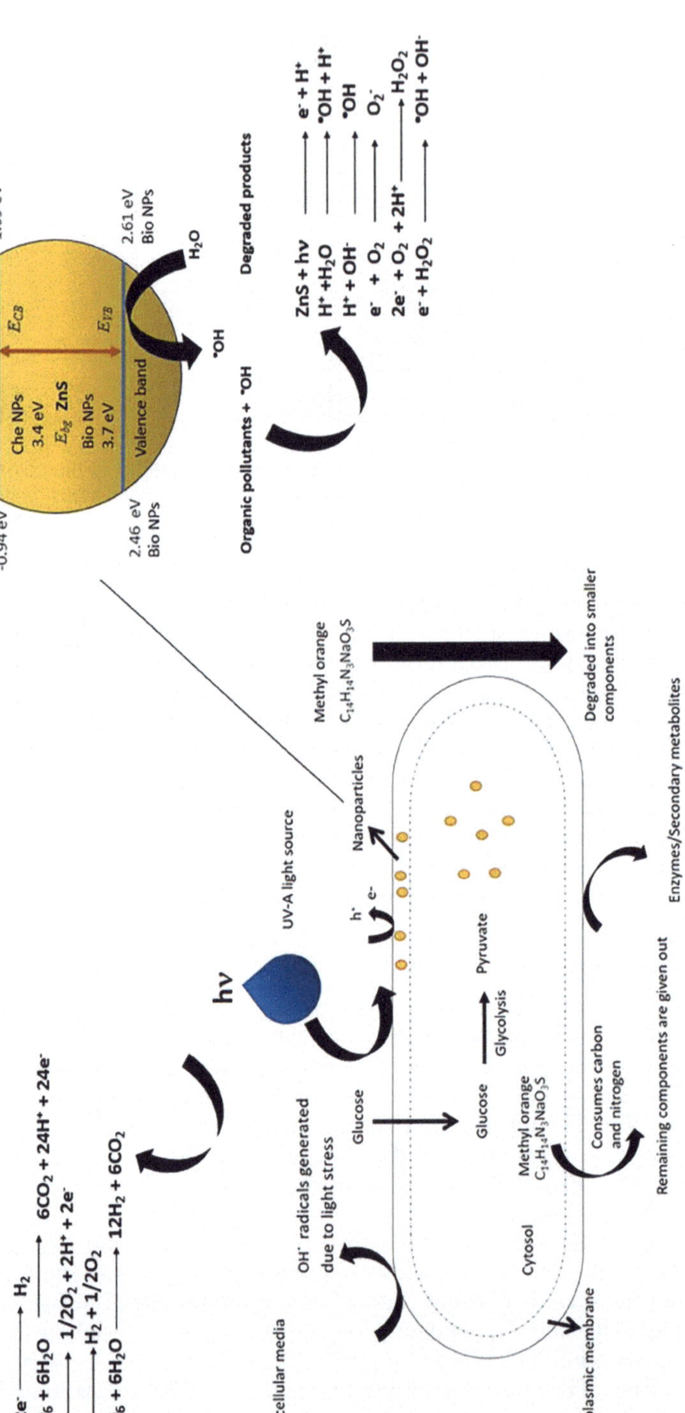

Figure 1.2 Illustration depicting the photocatalysis mechanism within *Aspergillus niger* cell-ZnS NBs. When the *Aspergillus niger* cells-ZnS NBs are exposed to light, the ZnS NPs absorb the light energy and generate electron–hole pairs. The electrons are excited to the conduction band, while the holes are left behind in the valence band. The overall effect is that the *Aspergillus niger* cell-ZnS NBs can degrade organic pollutants by a process called photocatalysis. (*Source*: Uddandarao & Lens 2022a).

Vasiljević *et al.*, 2021). Part 4 focuses on pollution control (Chapter 11) and photoelectrocatalysis (Chapter 12).

REFERENCES

Alabresm A., Chen Y. P., Decho A. W. and Lead J. (2018). A novel method for the synergistic remediation of oil-water mixtures using nanoparticles and oil-degrading bacteria. *Science of the Total Environment*, **630**, 1292–1297, https://doi.org/10.1016/j.scitotenv.2018.02.277

Andriukonis E., Celiesiute-Germaniene R., Ramanavicius S., Viter R. and Ramanavicius A. (2021). From microorganism-based amperometric biosensors towards microbial fuel cells. *Sensors*, **21**(7), 2442, https://doi.org/10.3390/s21072442

Cao X., Alabresm A., Chen Y. P., Decho A. W. and Lead J. (2020). Improved metal remediation using a combined bacterial and nanoscience approach. *Science of the Total Environment*, **704**, 135378, https://doi.org/10.1016/j.scitotenv.2019.135378

Chaudhary A., Dwivedi A. and Bhaskarwar A. N. (2021). Qualitative and quantitative analysis of water. In: Inamuddin, M. I. Ahamed, R. Boddula and T. A. Rangreez (eds), Applied Water Science Volume 1: Fundamentals and Applications, Wiley and Scrivener Publishing, Beverly, MA, pp. 469–502.

Chen H., Liu T., Meng Y., Cheng Y., Lu J. and Wang H. (2020). Novel graphene oxide/aminated lignin aerogels for enhanced adsorption of malachite green in wastewater. *Colloids and Surfaces A: Physicochemical and Engineering Aspects*, **603**, 125281, https://doi.org/10.1016/j.colsurfa.2020.125281

Daliran S., Oveisi A. R., Peng Y., López-Magano A., Khajeh M., Mas-Ballesté R. and Garcia H. (2022). Metal–organic framework (MOF)-, covalent-organic framework (COF)-, and porous-organic polymers (POP)-catalyzed selective C–H bond activation and functionalization reactions. Chemical Society Reviews, **51,** 7810–7882.

Das S. and Mahalingam H. (2020). Novel immobilized ternary photocatalytic polymer film based airlift reactor for efficient degradation of complex phthalocyanine dye wastewater. *Journal of Hazardous Materials*, **383**, 121219, https://doi.org/10.1016/j.jhazmat.2019.121219

Liang J. and Liang K. (2023). Nanobiohybrids: synthesis strategies and environmental applications from micropollutants sensing and removal to global warming mitigation. Environmental Research, **232**, 116317, https://doi.org/10.1016/j.envres.2023.116317

Liu J., Guo Z. and Liang K. (2019). Biocatalytic metal-organic framework-based artificial cells. *Advanced Functional Materials*, **29**(45), 1905321, https://doi.org/10.1002/adfm.201905321

Manikandan S., Subbaiya R., Saravanan M., Ponraj M., Selvam M. and Pugazhendhi A. (2022). A critical review of advanced nanotechnology and hybrid membrane based water recycling, reuse, and wastewater treatment processes. *Chemosphere*, **289**, 132867, https://doi.org/10.1016/j.chemosphere.2021.132867

Nasrollahzadeh M., Sajjadi M., Iravani S. and Varma R. S. (2021). Starch, cellulose, pectin, gum, alginate, chitin and chitosan derived (nano) materials for sustainable water treatment: A review. *Carbohydrate polymers*, **251**, 116986, https://doi.org/10.1016/j.carbpol.2020.116986

Safonov A. V., Babich T. L., Sokolova D. S., Grouzdev D. S., Tourova T. P., Poltaraus A. B. and Nazina T. N. (2018). Microbial community and in situ bioremediation of groundwater by nitrate removal in the zone of a radioactive waste surface repository. *Frontiers in Microbiology*, **9**, 1985, https://doi.org/10.3389/fmicb.2018.01985

Sakimoto K. K., Wong A. B. and Yang P. (2016). Self-photosensitization of nonphotosynthetic bacteria for solar-to-chemical production. *Science*, **351**(6268), 74–77, https://doi.org/10.1126/science.aad3317

Shen H., Shi H., Yang Y., Song J., Ding C. and Yu S. (2022). Highly efficient synergistic biocatalysis driven by stably loaded enzymes within hierarchically porous iron/cobalt metal–organic framework via biomimetic mineralization. *Journal of Materials Chemistry B*, **10**(10), 1553–1560, https://doi.org/10.1039/D1TB02596A

Tao M., Jin C., Lu H., Jin K., Yu L., Liu J. and Wu Y. (2022). Living and regenerative material encapsulating self-assembled *Shewanella oneidensis*-CdS hybrids for photocatalytic biodegradation of organic dyes. *Microorganisms*, **10**(12), 2501, https://doi.org/10.3390/microorganisms10122501

Teodor A. H., Sherman B. D., Ison Z. Y., Ooi E. J., Bergkamp J. J. and Bruce B. D. (2020). Green catalysts: applied and synthetic photosynthesis. *Catalysts*, **10**(9), 1016, https://doi.org/10.3390/catal10091016

Uddandarao P. and Lens P. N. L. (2022a). Enhanced removal of hydrocarbons BTX by light-driven *Aspergillus niger* ZnS nanobiohybrids. *Enzyme and Microbial Technology*, **157**, 110020, https://doi.org/10.1016/j.enzmictec.2022.110020

Uddandarao P. and Lens P. N. L. (2022b). Light driven *Aspergillus niger*-ZnS nanobiohybrids for degradation of methyl orange. *Chemosphere*, **298**, 134162, https://doi.org/10.1016/j.chemosphere.2022.134162

Vasiljević J., Jerman I. and Simončič B. (2021). Graphitic carbon nitride as a new sustainable photocatalyst for textile functionalization. *Polymers*, **13**(15), 2568, https://doi.org/10.3390/polym13152568

Wicker R. J., Kwon E., Khan E., Kumar V. and Bhatnagar A. (2023). The potential of mixed-species biofilms to address remaining challenges for economically-feasible microalgal biorefineries: a review. *Chemical Engineering Journal*, **451**, 138481, https://doi.org/10.1016/j.cej.2022.138481

Yadav J. and Chauhan P. (2022). Recent advances in synthesis, characterization, and application of nanotechnology in wastewater treatment – a review. *Nanoscience & Nanotechnology-Asia*, **12**(3), 16–33.

Yadav S. and Mondal S. S. (2022). A review on the progress and prospects of oxy-fuel carbon capture and sequestration (CCS) technology. *Fuel*, **308**, 122057, https://doi.org/10.1016/j.fuel.2021.122057.

Zuo W., Yu Y. and Huang H. (2021). Making waves: microbe-photocatalyst hybrids may provide new opportunities for treating heavy metal polluted wastewater. *Water Research*, **195**, 116984, https://doi.org/10.1016/j.watres.2021.116984

doi: 10.2166/9781789063592_0009

Chapter 2

Addressing the global water crisis: a comprehensive review of nanobiohybrid applications for water purification

I. Kalyan[1], *A. K. Nayak*[2] *and M. U. Khobragade*[3]

[1]*Indian Institute of Technology Kharagpur, School of Nano Science and Technology, Kharagpur, West Bengal 721302, India*
[2]*Prasad V. Potluri Siddhartha Institute of Technology, Department of Civil Engineering, Vijayawada, Andhra Pradesh 520007, India*
[3]*COEP Technological University (COEPTECH) Pune, Civil Engineering Department, Pune, Maharashtra 411005, India*

ABSTRACT

Systematic and organized handling of pre-existing and ever-emerging pollutants, which are continuously contaminating the freshwater resources and causing severe problems in aquatic bodies, aerial environments, and earthbound flora and fauna, is an absolute must. These pollutants contribute to one of the major global problems of scarcity of pristine water for daily use. In the last decades, different types of water cleaning and purification techniques have been embraced to address the issue. Traditional and standard water purification techniques have several shortcomings, like high time consumption, expensive operation and maintenance, and lower affinity with low efficiency toward newly emerging micropollutants. The circumstances demand that we adopt and develop a relatively new purification approach through 'nanobiohybrids.' Though for a long time, nanomaterials, nanohybrids, and enzymes have previously been used for water treatment, each of these materials has its unique challenges. Advance oxidation processes, production of toxic intermediates, lesser selectiveness and sensitivity toward dilute solutions, catalyst leaching and for enzymes shorter shelf life, extreme sensitiveness toward mechanical stress, and separation difficulties for reuse are some major problems. Now, the combination of nanomaterials with biomolecules (especially enzymes), known as 'nanobiohybrids,' is a powerful combination that holds the potential to overcome the abovementioned issues due to their synergistic properties. In the synthesis of nanobiohybrids, carbon nanotubes or graphene oxide, offer an excellent support matrix for enzymes, and play a significant role. This chapter discusses the background of nanobiohybrids, their recent development as a favorite candidate for novel water purification techniques, and their future.

Keywords: CNT, enzymes, nanomaterials, nanobiohybrids, pollutants, purification, water purification

2.1 INTRODUCTION

According to the United Nations report, with an increasing rate of over 0.84% per year, the world's population in the next 30 years is expected to increase by 2 billion (World Population Growth Rate 1950–2023 | MacroTrends). That means by 2050, it is estimated that there will be around 9.7 billion people in the world, and further, the world population will peak at 10.4 billion by 2080 (18 Surprising

Projections About the Future of Water | Seametrics). This information implies a great threat of freshwater scarcity to humankind. According to UNICEF, with our current water usage policies, at least two-thirds of the world's population is forced to experience severe water scarcity for at least a month each year (Water scarcity | UNICEF). Currently, at least 2 billion people are living in places where the water supply is inadequate for day-to-day livelihood. By 2025, around 1.8 billion people worldwide will live in areas tormented by extreme water scarcity (Das *et al.*, 2015). Not only that, but with around 1 billion additional mouths to feed by that time, worldwide agriculture will need 1 trillion cubic meters of water per year, which is comparable to 20 times the annual water flown in the great Nile River (18 Surprising Projections About the Future of Water | Seametrics). One UN study projected that around 2025 an estimated as low as 30 countries would globally undergo an extreme water crisis (Wheeler, 2012). Lack of sufficient water supply will force around 700 million people to be displaced by 2030. Another study predicts the global middle class will reach 4.9 billion by 2030, resulting in a significant increase in freshwater consumption. According to the International Energy Agency, by 2035, global energy consumption will increase by 35%, causing a 15% increase in water usage (Rodríguez & Sohns, 2014). By 2040 there will not be enough fresh water to quench the thirst globally, or enough water to keep the balance between energy generation and power supply. In 2050, it is predicted that around five times the earth's total land will be highly drought-prone compared to today.

These reports are quite paradoxical as the water covers 75% of the earth's surface. Interestingly, 2.5% of this is available as freshwater; two-thirds is locked up in glaciers and icecaps or present as permanent snow, making them impossible to be available for humans (Das, 2017). In addition to that, out of this accessible freshwater, 99% is available as underground water in aquifers. Some of these sources are non-replenishable, and some are renewable. On the contrary, the prediction of rising sea levels around 0.25–0.40 m by 2025 will cause salt waters to naturally protrude into the estuaries to increase the salinity of fresh river waters, further creating problems (Nicholls, 2002). The limited groundwater is slurped up by humans quickly with the growing population. So there is no wonder that the aforementioned facts will soon become prominent problems in human society. To overcome these scary problems appropriate water policies are needed. Overuse and needless exploitation of freshwater should be stopped, along with proper water treatment policies have to be implemented strongly to clean the different discharged polluted water and reuse them for different purposes other than drinking.

Several purifying techniques are available but no traditional technique is potent enough to handle every kind of polluted water. After implementing nanomaterials in water purification, significant improvements in efficiency and reaction time have been observed. Though there are some crucial drawbacks, like thermal instability, pollutant precipitation, and agglomeration of nanoparticles during the purification process. On the other hand, enzymes are highly efficient and can increase the reaction rate up to 1019 times. So because of their sustainable nature, eco-friendliness, cost-effectiveness, and low concentration use in any reaction makes them a perfect candidate to use as catalysts in the catalytic process of water purification. However, enzymes are also not perfect, and their application can often be limited by factors such as low stability, a shorter shelf life, and a lack of reusability. Materials that possess stability and performance similar to previously employed nanostructures for water purification, and exhibiting reaction rates and sustainability comparable to enzymes, have the potential to effectively address all types of water pollutants. These types of materials could be an effective alternative that may replace the available traditional water cleaning technologies. The idea of 'nanobiohybrids' (NBIOH) combines these two most powerful and novel technologies for purifying contaminated water. To generate NBIOH, immobilization of enzymes is essential, and for this a biocompatible molecular matrix like carbon nanotube (CNT) and graphene oxide (GO) is necessary. This chapter discusses the problems associated with conventional water purification techniques and how NBIOH can be implemented for water treatment along with their future potency as novel purification technique.

2.2 ROOT CAUSE BEHIND CONTINUOUS FRESHWATER SHRINKING

Illegal industrialization, ruthless deforestation, and unplanned urbanization are the main contributing factors to the consistent pollution of freshwater resources. The discharge of persistent pollutants to water bodies through point and nonpoint sources leads to a decrease in the availability of clean and safe water globally. The pollutants are comprising different types of inorganic species (arsenic, cobalt, cadmium, copper, chromium, selenium, and various other metals); organic compounds (various coloring agents, azo, non-azo dyes, pesticides, herbicides, fertilizers, and hydrocarbons); gas fumes (carbon monoxides, carbon dioxide, sulfur dioxide, nitric oxide, nitrous oxide, and nitrogen dioxides); and various biologically derived contaminants (viruses, bacteria, and fungi). These have been constantly defiling the clean water. Among all contaminants which are responsible for the dwindling of freshwater, phenolic water pollutants are the most notorious to deal with because of their wide availability, persistence, and bactericidal effects (Girelli *et al.*, 2006; Kolaczkowski *et al.*, 1997). These phenolic pollutants can be readily attached to natural and anthropogenic sources, causing rapid contamination in freshwater. For example, protocatechuic acid or 3,4-dihydroxybenzoic acid (3,4-DHBA), is an important pernicious phenolic water pollutant that can be largely traced in food processing wastewater effluents (Rivas *et al.*, 2005), especially in processing of olives in olive oil factories (Poulios *et al.*, 1999). Also various plant species secrete this chemical into water (Babich *et al.*, 2002). It is found that the concentration of 3,4-DHBA can be calculated as polyphenolic group, which is around 4–10 mg/L in the discharged wastewater of an olive oil mill (Hamdi, 1992). Therefore, effective and inexpensive regulatory tools should be developed to control the quantity of 3,4-DHBA and other commonly available toxic contaminants in public sewage systems.

2.3 METHODICAL HANDLING OF WATER POLLUTION

2.3.1 Treatment technologies

Over the years, different water treatment and purification techniques have been developed and adopted to tackle the variety of water pollutants. These technologies can be broadly classified into three segments. The first one is primary processes which include basic filtration, screening, sedimentation, coagulation, flocculation, and centrifugal separation. Secondary processes are often performed with aerobic and anaerobic treatments. Finally, tertiary processes involve complex purification techniques like crystallization, evaporation, distillation, solvent extraction, precipitation, ion-exchange method, oxidation, adsorption, microfiltration (MIF), ultrafiltration (ULF), nanofiltration (NAF), reverse osmosis (RO), electrolysis and electrodialysis (Gupta *et al.*, 2012).

2.3.2 Major drawbacks of current water purification techniques

No single technique is flawless or a complete solution for different types of water pollution. Often a combination of multiple processes is used to purify different types of contaminated water. Also very few have the capability of solving the water purification issue. Some methods are not commercially viable because of their high energy consumption and intensive operational issues (Das *et al.*, 2014). For example, the adsorption technique often used in water purification is easy and straightforward, though their water purification ability is unsatisfactory (Yang *et al.*, 2013). Figure 2.1 indicates some major shortcomings in water purification technologies that are currently available. The picture signifies that the water treatment process is a complex phenomenon and no single purifying technique is reliable enough to clean the water completely. The classical techniques available for water treatment become less sensitive as these methods are heavily dependent on the qualities of influent water, like temperature, pH, and turbidity. Thus, current technologies have to compromise with some of the water pollutants to let them pass through during the purification, which ultimately results in partially polluted water despite of purification. Excessive, unplanned urbanization and increasing industrialization also contribute to the introduction of a continually growing number of anthropogenic micropollutants. These pollutants ultimately find their way into water treatment plants and cannot be

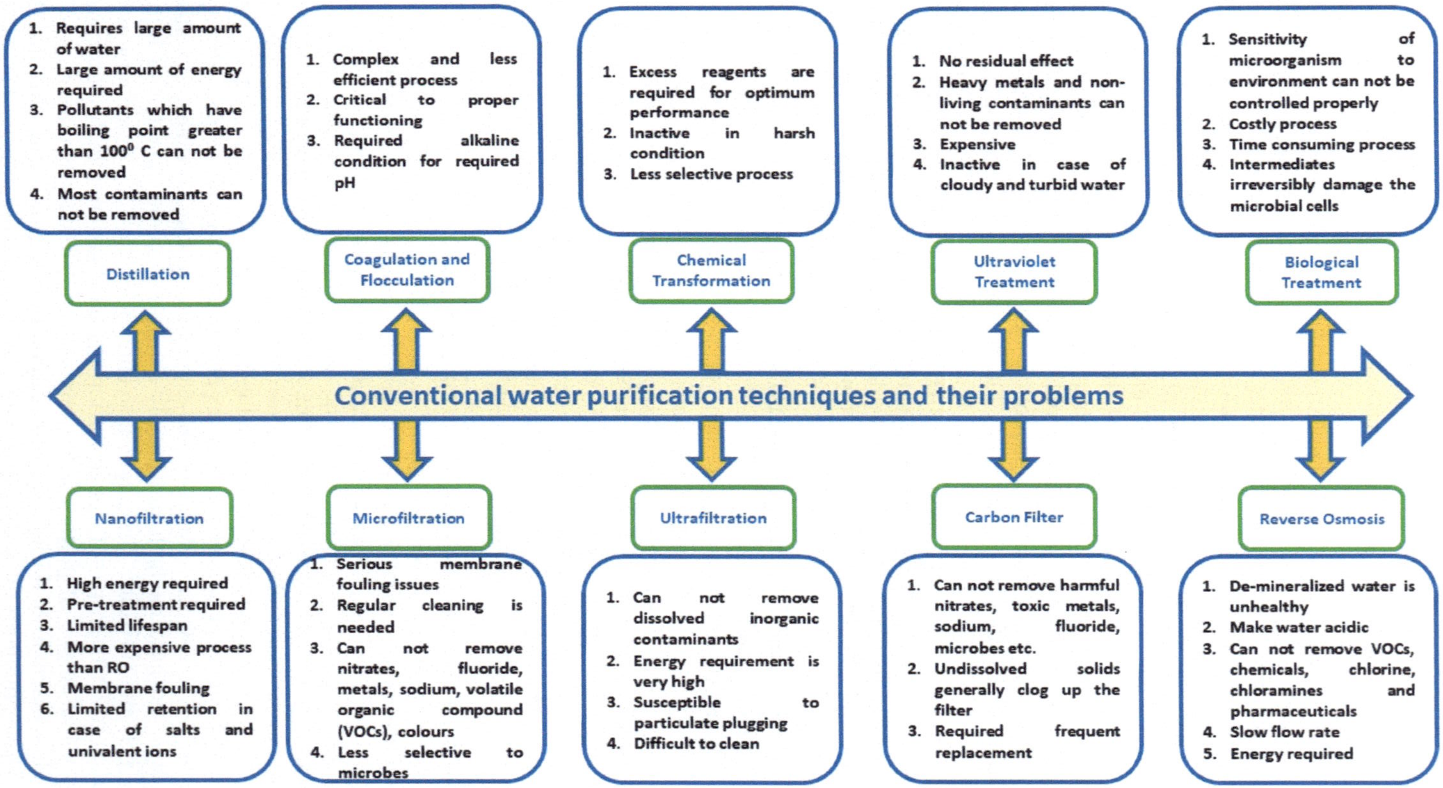

Figure 2.1 Major drawbacks of current water purification techniques.

effectively removed using our current technologies. Also, most effective purification techniques either have higher operating and maintenance costs or consume high amounts of energy. This results in the shrinking of effective technologies in developing countries (Upadhyayula *et al.*, 2009).

2.4 NANOBIOHYBRID (NBIOH) CATALYST IN WATER PURIFICATION

2.4.1 Use of nanoparticles in water purification and their problems

The advancement of nanoscience and technology has given immense opportunities to purify water even at ionic state. Different nanomaterials with different structures, like fullerenes, nanowires, zeolites, and nanofibers have been fabricated in such a way that these have a very high surface-to-volume ratio, tunable pore volume, and high reactivity so that these can take part in different electrostatic, hydrophobic, and hydrophilic interactions. These characteristics of nanomaterials make them very useful in adsorption, catalysis, sensing, and the mitigation of various water pollutants. TiO_2 nanoparticles have been utilized in the redox reactions transforming micropollutants (Kamat & Meisel, 2003). Photocatalytic nano-TiO_2 is used as a detoxifier for phenolic recalcitrant compounds and odorous chemicals (Canela & Jardim, 2008; San *et al.*, 2001). Silver, gold, iron, nanoparticles, and their oxides have been widely used in water purification. To remove inorganic heavy metals from waste waters, gold and iron nanoparticles have been frequently used (Khin *et al.*, 2012). Most of the nanomaterials improve salt rejection and avoid macrovoid formation (Goh *et al.*, 2013). CNTs have been used to make composites of RO, ULF, NAF, and MIF membranes, which enhance salt retention ability, curtails costs, land area, and energy for brackish and seawater desalination (Das *et al.*, 2014). Zeolite is a potent nanomaterial for water purification and has been mixed with a polymer matrix to form a thin film RO membrane (Fathizadeh *et al.*, 2011). Silica nanoparticles have been used in water desalination by doping into RO polymer matrices (Jadav & Singh, 2009). It improves pore diameters and networks in the polymer, which enhances the transport properties. Silver nanoparticles have been used as disinfecting agents for biological pollutants like bacteria, viruses, and fungi (Khin *et al.*, 2012).

Although several nanomaterials have been successfully utilized in water purification (Khin *et al.*, 2012), each has its own limitations. Some significant drawbacks include thermal instability, high-pressure requirements, fouling, pollutant precipitation, pore-clogging, low water influx, slower reaction rate, toxic intermediates, and aggregation on storage (Dai *et al.*, 2011). Low reusability and unknown risks to ecosystems are also major concerns (Khin *et al.*, 2012). There are other problems too, such as the unavailability of sufficient data on long-time operations, instability, and scalability that must be resolved before commercialization (Goh *et al.*, 2013). These drawbacks can be overcome by synthesizing and manipulating the nanomaterials in a manner that maximizes their surface reactivity, selectivity, and sensitivity toward a wide range of water pollutants. Therefore, proper fabrication and functionalization strategies are needed to produce these nano-structured materials for water purification processes.

2.4.2 Enzymes in water purification and their problems

With time, complex wastewater pollutants have emerged. The demand and the cost for specialized chemicals to purify these contaminants have also significantly increased. The circumstances set a perfect plot to introduce enzyme biocatalysts in water purification. Enzymes are crucial to reduce the generation of solid chemical wastes (Khin *et al.*, 2012; Xu *et al.*, 2013), since they are highly specific and efficient (Nelson *et al.*, 2008). Enzymes are an effective green catalyst and proteinacious in nature that are even active in milder reaction conditions. They have a broad substrate range, can operate across various pH and temperature ranges, exhibit low reaction times with no by-products, can function in harsh conditions, and are more eco-friendly compared to other physical and chemical treatments (Khin *et al.*, 2012). Though there are some disadvantages too. Free enzymes have poor stability and high sensitivity toward mechanical stresses (Brena *et al.*, 2013).

Various supports (Feng & Ji, 2011) are often used to immobilize enzymes to improve stability and reusability. For example, enzymes immobilized on mesoporous activated carbon (AC), graphene, and multi-wall carbon nanotube (MWCNT) have shown that the maximum loading capacity of the lipase enzyme was 3.69, 8.63 and 145 lg/mg of support, respectively (Ramani *et al.*, 2012; Tan *et al.*, 2012; Zhang *et al.*, 2014). The result clearly shows that the enzyme's effective loading is increasing in the order of MWCNT > graphene > AC which have good catalytic activity, higher mechanical stability, broad pH and temperature ranges compared with free enzymes (Brena *et al.*, 2013; Feng & Ji, 2011). In addition, to decrease its costs, enzyme catalyst recovery is necessary for multiple uses, which could only be possible upon immobilization rather than enzymes present in free state (Subrizi *et al.*, 2014). Therefore, the method is stable and commercially viable for developing novel water purification technology.

There are various reports available on the immobilization of enzymes, including tyrosinase (Subrizi *et al.*, 2014), horseradish peroxidase (HRP) (Lee *et al.*, 2006), laccase (Liu *et al.*, 2006), glucose oxidase (Feng & Ji, 2011; Guiseppi-Elie *et al.*, 2002), onto a carbon nanotube (CNT) matrix. But, these hybrid materials have only been used for sensing and detecting the permissible and non-permissible levels of water pollutants rather than eliminating them from the complex wastewater matrices. A few studies have shown effective degradation of various water pollutants using immobilized enzyme catalysts onto different supports (Dai *et al.*, 2011; Guiseppi-Elie *et al.*, 2002; Guzik *et al.*, 2014). Therefore, there is room for developing and optimizing a 'nanobiohybrid' catalyst, which could serve as a novel water purification technology suited for attenuating wastewater pollutants in the near future.

2.4.3 Use of NBIOH catalyst for water purification

2.4.3.1 Capacity of NBIOH to treat water

Among all available support matrixes to immobilize enzymes for water purification, CNT is the best. Single-wall carbon nanotubes (SWCNTs) as well as MWCNTs have also been utilized to attach enzymes for the purpose of decontaminating water. NBIOH have some important advantages over commonly used water purification processes like the chemical oxidation of water pollutants. First, it can be applicable to the decontamination process of wastewater, where no other chemical transformation of water pollutants is possible. Second, these catalysts show excellent efficiency toward disintegrating the electron resonance in the benzene ring containing aromatic water pollutants in contrast to photocatalysis. Third, recoverable NBIOH catalysts can be reused multiple times, making it a commercially viable process. Fourth, it can be utilized in the industry to produce commercially important compounds, which have been produced by mitigating organic wastewater contaminants. Finally, these nanohybrids can play three primary functions with very high sensitivity and selectivity in water purification: (1) binding and pre-concentration of the pollutants, (2) removal and degradation of pollutants (see Chapters 5–8), and (3) sensing as well as monitoring the pollutants in water (see Chapter 9). The concise information about various NBIOH systems for water purification has been presented in tabular form in Table 2.1.

Laccase was immobilized onto MWCNT and then utilized to treat water containing 2,2′-azino-bis-(3-ethylbenzthiazoline-6-sulfonic acid) diammonium salt (ABTS), bisphenol, and catechol (Suma *et al.*, 2012). The system successfully performed the catalysis. In this process, it was observed that enzyme loading was highest for O-MWCNT while it was lowest for C_{60}. The interesting aspect of this process is that no obvious structural changes were observed after enzyme immobilization (Pang *et al.*, 2015). Another system, a combination of SWCNTs with horseradish peroxidase (HPR) and n-dodecylβ-D-maltoside (DM) and concanavalin A (Con-A) was used to remove dibenzothiophene (DBT) from wastewater. A DBT removal rate of 66.46 and 94.65% for SWCNT-HRP and SWCNT-DM-Con A-HRP catalysts, respectively, was observed (Li *et al.*, 2013). Hydroxyquinol 1,2 dioxygenase (HQD) immobilized in SWCNT successfully removed catechol with a wide range of pH and temperature (Suma *et al.*, 2012). To remove phenol derivatives from wastewater, MWCNTs-tyrosinase NBIOH was used. These can selectively oxidize phenol and its derivatives and showed the highest stability (Subrizi *et al.*, 2014). Another bionanohybrid system, soybean peroxidase (SBP) immobilized into

MWCNT, exhibits an excellent removal efficiency toward *p*-cresol and retained 40–60% activity as well as stability after multiple use (Asuri *et al.*, 2006). Crude enzymes incorporated into SWCNT and MWCNT were used in water purification to remove aniline successfully from contaminated water. It was observed that enzyme adsorption increased on MWCNTs compared to SWCNTs, while the effective removal of aniline was higher on SWCNTs enzyme compared to MWCNTs enzyme. This indicates that the reaction between aniline and enzymes is favorable on the surface of SWCNTs. It plays a vital role in the rapid enzymatic biodegradation of aniline (Yan *et al.*, 2011). Super magnetic $FeCO_3$ nanorods combined with lipase were efficiently used to remove bisphenol-A (Benavente *et al.*, 2018). Othman *et al.* (2016) showed that MWCNT-laccase NBIOH can help to decolorize dyes. GO-HPR nanobiohybrids can successfully remove phenolic compounds (Zhang *et al.*, 2010). Another study showed that the organic–inorganic nanocomposites–laccase system is helpful in the removal of dichlorophenol. Protocatechuate 3,4-dioxygenase, or simply 3,4-POD, attached to CNT, has also been successful in removing 3,4-DHBA (Das, 2017).

2.4.3.2 Problems associated with nanobiohybrid

Pristine CNTs are often contaminated with metal catalysts, ash, and carbonaceous agents and often act as additional adsorbent sites toward the water pollutants. Nanotube's pore diameters and morphology are determined by its impurities. The impurities could influence or inhibit adsorption behavior (Agnihotri *et al.*, 2006). In catalysis, metal and amorphous carbon impurities can affect the observed catalytic behaviors and mislead total electrocatalytic activity (Banks *et al.*, 2006). Furthermore, CNT impurities could have several detrimental effects on enzyme immobilization. First, non-nanotube carbon materials can inhibit the maximum immobilization efficiency. Second, metals in pristine CNTs can affect the immobilized enzyme's reaction rates.

Moreover, different CNT geometries and impurities could complicate the procedures and compromise CNT performance in sensor devices. Separating metallic from semiconducting CNTs has remained a challenging job because of their complex carbon networks. Synthesis of homogenous CNTs with perfect geometry for high power density is also appreciable for upgrading this technology. One pot combination of CNTs with classical electrochemical techniques could be a nice electrical device for water pollution control. It would increase the overall catalysis rate and has fast mass transport and good control over the classical current flow in small geometry.

2.5 CONCLUSION

In the field of water purification, catalysts are the main workhorse. Working of a catalyst in water purification is performed by three main processes: (1) photocatalysis, (2) catalytic wet air oxidation, and (3) biocatalysis. Although an immense progress has been noticed in the field of the former two processes, biocatalysis has remained insufficiently discovered in the field of water purification. Enzymes are effective green biocatalysts with high reactivity, which has several quintessential properties, like multi-substrate specificity, high sensitivity, active in milder reaction conditions, wide pH, and temperature ranges, low reaction times, less by-product formation, and eco-friendly. However, enzymes in their free state are not stable and are highly sensitive to mechanical stresses. To improve their stability and reusability, enzymes if immobilized onto various supports can overcome the aforementioned obstacles. Conventional water purification technologies have some major demerits when newly emerging micropollutants are present in contaminated water. To overcome these issues, a novel approach toward water purification has emerged. This chapter shows a novel catalyst synthesized with nanomaterials and enzymes, which is a nanobiohybrid catalyst that can be successfully used in water purification. Different enzyme-immobilized nanomaterials and their applications in decontamination processes have been discussed. As of today, it can be observed that SWCNT and MWCNT-based nanobiohybrids are dominating in this field. Though ample reports are available for nanobiohybrids for detection and sensing purposes, there are few reports available

Table 2.1 Different NBIOH systems for water purification.

Nanomaterials	Enzyme	Water Pollutant	Observations	References
MWCNTs-Laccase	Laccase	ABTS, bisphenol and catechol	• Enzyme loading capacity was highest for O-MWCNTs and lowest for C_{60} • No obvious structural change of enzyme observed after immobilization	Pang *et al.* (2015)
SWCNTs-HRP SWCNT-DM-Con A- HRP	HPR, DM	Dibenzothiophene (DBT)	• Increased enzyme loading with high specific activity triggered efficient DBT degradation • DBT removal rate: 66.46 and 94.65% for SWCNT-HRP and SWCNT-DM-Con A-HRP catalysts, respectively	Li *et al.* (2013)
SWCNTs–1,2 HQD	HQD	Catechol	Removed catechol with wide pH and temperature ranges	Suma *et al.* (2012)
MWCNTs-tyrosinase	Tyrosinase	Phenol derivatives	Selectively oxidized phenol and its derivatives with high stability of the catalyst	Subrizi *et al.* (2014)
MWCNTs-SBP	SBP	p-Cresol	• Nanobiohybrid retained 40–60% activity after multiple uses • Highly stable and reusable than free enzyme in removing *p*-cresol	Asuri *et al.* (2006)
SWCNT, MWCNT, SWCNTs-crude enzyme, MWCNTs-crude enzyme	Crude enzyme	Aniline	• Enzymes adsorption increased in MWCNTs than SWCNTs • Effective aniline removal observed in SWCNTs enzyme compared with • MWCNTs enzyme indicated efficient reaction between aniline and enzymes on the surface of SWCNTs. This played a key role in the rapid enzymatic biodegradation of aniline	Yan *et al.* (2011)
Super magnetic $FeCO_3$ nanorod-lipase	*Candida antartica* B lipase (CAL-B)	Bisphenol-A	Efficiently removed toxic bisphenol-A	*Benavente et al.* (2018*)*
MWCNT-laccase	Laccase	Dye	Dye decolorization	Othman *et al.* (2016)
GO-HPR	HPR	Phenolic compound	Removed phenolic compound	Zhang *et al.* (2010)
Organic–inorganic nanocomposites-laccase	Laccase	Dichlorophenol	Dichlorophenol removal	Qiu *et al.* (2019)

MWCNT: multi-wall carbon nanotube; SWCNT: single-wall carbon nanotube; HPR: horseradish peroxidase; DM: n-dodecyl β-D-maltoside; HQD: hydroxyquinol 1,2 dioxygenase; SBP: soybean peroxidase.

regarding the purifying techniques for polluted water using nanobiohybrids. There is vast scope to explore these novel water purification techniques with 'nanobiohybrids' as catalysts to bring a sustainable world in the near future.

REFERENCES

18 Surprising Projections About the Future of Water | Seametrics. https://www.seametrics.com/blog/future-water/ (accessed 13 March 2023).

Agnihotri S., Mota J. P. B., Rostam-Abadi M. and Rood M. J. (2006). Adsorption site analysis of impurity embedded single-walled carbon nanotube bundles. *Carbon*, **44**(12), 2376–2383, https://doi.org/10.1016/j.carbon.2006.05.038

Asuri P., Karajanagi S. S., Sellitto E., Kim D., Kane R. S. and Dordick J. S. (2006). Water-soluble carbon nanotube-enzyme conjugates as functional biocatalytic formulations. *Biotechnology and Bioengineering*, **95**(5), 804–811, https://doi.org/10.1002/bit.21016

Babich H., Sedletcaia A. and Kenigsberg B. (2002). In vitro cytotoxicity of protocatechuic acid to cultured human cells from oral tissue: involvement in oxidative stress. *Pharmacology & Toxicology*, **91**(5), 245–253, https://doi.org/10.1034/j.1600-0773.2002.910505.x

Banks C. E., Crossley A., Salter C., Wilkins S. J. and Compton R. G. (2006). Carbon nanotubes contain metal impurities which are responsible for the 'electrocatalysis' seen at some nanotube-modified electrodes. *Angewandte Chemie International Edition*, **45**(16), 2533–2537, https://doi.org/10.1002/anie.200600033

Benavente R., Lopez-Tejedor D. and Palomo J. M. (2018). Synthesis of a superparamagnetic ultrathin $FeCO_3$ nanorods–enzyme bionanohybrid as a novel heterogeneous catalyst. *Chemical Communications*, **54**(49), 6256–6259, https://doi.org/10.1039/C8CC02851F

Brena B., González-Pombo P. and Batista-Viera F. (2013). Immobilization of enzymes: a literature survey. In: Immobilization of Enzymes and Cells. Methods in Molecular Biology, J. Guisan (ed.), Humana Press, Totowa, NJ, Vol. **1051**, pp. 15–31.

Canela M. C. and Jardim W. F. (2008). Identification and photocatalytic destruction of malodorous compounds in sewage. *Environmental Technology*, **29**(6), 673–679, https://doi.org/10.1080/09593330801984936

Dai Y., Yin L. and Niu J. (2011). Laccase-carrying electrospun fibrous membranes for adsorption and degradation of PAHs in shoal soils. *Environmental Science & Technology*, **45**(24), 10611–10618, https://doi.org/10.1021/es203286e

Das R. (2017). Nanohybrid Catalyst Based on Carbon Nanotube: A Step-by-Step Guideline From Preparation to Demonstration. Springer, Cham.

Das R., Abd Hamid S. B., Ali M. E., Ismail A. F., Annuar M. S. M. and Ramakrishna S. (2014). Multifunctional carbon nanotubes in water treatment: the present, past and future. *Desalination*, **354**, 160–179, https://doi.org/10.1016/j.desal.2014.09.032

Das R., Abd Hamid S. B. and Ali M. E. (2015). Nanobiohybrid: a favorite candidate for future water purification technology. *Advanced Materials Research*, **1131**, 193–197, https://doi.org/10.4028/www.scientific.net/AMR.1131.193

Fathizadeh M., Aroujalian A. and Raisi A. (2011). Effect of added NaX nano-zeolite into polyamide as a top thin layer of membrane on water flux and salt rejection in a reverse osmosis process. *Journal of Membrane Science*, **375**(1–2), 88–95, https://doi.org/10.1016/j.memsci.2011.03.017

Feng W. and Ji P. (2011). Enzymes immobilized on carbon nanotubes. *Biotechnology Advances*, **29**(6), 889–895, https://doi.org/10.1016/j.biotechadv.2011.07.007

Girelli A. M., Mattei E. and Messina A. (2006). Phenols removal by immobilized tyrosinase reactor in on-line high performance liquid chromatography. *Analytica Chimica Acta*, **580**(2), 271–277, https://doi.org/10.1016/j.aca.2006.07.088

Goh P. S., Ismail A. F. and Ng B. C. (2013). Carbon nanotubes for desalination: performance evaluation and current hurdles. *Desalination*, **308**, 2–14, https://doi.org/10.1016/j.desal.2012.07.040

Guiseppi-Elie A., Lei C. and Baughman R. H. (2002). Direct electron transfer of glucose oxidase on carbon nanotubes. *Nanotechnology*, **13**(5), 559, https://doi.org/10.1088/0957-4484/13/5/303

Gupta V. K., Ali I., Saleh T. A., Nayak A. and Agarwal S. (2012). Chemical treatment technologies for waste-water recycling – an overview. *RSC Advances*, **2**(16), 6380–6388, https://doi.org/10.1039/c2ra20340e

Guzik U., Hupert-Kocurek K., Krysiak M. and Wojcieszyńska D. (2014). Degradation potential of protocatechuate 3, 4-dioxygenase from crude extract of *Stenotrophomonas maltophilia* strain KB2 immobilized in calcium alginate hydrogels and on glyoxyl agarose. *BioMed Research International*, **2014**, 1–8, https://doi.org/10.1155/2014/138768

Hamdi M. (1992). Toxicity and biodegradability of olive mill wastewaters in batch anaerobic digestion. *Applied Biochemistry and Biotechnology*, **37**, 155–163, https://doi.org/10.1007/BF02921667

Jadav G. L. and Singh P. S. (2009). Synthesis of novel silica-polyamide nanocomposite membrane with enhanced properties. *Journal of Membrane Science*, **328**(1–2), 257–267, https://doi.org/10.1016/j.memsci.2008.12.014

Kamat P. V. and Meisel D. (2003). Nanoscience opportunities in environmental remediation. *Comptes Rendus Chimie*, **6**(8–10), 999–1007, https://doi.org/10.1016/j.crci.2003.06.005

Khin M. M., Nair A. S., Babu V. J., Murugan R. and Ramakrishna S. (2012). A review on nanomaterials for environmental remediation. *Energy & Environmental Science*, **5**(8), 8075–8109, https://doi.org/10.1039/c2ee21818f

Kolaczkowski S. T., Beltran F. J., McLurgh D. B. and Rivas F. J. (1997). Wet air oxidation of phenol: factors that may influence global kinetics. *Process Safety and Environmental Protection*, **75**(4), 257–265, https://doi.org/10.1205/095758297529138

Lee Y.-M., Kwon O.-Y., Yoon Y.-J. and Ryu K. (2006). Immobilization of horseradish peroxidase on multi-wall carbon nanotubes and its electrochemical properties. *Biotechnology Letters*, **28**, 39–43, https://doi.org/10.1007/s10529-005-9685-8

Li Y., Huang X. and Qu Y. (2013). A strategy for efficient immobilization of laccase and horseradish peroxidase on single-walled carbon nanotubes. *Journal of Chemical Technology & Biotechnology*, **88**(12), 2227–2232, https://doi.org/10.1002/jctb.4091

Liu Y., Qu X., Guo H., Chen H., Liu B. and Dong S. (2006). Facile preparation of amperometric laccase biosensor with multifunction based on the matrix of carbon nanotubes–chitosan composite. *Biosensors and Bioelectronics*, **21**(12), 2195–2201, https://doi.org/10.1016/j.bios.2005.11.014

Nelson D. L., Lehninger A. L. and Cox M. M. (2008). Lehninger Principles of Biochemistry, New York.

Nicholls R. J. (2002). Case study on sea-level rise impacts. OECD Workshop on the Benefits of Climate Policy: Improving Information for Policy Makers, 12–13 December, Paris. Vol. **9**, pp. 69–86.

Othman A. M., González-Domínguez E., Sanromán Á., Correa-Duarte M. and Moldes D. (2016). Immobilization of laccase on functionalized multiwalled carbon nanotube membranes and application for dye decolorization. *RSC Advances*, **6**(115), 114690–114697, https://doi.org/10.1039/C6RA18283F

Pang R., Li M. and Zhang C. (2015). Degradation of phenolic compounds by laccase immobilized on carbon nanomaterials: diffusional limitation investigation. *Talanta*, **131**, 38–45, https://doi.org/10.1016/j.talanta.2014.07.045

Poulios I., Makri D. and Prohaska X. (1999). Photocatalytic treatment of olive milling waste water: oxidation of protocatechuic acid. *Global Nest: The International Journal*, **1**(1), 55–62.

Qiu X., Qin J., Xu M., Kang L. and Hu Y. (2019). Organic-inorganic nanocomposites fabricated via functional ionic liquid as the bridging agent for laccase immobilization and its application in 2, 4-dichlorophenol removal. *Colloids and Surfaces B: Biointerfaces*, **179**, 260–269, https://doi.org/10.1016/j.colsurfb.2019.04.002

Ramani K., Karthikeyan S., Boopathy R., Kennedy L. J., Mandal A. B. and Sekaran G. (2012). Surface functionalized mesoporous activated carbon for the immobilization of acidic lipase and their application to hydrolysis of waste cooked oil: isotherm and kinetic studies. *Process Biochemistry*, **47**(3), 435–445, https://doi.org/10.1016/j.procbio.2011.11.025

Rivas F. J., Frades J., Alonso M. A., Montoya C. and Monteagudo J. M. (2005). Fenton's oxidation of food processing wastewater components. Kinetic modeling of protocatechuic acid degradation. *Journal of Agricultural and Food Chemistry*, **53**(26), 10097–10104, https://doi.org/10.1021/jf0512712

Rodríguez D. J. and Sohns A. (2014). Will Water Constrain Our Energy Future?. https://www.worldbank.org/en/news/feature/2014/01/16/will-water-constrain-our-energy-future/ (accessed 13 March 2023).

San N., Hatipoğlu A., Koçtürk G. and Çınar Z. (2001). Prediction of primary intermediates and the photodegradation kinetics of 3-aminophenol in aqueous TiO_2 suspensions. *Journal of Photochemistry and Photobiology A: Chemistry*, **139**(2–3), 225–232, https://doi.org/10.1016/S1010-6030(01)00368-9

Subrizi F., Crucianelli M., Grossi V., Passacantando M., Pesci L. and Saladino R. (2014). Carbon nanotubes as activating tyrosinase supports for the selective synthesis of catechols. *ACS Catalysis*, **4**(3), 810–822, https://doi.org/10.1021/cs400856e

Suma Y., Kim D., Lee J. W., Park K. Y. and Kim H. S. (2012). Degradation of catechol by immobilized hydroxyquinol 1, 2-dioxygenase (1, 2-HQD) onto single-walled carbon nanotubes. Proceedings of the International Conference on Chemical, Environmental Science and Engineering (ICEEBS), 28–29 July, Pattaya, Thailand.

Tan H., Feng W. and Ji P. (2012). Lipase immobilized on magnetic multi-walled carbon nanotubes. *Bioresource Technology*, **115**, 172–176, https://doi.org/10.1016/j.biortech.2011.10.066

Upadhyayula V. K. K., Deng S., Mitchell M. C. and Smith G. B. (2009). Application of carbon nanotube technology for removal of contaminants in drinking water: a review. *Science of the Total Environment*, **408**(1), 1–13, https://doi.org/10.1016/j.scitotenv.2009.09.027

Water scarcity | UNICEF. https://www.unicef.org/wash/water-scarcity/ (accessed 13 March 2023).

Wheeler W. (2012). Global water crisis: too little, too much, or lack of a plan?. https://www.csmonitor.com/World/Global-Issues/2012/1202/Global-water-crisis-too-little-too-much-or-lack-of-a-plan/ (accessed 13 March 2023)

World Population Growth Rate 1950–2023 | MacroTrends. https://www.macrotrends.net/countries/WLD/world/population-growth-rate/ (accessed 13 March 2023).

Xu R., Chi C., Li F. and Zhang B. (2013). Laccase-polyacrylonitrile nanofibrous membrane: highly immobilized, stable, reusable, and efficacious for 2,4,6-trichlorophenol removal. *ACS Applied Materials & Interfaces*, **5**(23), 12554–12560, https://doi.org/10.1021/am403849q

Yan H., Yang X., Chen J., Yin C., Xiao C. and Chen H. (2011). Synergistic removal of aniline by carbon nanotubes and the enzymes of *Delftia* sp. XYJ6. *Journal of Environmental Sciences*, **23**(7), 1165–1170, https://doi.org/10.1016/S1001-0742(10)60531-1

Yang H. Y., Han Z. J., Yu S. F., Pey K. L., Ostrikov K. and Karnik R. (2013). Carbon nanotube membranes with ultrahigh specific adsorption capacity for water desalination and purification. *Nature Communications*, **4**(1), 2220, https://doi.org/10.1038/ncomms3220

Zhang F., Zheng B., Zhang J., Huang X., Liu H., Guo S. and Zhang J. (2010). Horseradish peroxidase immobilized on graphene oxide: physical properties and applications in phenolic compound removal. *The Journal of Physical Chemistry C*, **114**(18), 8469–8473, https://doi.org/10.1021/jp101073b

Zhang G., Ma J., Wang J., Li Y., Zhang G., Zhang F. and Fan X. (2014). Lipase immobilized on graphene oxide as reusable biocatalyst. *Industrial & Engineering Chemistry Research*, **53**(51), 19878–19883, https://doi.org/10.1021/ie503596j

doi: 10.2166/9781789063592_0021

Chapter 3

Biological production of nanoparticles and their application in photocatalysis

Priyanka Uddandarao[1,2]* *and Piet N. L. Lens*[1]

[1]Department of Microbiology, National University of Ireland, Galway, University Road, Galway, Ireland
[2]Department of Biotechnology, National Institute of Technology, Andhra Pradesh
**Corresponding author: uddandaraopriyanka@gmail.com*

ABSTRACT

With the advancement in science and technology, the demand for nanotechnology has increased significantly in terms of research and developing strategies for their manufacturing. Nanotechnology is an interdisciplinary field of science where the size of the material is designed such that at least one of its dimensions is in this nanometer (nm) range. It combines various disciplines of biology, chemistry, physics and material science. The exceptional size-dependent fluorescence and quantum confinement properties of nanomaterials have allured their utility in diverse sectors. They are potential tools for various photocatalytic devices. Photocatalysis is an advanced oxidation process driven by a light source and is a promising technology for degradation of hazardous chemical compounds in water. The vast literature indicates that the methods followed for synthesis of nanoparticles are mostly chemically based approaches and the produced nanoparticles properties depend on the reaction conditions. On the other hand, use of microbial sources for biosynthesis of nanoparticles is an ecofriendly and cost-effective strategy for nanomaterials. In this perspective, this chapter presents an overview of synthesis of nanomaterials by plants, bacteria, fungi and algal sources and their role in photocatalysis. Furthermore, some of the latest advances in these technologies are highlighted and an outlook at the major challenges for further developments is provided.

Keywords: algae, bacteria, fungi, nanomaterials, plants, photocatalysis, photobioreactor

3.1 INTRODUCTION

With the development of society, various industries are being setup to captivate human needs. In this scenario, various persistent organic pollutants are discharged into the environment harming the ecosystems (Cabral *et al.*, 2019; des Ligneris *et al.*, 2018; Rajaram & Das, 2008; Samanta *et al.*, 2019). The persistence of these pollutants in the environment is a serious concern due to their long-range transport ability to bio-magnify and bio-accumulate in ecosystems (Islam *et al.*, 2018; Yu *et al.*, 2018). These significant amounts of wastewaters are discharged into the water bodies directly or indirectly from various industries, which include pharmaceuticals, personal care products, endocrine disruptors, pesticides, detergents, organic dyes and common industrial organic wastes (Alharbi *et al.*, 2018; Han & Currell, 2017; Olatunji, 2019). The increase in the production, use and disposal of numerous chemicals frequently employed in medicine, industry, agriculture and households has led to an uncontrolled discharge of these substances into the aquatic compartments, even at trace concentrations, that is, ng/L–μg/L (Griffero *et al.*, 2019; Murgolo *et al.*, 2021; Rasheed *et al.*, 2020). Although various methods have been developed for degrading these pollutants, they encounter various limitations because of their complex and stable chemical structures, making them resistive toward traditional treatment techniques (Crini, 2005; Heng *et al.*, 2021; Kannan *et al.*, 2020; Singh & Arora, 2011).

The common treatment methods include biological treatment techniques, coagulation/precipitation, Fenton oxidation technology and photocatalytic oxidation technology. Among them, photocatalysis is a promising technology for degradation and inactivation of hazardous chemical compounds (Han *et al.*, 2015; Saharan *et al.*, 2015). Photocatalysis falls in the category of advanced oxidation processes. They can be driven by different light sources such as light-emitting diodes, ultraviolet, visible or sunlight. The mechanism of photocatalysis relies on the formation of reactive oxygen species, activated by a light source, leaving holes in the valence band as they tend to move from the valence band to the conduction band forming electron–hole pairs and inducing a series of chemical reactions. These reactions form free radicals (•OH) and are capable of undergoing secondary reactions where the electrons can be absorbed by oxygen to form superoxide or they can react with oxygen along with hydrogen to create hydrogen peroxide. Subsequently, these reactants help in the degradation of the pollutants (Bajorowicz *et al.*, 2018; Koe *et al.*, 2020; Murgolo *et al.*, 2021; Soleimani *et al.*, 2021).

The amalgamation of nanotechnology helps in enhancing the degradation process. The nanoparticles possess quantum confinement properties and have a large surface area-to-volume ratio (Bajorowicz *et al.*, 2018; Xue *et al.*, 2012; Yanhong *et al.*, 2004). These nanoparticles show a diverse behavior in their electronic, optical and magnetic properties by size tuning. When the size of the material reduces to the atomic level, they tend to breakdown into discrete levels and the size of the nanoparticle is a function of the band gap of the material (Fernández-García *et al.*, 2004; Hodes, 2007). With the advent of nanobiotechnology, microbial synthesis of nanoparticles has become a green chemistry approach, which is environmentally viable and economically feasible. Therefore, biologically driven processes have gained importance in the synthesis of nanoparticles, the natural sources comprise of plants, bacteria, fungi and algae (Das *et al.*, 2017; Gahlawat & Choudhury, 2019; Thakkar *et al.*, 2010). This chapter discusses (1) the biological synthesis of nanoparticles by various biological sources, (2) approaches based on photocatalysis and (3) challenges encountered in the application of nanomaterials.

3.2 GREEN SYNTHESIS OF NANOPARTICLES

The variability of nanofactory systems requires standardization protocols to obtain nanomaterials with increasingly uniform and reproducible physical–chemical characteristics (Grasso *et al.*, 2020). In some cases, green synthesis of the nanoparticles involves the use of biological entities for the chemically synthesized nanoparticles (Khatami *et al.*, 2018; Nadeem *et al.*, 2018; Shah *et al.*, 2015). This includes the use of biological agents as capping agents to attain stability, use of amino acids/biomolecules or polysaccharides as reducing agents for chemical synthesis Saratale *et al.* (2018). Stabilization of nanomaterials is important because, as explained by the classical theory of the stability of colloids, the balance between two opposing forces: electrostatic repulsion and van der Waals attraction. In most of the systems during synthesis of nanoparticles they tend to agglomerate as the van der Waals force of attraction becomes significant when two colloids approach each other and their double layers begin to interfere. This agglomeration can be avoided by using capping agents such as surfactants (Dzimitrowicz *et al.* 2019; Franco-Ulloa *et al.*, 2020; Polte, 2015; Segets *et al.*, 2011). Researchers have reported biologically capped nanoparticles which include pectin-capped Ag (Devasvaran & Lim, 2021; Durán *et al.*, 2011; Hileuskaya *et al.*, 2020), levan-capped gold (Akturk, 2020; González-Garcinuño *et al.*, 2019), gum arabic stabilized copper (Chawla *et al.*, 2020, 2021; Yao *et al.*, 2013), alanine-capped CuS (Mofokeng *et al.*, 2018, 2019; Nelwamondo *et al.*, 2012) and chitosan capped zinc oxide nanoparticles (AbdElhady, 2012; El-Saied & Ibrahim, 2020; Gong *et al.*, 2019).

Amino acids act as efficient reducing as well as capping agents to synthesize nanoparticles with uniform structure. Primary amine and hydroxyl groups have a powerful affinity to metal ions to decrease particle size and stop agglomeration (Basnet & Chatterjee, 2020; Dumur *et al.*, 2011). These include the Fe- and Cu-containing humic acids extracted from coal waste which was further used for the solar-based disinfection processes, namely solar light (*hv*), hv/H_2O_2 and $hv/H_2O_2/Fe^{2+}$ (photo-Fenton) processes and L-histidine was found to reduce tetrachloroauric acid to Au nanoparticles (Liu *et al.*, 2019; Porras *et al.*, 2018). Polysaccharides are very attractive biopolymers for synthesis of nanoparticles. Also polysaccharides are a class of polymeric carbohydrate molecules with repeating units of mono or disaccharides linked together by glycosidic linkages. Biocompatibility, biodegradability, low toxicity, mucoadhesive properties and inherent antimicrobial activity are the advantages favoring the broad use of polysaccharides such as dextran, pectin, chitosan and alginate in fabrication of novel materials (Jones & McClements, 2011; Joye & McClements, 2014; Zafar *et al.*, 2016).

3.3 BIOLOGICAL NANOPARTICLES

3.3.1 Plants

Plant-mediated nanoparticles synthesis, which is often called phytonanotechnology, is usually single step. Moreover, these nanoparticles are biocompatible and scalable. Further, it excludes the steps included in microbial synthesis such as microbial isolation, cell culture maintenance and downstream processing (Abdelghany *et al.*, 2018; Agarwal *et al.* 2017; Ahmed *et al.*, 2016a, 2016b; Akhtar *et al.*, 2013; Banerjee *et al.* 2014; Dzimitrowicz *et al.* 2019; Ebrahiminezhad *et al.*, 2018; Elia *et al.* 2014; Mashwani *et al.*, 2015; Singh *et al.*, 2016). Phyto nanotechnology is majorly divided into three classifications: in-vivo, in-vitro and phyto constituents mediated (Figure 3.1). The in-vivo-mediated synthesis involves the growth of plants in metal-rich media (Rana *et al.*, 2023), whereas in-vitro synthesis involves the use of leaf, stem, root, fruit and seed extracts (Balachandar *et al.* 2019; Dinga *et al.*, 2022). Photochemicals-mediated synthesis comprises the addition of the precursor solution for the reduction process (Fahimirad *et al.*, 2019; Sharma *et al.*, 2018). A summary on the recent advancements in the phytonanotechnology is reported in Table 3.1.

The components present in plant crude extracts contain phytochemicals/secondary metabolites consisting of phenolic acid, flavonoids, alkaloids and terpenoids responsible for the reduction of metal ions. Likewise, the proteins, amino acids, enzymes, polysaccharides and organic acids serve as a stabilizing material around the nanoparticles (Dauthal & Mukhopadhyay, 2016; Matussin *et al.*, 2020; Verma & Bharadvaja, 2021; Vanti *et al.*, 2020). The role of phenolics, flavonoids and terpenoids is elucidated by the phenylpropanoid/shikimate pathway responsible for the acetate/malonate pathway Shojaei Yeganeh *et al.* (2019), in which *p*-coumaroyl-coenzyme A is produced by a phenylpropanoid alleyway utilizing three units of malonyl-coenzyme A residues (initiated from the malonic acid pathway) and isopentenyl diphosphate (produced by the mevalonate and methylerythritol phosphate pathway, respectively) (Logeswari *et al.*, 2015; Ovais *et al.*, 2018a, 2018b).

In-vivo synthesis of nanoparticles was carried out in 2002 by Gardea-Torresdey, by using the plant alpha alpha Gardea-Torresdey *et al.* (2002), by growing it in jars with a medium containing gold $AuCl_2$, further converting them into zero valence gold nanoparticles of 4 nm with icosahedron structure. In 2007, Sharma *et al.* reported intracellular formation and growth of spherical gold nanoparticles of size 6–20 nm during the in vitro growth of seedlings in a gold solution. Over the years, the synthesis

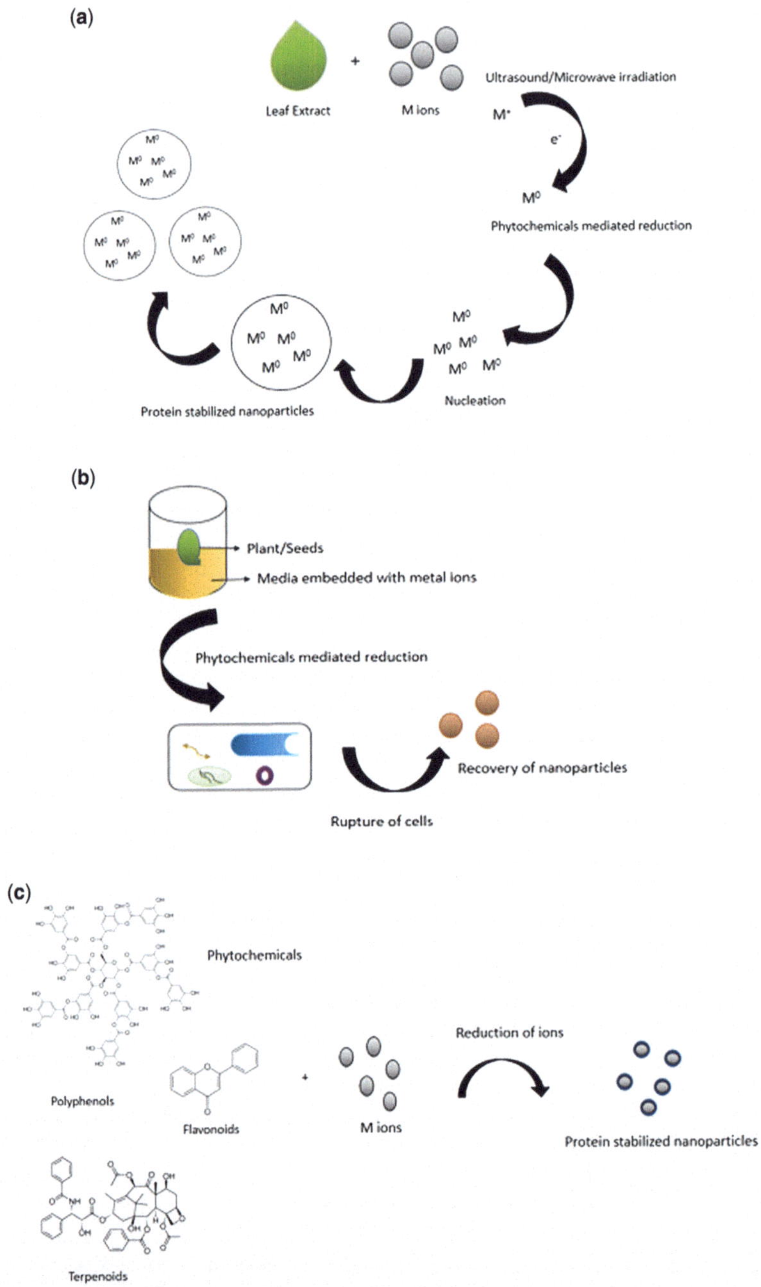

Figure 3.1 Pictorial representation of the mechanism for (a) in-vitro, (b) in-vivo, and (c) phytochemical nanoparticles synthesis from plants.

Table 3.1 Plant-mediated synthesis of nanoparticles.

Plant Species	Part of Plant	Nanoparticles	Category	Morphology	Size (nm)	UV–Vis (nm)	Experimental Conditions	Plant Metabolites Involved in Bioreduction	References
Mentha piperita	Leaf extract	Cobalt aluminate	Semiconductor	Rhombic and rectangular	35	200–1800	Heating (10 min)	Phenolic acids, menthol and flavonoids	Gingasu *et al.* (2021)
Saponaria officinalis	Roots	ZnO	Semiconductor	Quasi-spherical to flower-like	–	400	Microwave-assisted hydrothermal method (5 min at 150°C)	Glycoside	Tănase *et al.* (2021)
Leea grandifolia	Leaf extract	ZnO	Semiconductor	Nanorods	20–200		Heating (80°C for 3 h)	Phytochemical compounds	Ali *et al.* (2021)
Catharanthus roseus	Leaf extract	Silver and copper oxide	Metallic and semiconductor	Spherical	58.4–97.4 and 70.1 –99.3	300–700 and 200–700	Incubater shaker (45°C for 24 h)	Phytochemicals	Verma and Bharadvaja (2021)
Gomphrena globosa and *Gomphrena serrata*	Leaf extract	Copper oxide	Semiconductor	Rods and spheres	345 and 380	200–800	Heating (80°C for 60 min)	Phyto-compounds	Chandrasekar *et al.* (2021)
Salvadora persica	Root extract	Silver	Metallic	Spherical	37.5	420	Heating (70°C for 24 h)	Phenolics, proteins or nitrogenous compounds	Arshad *et al.* (2021)
Aloe vera and flax	Leaf extract and seed extract	Iron oxide	Superpara magnetic	Spheres	<50 nm	250 and 230	Co-precipitation	Phenolics and phenolic glycosides	Rahmani *et al.* (2020)
Drosera indica, *Drosera binata*, *Drosera spatulata*, and *Dionaea muscipula*	Plant extract	Silver	Metallic	Spherical	5–10	420	Microwave method (2 h at 70°C)	–OH and the C = O groups of appropriate polyphenolic compounds	Banasiuk *et al.* (2020)
Amaranthus blitum	Leaf extract	Silver ferrite	Magnetic	Spherical	–	356	Co-precipitation method	Phytochemical compounds	Muthukumar *et al.* (2020)
Elaeis guineensis	EGL powder	Gold	Metallic	Spherical	16.26 ± 5.84	500–550	Heating (30–90°C for 10 min)	Polyphenolic compounds	Ahmad *et al.* (2020)

(Colntinued)

Table 3.1 Plant-mediated synthesis of nanoparticles (*Continued*).

Plant Species	Part of Plant	Nanoparticles	Category	Morphology	Size (nm)	UV–Vis (nm)	Experimental Conditions	Plant Metabolites Involved in Bioreduction	References
Stevia rebaudiana Bertoni	Seed	Iron oxide@ glycoside	Core@shell	Spherical	25	–	Hydrothermal method (13 h at 170°C)	Glycoside	Khatami *et al.* (2019)
Impatiens Balsamina and *Lantana camara*	Aqueous extracts of leaves	Silver	Metallic	Spherical	24	450–420 and 450–440	Heating (5 h at 60°C)	Proteins and secondary metabolites	Aritonang *et al.* (2019)
Matricaria chamomilla, Olea europaea and *Lycopersicon esculentum*	Chamomile flower, Olive leave and Red tomato fruit	Zinc oxide	Semiconductor	Cubic structures	41.0, 51.2 and 51.6	384, 380, and 386	Heating (4 h at 60–70°C)	Polyphenols such as the tannins, glycosides, and flavonoids	Ogunyemi *et al.* (2019)
Brassica oleracea	Aqueous extracts of leaves	Zinc oxide	Semiconductor	Spherical	14 and 17	303 and 320	Heating (70°C for 20 min)	Hydroxyl group of phenol overlapped with the N-H of the amines C = C group of the aromatic conjugates of the biomolecules	Osuntokun *et al.* (2019)
Andrographis paniculata, Phyllanthus niruri, and Tinospora cordifolia	Stem and leaves	Silver	Metallic	Spherical	70–95,70–120 and 50–70	425, 424 and 452	Stirring (incubated at room temperature for 24 h)	Alcoholic/amino/ aldehydic groups	Sharma *et al.* (2019a, 2019b, 2019c)
Fritillaria	Flower plant extract	Silver	Metallic	Spherical	10	430	Stirring (incubated at room temperature for 24 h)	Polyphenols and triterpenes/C = O, C-N, C-H and O-H corresponds to hydroxyl, amid and carbonyl groups	Hemmati *et al.* (2019)
Salvia spinosa	Plant extract	Silver	Metallic	Spherical	19–125	450	Stirring (27°C for 6 h)	Phenols and alcohols with free OH group NH and C = C in aromatic compounds and proteins.	Pirtarighat *et al.* (2019)

Parkia speciosa	Leaf aqueous extract	Silver	Metallic	Spherical	35	410.5	Heating (pH 11.0, 60°C and 2 min reaction time)	–OH and –NH_2 groups of flavonoids and phenolic compounds	Ravichandran *et al.* (2019)
Origanum vulgare	Plant extract	Silver	Metallic	Spherical	2–25	430	Stirring (90°C for 120 min)	Phytomolecules Oxygen comprising functional groups	Shaik *et al.* (2018)
Cleome viscosa	Fruit extract	Silver	Metallic	Spherical	20–50	410–430	Stirring (incubated at room temperature for 24 h)	C–H stretching proteins by the amine or amide I band OH phenolic group alkaloids, phenolic compounds, amino acids carbohydrates and tannins	Lakshmanan *et al.* (2018)
Ginger	Rhizome powder	Copper-silver, copper-nickel	Bimetallic	Spherical	59 and 68	–	Precursors added to sodium borohydride for reduction of ions	–OH functional groups of shogaol, paradols and gingerol-like phenolic compounds	Ismail *et al.* (2018)
Ocimum sanctum (tulsi)	Leaf extracts	Silver	Metallic	Spherical	10–20	400–450	Precipitation method (pH 10, 35°C for 30 min)	Quercetin (molecular formula $C_{15}H_{10}O_7$) is a polyphenolic flavonoid	
Rosa damascene, Thymus vulgaris, and *Urtica dioica*	Plant extracts	Iron	Zerovalent metal	Irregular in shape	100	–	Precipitation method (pH 6, 80°C for 1 h)	C = C in alkene groups of no-saturated hydrocarbon compounds. Polyphenols proteins and organic acids	Fazlzadeh *et al.* (2017)
Passiflora caerulea	Leaf extract	Zinc oxide	Semiconductor	Spherical	30–50	380	Precipitation method (60°C for 15 min)	C = C stretch in aromatic ring and C = O stretch in polyphenols and C–N stretch of amide-I in protein.	Santhoshkumar *et al.* (2017)

of the plant-mediated nanoparticles has moved from intracellular to extracellular as the method is simple and less time consuming. Bali and Harris (2010) reported the use of the vascular plants *Brassica juncea* and *Medicago sativa* with aqueous solution of $KAuCl_4$ both in in vivo and ex vivo. This resulted in the formation of nanoparticles of 5–10 nm and 10–20 nm by the reduction of Au (III) to Au (0) Teimouri *et al.* (2018). Formation of silver nanoparticles of different shapes and sizes were found in the cell wall, cytoplasm and chloroplasts of *Brassica juncea*, *Festuca rubra* and *Medicago sativa* as per the work of Marchiol *et al.* (2014). Chandran *et al.* (2006) reported the synthesis of gold nanotriangles of dimensions 50–350 nm and spherical silver nanoparticles with a size of 15.2 nm using *Aloe vera* extract by adding aqueous $HAuCl_4$ solution and incubating it for 5 h.

3.3.2 Bacteria

Another biological approach of nanoparticles synthesis is by bacteria (Figure 3.2). Aquatic motile bacteria that align and swim along magnetic field lines were first reported by Salvatore Bellini in 1963. He referred to these microorganisms as 'batteri magnetosensibili' or magnetosensitive bacteria Fang *et al.* (2014). These organisms were later rediscovered by R. P. Blakemore, in 1975, who called them magnetotactic bacteria and elucidated the reason for their magnetotactic behavior Hamouda *et al.* (2019). Initial evidence of bacterial synthesis is found in magnetotactic bacteria where they synthesize a chain of magnetic nanoparticles inside their cells (Elfick *et al.*, 2017; Fang *et al.*, 2019; Kang *et al.*, 2008; Khan *et al.*, 2019). Magnetic nanoparticles were covalently immobilized on the surface of *Acetobacter xylinus* and the location of the bacteria was controlled to manipulate bacterial bioactivation (Patil and Kim, 2017; Sumanth *et al.*, 2020). Bacterial magnetosomes are organic-coated intracellular nanocrystals of Fe_3O_4 and Fe_3S_4, biosynthesized by both magnetotactic and non-magnetotactic bacteria Iravani and Varma (2020). The composition of the magnetic inorganic part is species-specific and the external organic coating layer is derived from the bacterial phospholipid bilayer membrane (Bharde *et al.* 2005; Bereczk-Tompa *et al.* 2017). The putative functions of the protein component of the external organic coating layer in the magnetosome biomineralization process have been hypothesized. Magnetic nanoparticles into ordered structures control the motion of magnetotactic bacteria with microelectromagnets. *Magnetospirillum magnetotacticum* (MS-1), a variety of magnetotactic bacteria, produce a single chain of intracellular magnetite (Fe_3O_4) nanoparticles López and Antuch (2020).

Bacterial species have evolved various natural defence mechanisms such as intracellular sequestration, efflux pumps, change in metal ion concentration and extracellular precipitation, to cope up with metal-induced stress conditions (Ali *et al.* 2019; Asghar *et al.* 2020; Mitra *et al.*, 2018). The transformation of metals occurs via several specific or unspecific reductases, for example, thiol-containing or glutathione reductases, nitrite reductases and a vast variation of small molecules and enzymatic activities, in different bacterial species (Chen *et al.* 2015; Chwalibog *et al.* 2010; Dahoumane *et al.* 2017; Fernández-Llamosas *et al.*, 2017; Gahlawat *et al.*, 2019). There are two common methods involved in the synthesis of nanoparticles: intracellular synthesis and extracellular synthesis. Intracellular methods involve the reduction of nanoparticles inside the cell Singh *et al.* (2018). Furthermore, nanoparticles are recovered from the cell by cell lysis methods such as sonication. In the case of extracellular synthesis, the synthesis is by using biomass followed by recovery of the nanoparticles from the supernatant. A summary on the recent advancements of the bacterial synthesis of nanoparticles is reported in Table 3.2.

3.4 FUNGI

Myconanotechnology includes synthesis of nanoparticles from microfungi, macrofungi and yeast (Figure 3.3). Their advantages over bacterial synthesis include easy culturing and fast growth, further extracellular discharge of proteins has an additional advantage in the downstream processing and treatment of biomass, when compared with the bacterial biosynthesis of nanoparticles Feroze *et al.* (2020). Enzyme components such as nathoquinones, anthraquinones and nitrate reductase act as

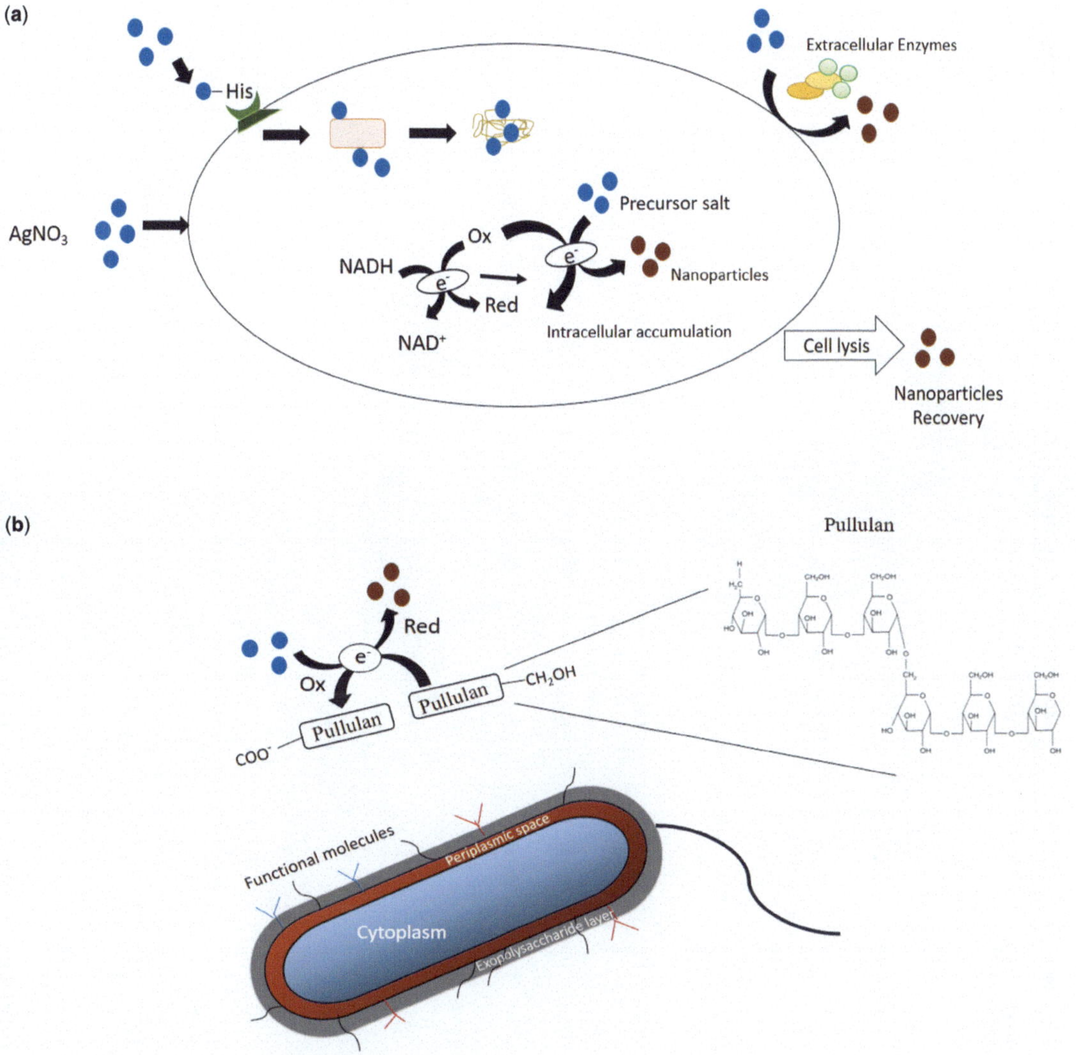

Figure 3.2 Pictorial representation of the mechanism for nanoparticle synthesis from bacteria. (a) Cellular uptake of silver ions and activation of silver reduction machinery; electron shuttle system, intra or extracellular localization of AgNPs; and reduction through extracellular enzymes and other organic molecules released in solution (Xu *et al.*, 2020) and (b) proposed mechanistic scheme of the biomineralization of gold ions by exopolysaccharide pullulan. (Source: Gahlawat and Choudhury 2019).

reducing agents in the synthesis of nanoparticles (Figure 3.3a) Subashini and Bhuvaneswari (2018). Not only enzymes, also electron shuttle compounds play a key role in the fungal formation of nanoparticles (Siddiqi & Husen, 2016).

The summary on the recent advancements of the fungal synthesis of nanoparticles is reported in Table 3.3. The most common method in case of microfungus is the use of fungal hyphae growth in the

Table 3.2 Bacterial-mediated synthesis of nanoparticles.

Bacteria	Source	Nanoparticles	Category	Morphology	Size (nm)	UV–Vis	Conditions	References
Alcaligenes aquatilis	Cell free supernatant	Ag_2O/ $AgO\text{-}TiO_2$	Nanocomposite	Spherical	39.6	One at 326 nm and a broad peak at 439 nm	50 ml of 100 mM $AgNO_3$ solution pH 11 and incubated for 24 h at 28 ± 2°C	Kulal and Shetty Kodialbail (2021)
Paenarthrobacter nicotinovorans MAHUQ-43	Soil sample/ extracellular synthesis	Silver	Metal	Spherical	13 to 27	466	1 mM $AgNO_3$ solution and incubated at 48 h at 30°C	Huq and Akter (2021)
Raoultella planticola and *Pantoea agglomerans*	Cell-free culture supernatant	Ag and AgCl	Metal/ semiconductor	Spherical	10–50	440	45 mL of 1 mM $AgNO_3$ solution and incubated at 48 h at 33°C	Ghiuta *et al.* (2021)
Lactobacillus spp	Intracellular synthesis	ZnO	Semiconductor	–	–	379	Zn^{2+} at concentration of 500 mM and incubated at 37°C for 24 h	Suba *et al.* (2021)
Nocardiopsis dassonvillei- DS013	Marine sample/ culture supernatant	Silver	Metal	Circular and non-uniform distributions	30–80	408 nm	$AgNO_3$ solution	Dhanaraj *et al.* (2020)
Bacillus sp. KFU36	Marine strain/ extra/supernatant	Silver	Metal	Spherical	5–15	430	50 mL of 1 mM $AgNO_3$ solution pH 8.5	Almalki and Khalifa (2020)
Cupriavidus sp.	Urban soil/ extracellular supernatant nitrate reductase	Silver	Metal	Spherical	10–50 nm	420	Cell supernatant 1 mM $AgNO_3$	Ameen *et al.* (2020)
Escherichia coli, *Exiguobacterium aurantiacumm*, and *Brevundimonas diminuta* MF754138, MF754139, and MF754140	Soil samples	Silver	metal	Spherical	5–50	420	LB broth and incubated at 37°C for 24 h. 1 mM $AgNO_3$ cell free supernatant	Saeed *et al.* (2020)
Halobacillus sp.	Intracellular/extra polyextremophile halophilic bacteria isolated from Atacama Salt Flat (Chile), Uyuni Salt Flat (Bolivia) and the Dead Sea (Israel)	CdS	Semiconductor	Spherical	2–5 nm	360	0.33 mM $CdCl_2$. Then cultures were grown at 28°C for 24 h. Cell pellets and supernatants	Bruna *et al.* (2019)
Magnetospirillum gryphiswaldense MSR-1	Intracellular proteins (MamCHAF) Magnetosomes	Nd–Fe–B magnets	Bimetallic	Spherical	35–42 nm	-	Culture medium. Ferric citrate was added to the fermentation medium at a final concentration of 30 μM	Zhang *et al.* (2019)

Azospirillum brasilense Sp245	Intra/culture	Selenium	Metal	Spherical	40–200 nm	-	Aqueous 1 mM Na_2SeO_3/ Na_2SiO_3 solution and a cell-free filtrate of the culture liquid with an intracellular bacterial extract	Vetchinkina *et al.* (2019)
	Extra/cell-free filtrate			nanowires	width of 20–150 nm 10 μm			
	Cell-free filtrate after autoclaving	Silicon		Nanospheres	5–15 nm to 250 nm			
Exopolysaccharides (EPS) from the probiotic strain Bacillus licheniformis Dahb1	Exopolysaccharides EPS-coated zinc oxide nanoparticle (EPS-ZnONPs)	ZnO	Semiconductor	Hexagonal	10–100 nm	375 nm	0.5 g of EPS crude powder was added to the aqueous solution (0.02 M) of zinc acetate and NaOH (2 M) under vigorous stirring at 37°C.	Abinaya *et al.* (2018)
Bacillus brevis (NCIM 2533)	Extracellular	Silver	Metal	Spherical	41–68 nm	420 nm	1 mM $AgNO_3$ was added into 100 mL of cell filtrate, agitated at 200 rpm for 24 h, 37°C	Saravanan *et al.* (2018)
Salmonella	External shell surface of MCP as a scaffold for the fabrication of gold nanoparticles displayed in 3D protein scaffolded gold nanoparticle shell enclosing an active enzyme cluster	Gold	Metal	Quantum dots	Size below 3 nm	540 nm	MCP and $AuCl_3$ more than 0.1 mg and 250 μM, respectively, leads to MCP-AuNPs aggregate formation	Bari *et al.* (2018)
Shewanella oneidensis	Anaerobic bioreductive deposition	Pd	Metal	Spherical	10 nm	400 nm	Aqueous Na_2PdCl_4 was added to a concentration of 400 μM and the serum bottles were incubated for 3 h at 30°C bioreductive convention of $PdCl_4{}^{2-}$ into the Pd NPs	Xiong *et al.* (2018)
Leuconostoc lactis	Bacterial exopolysaccharide	Silver	Metal	Spherical	35 nm	425 nm	Cell suspension was heated to 100°C for 10 min to inactivate the enzymes. EPS (10 mg) was dissolved in 10 mL of Milli Q water to form a uniform dispersion and 9 mM $AgNO_3$	Saravanan *et al.* (2017)
Vibrio natriegens	Aerobic/ intracellular SefA (selenium factor A) protein genetically modified	Selenium	Metal	Spherical	100–400 nm	500	20 ml of cell culture grown at 30°C on LB 1 mM selenite	Fernández-Llamosas *et al.* (2017)

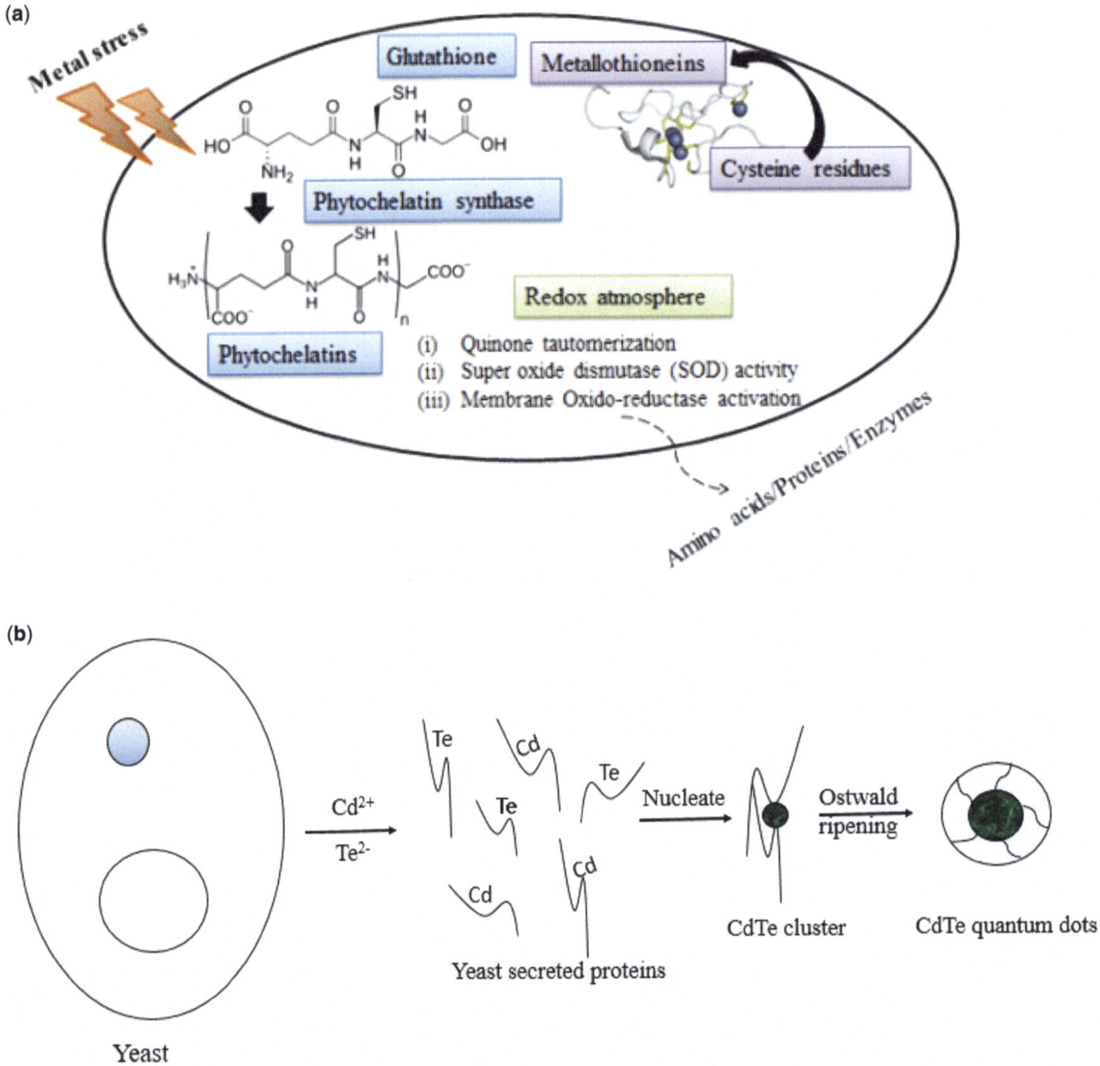

Figure 3.3 Pictorial representation of the mechanism of nanoparticle synthesis by fungi. (a) Fungal phytochelatins synthesis and (b) biosynthesis of protein-capped CdTe quantum dots in yeast.

liquid media incubated on a shaker for 24–48 h. After attaining growth they are washed with distilled water to remove growth media and added to the solution containing the metal precursor. Based on the color change after 2–3 days, the formation of nanoparticles is known (Meena *et al.*, 2021; Sumanth *et al.*, 2020). Fungal phenol oxidases such as laccases, tyrosinases, and Mn-peroxidases have also been reported in *Lentinus edodes*, *Pleurotus ostreatus*, *Ganoderma lucidum* and *Grifola frondosa*. For macrofungus, both the fruiting body and the mycelium of the mushrooms (basidiocarps) are subjected to blending or crushing of the cleaned fruit bodies. Then suspending the resulting paste in a known volume of distilled water, the resultant filtrate/supernatant can be utilized for generation of nanoparticles by supplementation of a metal salt solution followed by incubation. In 2004,

Numata *et al.* produced nanofibers from the purified polysaccharides (β-1,3-glucan), the so-called *Schizophyllan*, from the mushroom *Schizophyllum* commune. In case of yeast, eukaryotic systems namely, *Candida glabrata* and *Schizosaccharomyces pombe* have been employed for nanoparticles production (Table 3.3) Eugenio *et al.* (2016). Moreover, intracellular synthesis of sulfide nanoparticles by *Schizosaccharomyces pombe* has also been reported.

3.5 ALGAE

Algae are known to hyperaccumulate heavy metal ions and possess an exceptional capability to remodel them into more malleable forms (Mukherjee *et al.*, 2021). Because of these alluring attributes, algae have been foreseen as model organisms for fabricating bio-nanomaterials (AlNadhari *et al.* 2021; Khanna *et al.*, 2019). Intracellular biosynthesis requires no pre-treatment of microalgae because the process relies on metabolic pathways responsible for synthesis such as photosynthesis, respiration and nitrogen fixation Yılmaz Öztürk *et al.* (2020). The reducing agents may be NADPH or NADPH-dependent reductases in the energy generating steps during photosynthesis *via* the electron transport system (ETS) or may be respiratory ETS at thylakoid membranes. Recent advancements of the algal synthesis of nanoparticles are reported in Table 3.4.

From the available reports, extracellular algae-mediated synthesis of nanoparticles involves preparation of algal extract, metal precursor solution and incubation of the algal extract with the metal precursor solution (Figure 3.4). The algal biomass is subjected to rudimentary pre-treatments such as washing and blending, followed by mixing the liquid algal extract with the molar solution of the metal precursor. Typically, the color change of the reaction mixture demarcates as a visible signature for the initiation of reaction, which involves nucleation, followed by growth of nanoparticles in which the adjoining nucleonic particles aggregate together, thus forming thermodynamically stable nanoparticles of different morphologies. Algal extracts consist of carbohydrates, proteins, minerals, oil, fats and polyunsaturated fatty acids along with bioactive compounds such as antioxidants (polyphenols and tocopherols) and pigments such as carotenoids (carotene and xanthophyll), chlorophylls, and phycobilins (phycocyanin and phycoerythrin) Chen *et al.* (2019b). They further comprise metabolites, ions, pigments, various proteins (enzymes) and non-protein entities such as DNA, RNA, microbial by-products (hormones and antioxidants) and lipids.

Diatoms, single-celled photosynthetic algae, are a great source of intricate natural porous silica materials. The whole diatom cell is confined within a three-dimensional ordered nanopatterned silica cell wall, called a frustule Bozarth *et al.* (2009). Diatoms are robust and can inhabit virtually all photic zones from the equator to the seemingly inhospitable sea ice where they are highly useful indicators of environmental conditions in their rapid response to environmental changes, including their capacity to react to sea ice freezing around them with their natural antifreeze ice-binding proteins (Jamali *et al.*, 2012).

3.6 PHOTOCATALYSIS

Photocatalysis is a light-driven process. When the nanoparticles are exposed to a light source, the electrons jump from the valance band to the conduction band creating electron–hole pairs (Li *et al.*, 2021). The confinement of an electron and hole in nanocrystals significantly depends on the material properties, namely on the Bohr radius a_B (Sahoo, 2019). When the radius of a microcrystalline semiconductor (R) is reduced relative to the Bohr radius (a_B), one of the most important consequences of the spatial confinement effect is a blue shift in the energy of the band-to-band excitation peaks. This is because of the main electron–hole interaction energy (which includes the Coulomb term), the confinement energy of the electron and hole, and kinetic energy changes in different regimes of the quantum confinement (Koole *et al.*, 2014; Mirza *et al.*, 2019). The surface of a semiconductor is unstable and has dangling bonds, which can trap charge carriers and degrade the electrical and optical

Table 3.3 Fungal-mediated synthesis of nanoparticles.

Fungi	Source	Nanoparticles	Category	Morphology	Size (nm)	UV–Vis	Conditions	References
Colletotrichum plurivorum	Endophytic fungus from a *Citrus pseudolimon* plant	Silver oxide	Metal oxide	Cubic	200–250 nm in length and 80–150 nm in width	560 nm	0.1–1.0 mL of the endophytic extract was mixed with 0.3 mL of metal precursors, that is, $AgNO_3$ solution (1 mM) and $HAuCl_4$ solution (1 mM) and incubated at 25°C for 16 h.	Kumar *et al.* (2021)
Acremonium potronii	Jubail coast in the Eastern province on the Arabian Gulf coast of Saudi Arabia	ZnO	Metal oxide	Spherical	13 and 15	320 and 350 nm	100 ml of the fungal filtrate and zinc acetate hexahydrate were mixed and incubated at 35°C in orbital shaker	Ameen *et al.* (2021)
Ganoderma enigmaticum and *Trametes ljubarskyi*	White-rot fungi, in the form of fruiting bodies	Silver	Metal	–	–	–	1 mm $AgNO_3$ aerobically in liquid medium and incubated at 30°C	Krishna *et al.* (2021)
Penicillium notatum, P. purpurogenum and *P. aculeatum*	-	Zirconium	Magnetic	Spherical	Size below 100 nm	–	50 mL of each supernatant was added to 50 mL $ZrCl_4$ solution	Golnaraghi-Ghomi *et al.* (2021)
Trichoderma sp.	*Bertholletia excelsa* (Brazil-nut) seeds and the soil	Silver	Metal	Spherical	200 nm	420 nm	$AgNO_3$ solution at 1.0 mmol/L and pH 5 over 9 days	Ramos *et al.* (2020)
Aspergillus niger BSC-1	Isolated from sediments sample of Bali Island, Sundarban Mangrove Biosphere, India	Iron oxide	Metal oxide	Nanoflakes	20–40 nm	296 nm	Cell free filtrate added to iron precursor salt mixture incubated for 3 h at 28°C at 120 rpm	Chatterjee *et al.* (2020)
Periconium sp.	*Balanites aegyptica* plant leaves endophyte	ZnO	Metal oxide	Spherical	16–78 nm	360	20 g dried fungal biomass boiled at 70°C for 2 h	Ganesan *et al.* (2020)
Reishi mushroom (*Ganoderma lucidum*)	-	Ag	Metal	Spherical	15–22 nm	400–460 nm	20 mL of Reishi mushroom extract 15 mg of $AgNO_3$ was added to the mixture and stirred for reduction of ions – microwave-assisted process	Aygün *et al.* (2020)
Dry baker's yeast	AB/MAURI Co., Ltd	Silver	Metal	Spherical	13.8	418	Cell-free yeast extract was collected for the biosynthesis of Ag NPs at pH 10 by centrifugation at 2000 rpm for 5 min	Shu *et al.* (2020)

Pleurotus tuber-regium, Polyporus rhinoceros, Coriolus versicolor, Ganoderma lucidum	-	Selenium	Metal	Spherical	91–102 nm	–	Polysaccharides stock solution of different mushrooms were mixed with sodium selenite solution	Zeng *et al.* (2019)
Trichoderma longibrachiatum	Fungal isolates were collected from plants from around Alryadah city, Saudi Arabia	Silver	Metal	Spherical	10 nm	385 nm	10 g fungal biomass, a reaction temperature of 28°C, a 72 h incubation time	Elamawi *et al.* (2018)
Penicillium polonicum	Marine endophytic fungus algal samples *Chetomorpha antennina*	Silver	Metal	Spherical	10–15 nm	430 nm	Fungal cell filtrate was added to $AgNO_3$ solution at 200 rpm	Neethu *et al.* (2018)
Rhizomucor pusillus, Sporotrichum thermophile, Thermoascus thermophilus, Thermomyces lanuginosus	Thermophilic fungi	Gold	Metal	Spherical	6–40 nm	555	Secreted biomolecules autolysate intracellular biomolecules extracellular liquid	Molnár *et al.* (2018)
Cladosporium cladosporioides	Seaweed *Sargassum wightii* was collected from the coastal region of Kanyakumari	Gold	Metal	Spherical	100 nm	540 nm	Mycelial biomass of *C. cladosporioides* grown 48 h aqueous solution of 1 mM $HAuCl_4$ with the fungal suspension filtrate	

Table 3.4 Algae-mediated synthesis of nanoparticles.

Algae	Source	Nanoparticles	Category	Morphology	Size (nm)	UV–Vis	Conditions	References
Padina boryana	Brown seaweed of *P. boryana* was collected from the Saudi coastal region of the Arabian Gulf	Palladium	Metal	Pleomorphic	8.7 nm 11.16 nm	293	Pd-NPs prepared by facile one-pot fabrication method using green *Padina boryana* extract.	Sonbol *et al.* (2021)
Avrainvillea amadelpha and *Codium edule*	*A. amadelpha* were collected from the Salman Gulf while *C. edule* was collected from the Kharar region Red Sea coast in Saudi Arabia	Silver	Metal	Spherical	–	400 nm and 420 nm	Algae extracts 1 mL of each extract and 1 mL of silver nitrate $AgNO_3$ 10^{-3} M to 97.5 mL of distilled water	Mahyoub (2021)
Sargassum longifolium	Seaweed *S. longifolium* was obtained from the coastal environment of the Gulf of Mannar, South India	Copper oxide	Metal oxide	Irregular shape	40–60 nm	350 nm	$CuSO_4$ solution is added to algal extract of 10 mL. After 6 h color change is observed	Rajeshkumar *et al.* (2021)
Cystoseira myrica, *Sargassum latifolium* and *Padina australis*	Brown algae extracts collected from the Bushehr coast Persian Gulf	Cuprous oxide	Metal oxide	Irregular shapes	12 and 26 nm	475 nm and 491 nm	1 g $CuSO_4 5H_2O$, 5 g sodium citrate and 2.5 g of sodium carbonate in 200 mL distilled water. 6 mL of algal extract was added dropwise to it under stirring at 100°C	Taherzadeh Soureshjani *et al.* (2021)
		$CaCO_3$		Spindle-shaped	1–2 μm in length and 300–500 nm in width		Calcinated dolomite fine powder was dissolved in water for hydration, was mixed with aqueous solution of magnesium chloride at 80°C and then CO_2 was bubbled for 3 h	
Gelidium corneum	Marine red alga *G. corneum* was collected from the upper infralittoral of Hatay coasts (Mediterranean coasts of Turkey)	Silver	Metal	Spherical	20–50 nm	420–430 nm	5 g of dried algae powder extract added to silver nitrate due to reduction reddish to dark brown indicating silver nanoparticles formation	
Portieria hornemannii	Red algae (*P. hornemanni*) were collected from Gulf of Mannar	Silver	Metal	Spherical	60–70 nm	418	1 mM silver nitrate concentration and 5 mL of algal extract continuous stirring 24–48 h the color changes to dark brown	Fatima *et al.* (2020)
Stephanopyxis turris	*S. turris* was isolated from the North Sea	Gold	Metal	Spherical	19 nm, 31 nm, 51 nm, 75 nm, 125 nm	-	5 mM gold precursor added to extract along with ascorbic acid and trisodium citrate boiling temperature for 30 min	Pytlik *et al.* (2019)

Botryococcus braunii	Green algae *B. braunii* was isolated from Udaisagar Lake, Udaipur, Rajasthan, India	Silver	Metal	Cubical, spherical and truncated triangular	2–100 nm	490 nm	5 ml algal extract mixed with 1 mM silver nitrate aqueous solution and put on magnetic stirrer at room temperature for 3 h	Arya *et al.* (2019)
Trichodesmium erythraeum	Dense *T. erythraeum* bloom was observed at Tuticorin harbor southeast coast of India	Silver	Metal	–	26.5 nm	438	10 g of dried *P. oedogonia* is boiled for 15 min 60°C aqueous 1 mM silver nitate solution brownish yellow solution	Sathishkumar *et al.* (2019)
Padina pavonia	The brown alga *Padina pavonia* was collected from marine water seashore, Umluj coast of Saudi Arabia	Silver	Metal	Spherical, triangular, rectangle, polyhedral and hexagonal	49.58–86.37 nm	401 nm	1 g of *P. pavonia* dry matter 10^{-3} M silver nitrate aqueous solution within 3 h stirring under light at room temperature	Abdel-Raouf *et al.* (2019)
Sargassum muticum	-	Zinc oxide	Metal oxide	–	–	–	*S. muticum* algae extract is mixed with 2 mM of zinc acetate di-hydrate solution and stirred at 70°C for 3–4 h	Sanaeimehr *et al.* (2018)
Botryococcus braunii	Green alga *Botryococcus braunii* was collected from Udaisagar Lake Udaipur Rajasthan, India	Copper and silver	Metal	–	2–100 nm	490 and 258 nm	5 mL algal extract was added to 1 mM aqueous copper acetate/silver nitrate in with vigorous stirring at 100°C for 24 h.	Arya *et al.* (2018)
Navicula	Diatom species *Navicula* is collected from local fresh water bodies of Assam, India	Silver	Metal	–	70–80 nm	200–260 nm	10 mM silver nitrate solution and the diatom frustules are added to precursor solution in the presence of light to initiate the reaction at room temperature	Chetia *et al.* (2017)
Nitzschia	Water samples were obtained from river and pond biofilm near Jalgoan, India	Gold	Metal	Spherical	15–90 nm	529 nm	Diatoms were centrifuged at 1042 g pellet after incubation, 0.1 mM gold chloride salt was added and further incubated for 24 h.	Borase *et al.* (2017)
Caulerpa serrulata	*C. serrulata* was collected from the Red Sea coast, Egypt	Silver	Metal	Spherical	10±2 nm	412 nm	*C. serrulata* extract 5–25 mL was added to 10^{-3} M of silver nitrate solution with color change from yellow to reddish brown	Aboelfetoh *et al.* (2017)
Stephanopyxis turris	*S. turris* was isolated from the North Sea	Gold	Metal	Spherical	30–10 nm	100 nm	*S. turris* was incubated in gold precursor containing artificial sea water for 1–3 days	Pytlik *et al.* (2017)

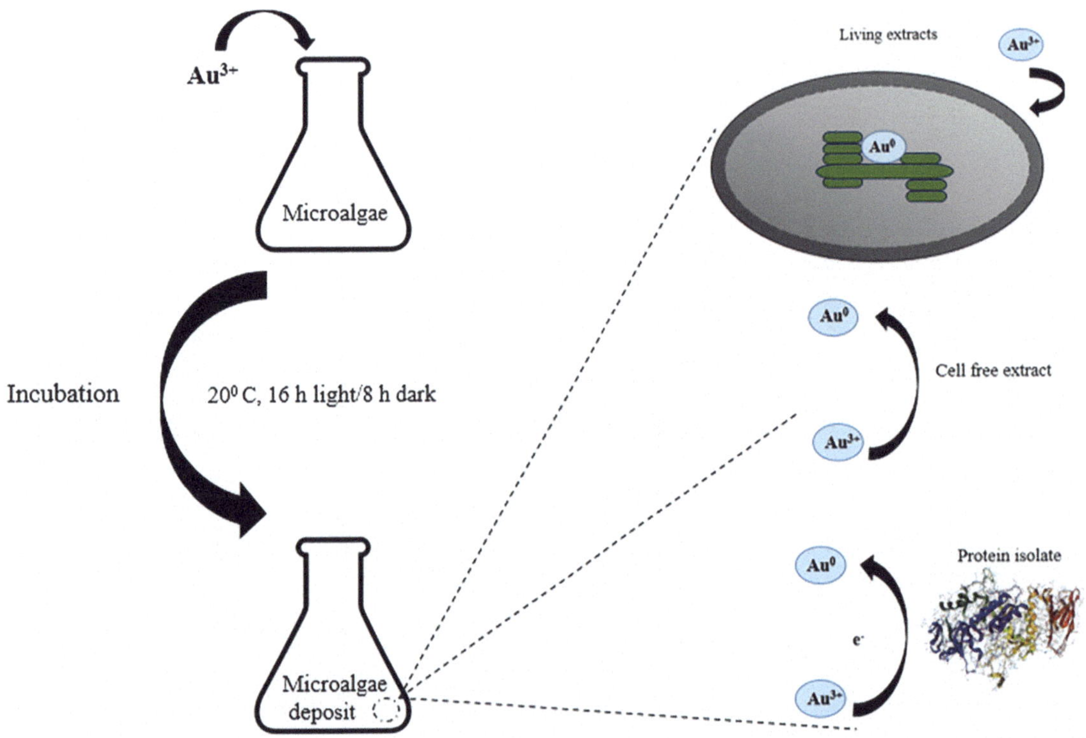

Figure 3.4 Experimental procedure for the biosynthesis of Au nanoparticles using a living culture of a microalga or its cell free extract or protein isolate. (Source: Shen *et al.* 2012).

properties of the material. To reduce the surface energy, the surface atoms rearrange themselves, forming a new structure called a reconstructed surface. This reconstruction can lead to the formation of energy levels in the forbidden gap of the semiconductor, which can also act as traps for charge carriers. In an ideal semiconductor nanocrystal, the surface atoms are bonded to other materials in a way that removes these defects, resulting in a high-quality material with superior electrical and optical properties (Pradeep, 2007).

The efficiency of the nanomaterials depends on the morphology and the band gap of the material. In 1839, Becquerel discovered that when a silver chloride electrode was immersed in an electrolyte solution and connected to a counter electrode, a voltage and an electric current were produced when the electrode was illuminated with sunlight. As early as 1929, it was known that the pigment titanium white (TiO_2) was responsible for fading in paints and several major studies were carried out on the photosensitizing action of TiO_2 (Kansal, 2013). Later in 1955, Brattain and Garret explained the Becquerel effect on germanium semiconductor electrodes, which is the origin of this photovoltaic phenomenon. According to this phenomenon, the semiconductor zinc oxide had attracted a great deal of attention as a photosensitizer for the decomposition of organic compounds and as a sensitizer for a number of inorganic photoreactions. Fujishima and Honda in 1972 reported the capability of photocatalytic properties of certain materials to convert solar energy into chemical energy to oxidize or reduce materials to obtain useful materials including hydrogen/hydrocarbons and to remove/degrade pollutants.

One of the parameters influencing the photocatalysis phenomenon is the morphology of the nanoparticles as it plays a key role in the light absorption based on localized surface plasmon resonance

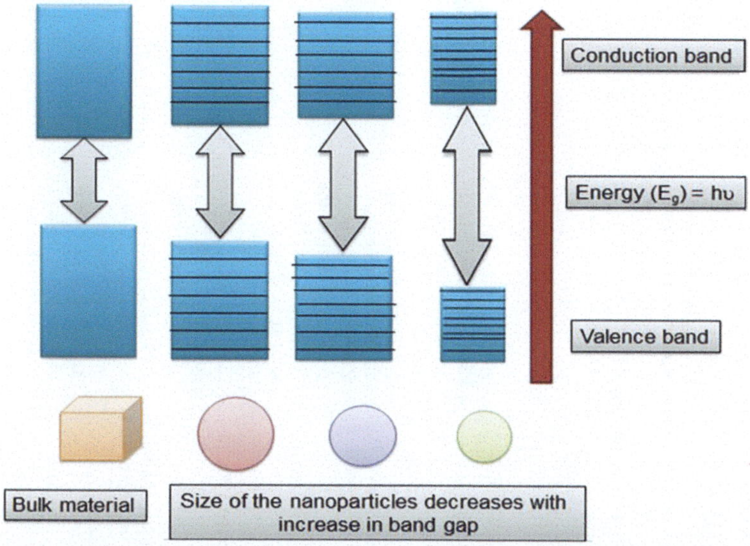

Figure 3.5 Schematic representation of decreasing size with the increasing band gap.

(LSPR). The localized surface plasmon resonance effect is the resonant photon-induced coherent oscillation of charge at the metal–dielectric interface, established when the photon frequency matches the natural frequency of metal surface electrons oscillating against the restoring force of their positive nuclei. The other important parameter in photocatalysis is the photonic band gap of the materials, at nanosize material is reduced in size to the atomic level, the energy band is broken into discrete levels so that the band gap is changed (Figure 3.5). The electrons are confined to a point in space that they have no freedom in any dimension and electrons are said to be localized at a point, implying that a change in all directions changes the properties. Thus, by altering the size and composition of materials, we can change the band gap energy, for example, the color of light emitted can be changed because emission color depends on band gap energy of the material. This implies that the absorption and emission wavelength both become shorter when they have larger band gap energy and vice versa. This is due to the fact that the specific wavelength of light emitted is a function of the band gap energy (Koole *et al.*, 2014).

3.6.1 Batch degradation of organic pollutants using NPs

In a typical photoreactor, the system is equipped with illumination, which fundamentally affects the design of any photocatalytic reactor. As a rule of thumb, indicatively the best choices are the usage of a linear fluorescent/light-emitting diodes (LED)/ultraviolet (UV)/visible lamp with the reactor on the z-axis. In practice, once the light source is chosen, the reactor must be designed to get the maximum utilization of both the emitted light and the volume of the reactor. Both depend on the capacity of light absorption of the system, which in a reactor (i.e., with suspended photocatalytic powders) is given by the photocatalyst concentration (Dolat *et al.*, 2012). The radiation inside the photocatalytic system depends on the power and geometry of the irradiation source, the distance to the photocatalytic system and the thickness of the suspension layer. These should be constant over all the experiments to get results with significance. If one increases the photocatalyst concentration, a high absorption of the light is obtained. Thus, the amount of utilized radiant energy increases, but the system is nevertheless less effective since large parts of the reactor work at lower level of illumination. This means that any design represents a compromise between conflicting requirements and the compromise choice ultimately is affected by economic evaluations, which depend on the cost of the radiant energy and

Table 3.5 Overview of NPs usage for photocatalytic degradation of pollutants.

Source	Nanoparticles	Reactor Type	Target Pollutant	Light Illumination/ Band Gap	Degradation Efficiency	Parameters Analyzed	References
Sol–gel method	Immobilized-TiO_2 NPs borosilicate glass (BG)	Photocatalytic reaction	Methyl orange	Solar	97.8%	Isopropoxide: C_2H_5OH:H_2O = 1:7.14:4 were mixed in a molar ratio and stirred at 600 rpm on a magnetic stirrer for 180 min at room temperature followed by aging for about 12 h to obtain the gel.	Al-mamun *et al.* (2021)
	Immobilized-TiO_2 NPs cement coated borosilicate glass (CCBG)				88.6%		
	Immobilized-TiO_2 NPs steel wire mesh (SWM)				21.5%		
In-situ ion exchange method	Fe_3O_4/ZnO/ ZnSe	Magnetically fixed-bed reactor	Rhodamine B	Solar	97.68% after 80 min	30 mmol of $NaBH_4$ was dissolved in 60 mL of deionized water and then 3 mmol of Se powder and stirred for 10 min. After Fe_3O_4/ ZnO was placed in the above solution, the resulting mixture was kept at 55°C for 1, 2 and 3 h	Xu *et al.* (2020)
Ultrasonic method	Gold	Membrane reactor	Methylene blue and 4-nitrophenol	Sunlight irradiation	Above 85%	20 mL $HAuCl_4$ solution (1 mM) for 3 h ultrasonic for 30 min	Yu *et al.* (2021)
Co-reduction	Fe/Ni-NPs on silica, bentonite and chitosan beads	Fluidized-bed reactors	Direct Red 80	Visible	Higher than 60% at the end of 22 h	9.85 g of iron(II) sulfate heptahydrate, 0.996 g of nickel(II) nitrate hexahydrate and 4.6 g of the support (silica or bentonite or chitosan beads) were added to 50 mL of an ethanol-deionized water solution (4:1, v/v) in a 250-mL Erlenmeyer flask	Puiatti *et al.* (2020)

Hydrothermal method	TiO_2	Membrane bioreactor/ hollow fiber UF membrane	Petroleum refinery effluent	Low-pressure mercury vapor lamp (6 W)/5.5×10^{-5} Einstein/min	TOC and TN removals were 32% and 67%	Reaction time of 90 min, pH 10 and catalyst dose of 100 mg/L Electrical energy per order (EE_0) was 356.29 kWh/m^3/order	Oliveira *et al.* (2020)
Shanghai Macklin Biochemical Company of China	Fe_3O_4 nanoparticle-immobilized *Pseudomonas* sp.	Batch	Di-n-butyl phthalate/ plasticizers	–	99.69%	Polydopamine films – Fe_3O_4 iron nanoparticles, suspended in liquid mineral salt medium (1 mg/mL), were added to the bacterial solution at a volume ratio of 1:10 and then immobilized for 2 h at 4°C and 150 rpm	Wang *et al.* (2020)
Co-reduction	Fe/Ni	Fluidized-bed reactor	Direct Red 80 dye	Visible	Higher than 60% at the end of 22 h	10 mg/L DR80 solution, initially contained in a cylindrical tank 7.5 g of CB-Fe/Ni flow rate of 5.7 mL/s bed height of 11.5 cm	Puiatti *et al.* (2020)
Chemical and laser-assisted reduction technique	Au-Ag and Ag-Au	Membrane reactor	Ofloxacin and methylene blue	405 nm laser 532 nm laser Xenon	71% Au-Ag 75% Ag-Au 65% and 59% Au-Ag and Ag-Au	30 mL of aqueous solutions containing OFX with the concentration range of 5–25 ppm or MB with the concentration range of 5–15 ppm along with 0.3–1.7 g/L of each photocatalyst	Amoli-Diva *et al.* (2020)
Wet impregnation method followed by hydrothermal treatment	TiO_2 NPs/ activated carbon fiber	Continuous flow designed photochemical reactor.	Methylene blue	UV-C	75%	50 mg of TiO_2/ ACF were dispersed with methylene blue solutions and continuously stirred at 300 rpm	Jain *et al.* (2019)
Cyanobacteria (*Microchaete NCCU-342*)	Silver	Batch	Azo dye methyl red	UV-light/2000 ± 200 lux	84.60%	Biomass quantity of 80 μg/mL, pH 5.5, 60°C temperature, duration of 60 min UV light exposure and 1 mM $AgNO_3$ concentration	Husain *et al.* (2019)

(*Continued*)

Table 3.5 Overview of NPs usage for photocatalytic degradation of pollutants (*Continued*).

Source	Nanoparticles	Reactor Type	Target Pollutant	Light Illumination/ Band Gap	Degradation Efficiency	Parameters Analyzed	References
Sol–gel and Hummers method	Fe-doped ZnO/rGO nanocomposite	Photocatalytic membrane reactor	Congo red	Solar	58–87%	Feed solution was prepared by mixing fixed amount of photocatalyst (1 g/L) in CR (100 mg/L) solution with working volume 8 L and feed flow rate 12 L/h	Ong *et al.* (2019)
–	Nano MgO_2 and *Rhodococcus zopfii*	Small bioreactor platform capsules	17α-Ethynylestradiol	UV-light	100% in 25 min	MgO_2 (25 mg) was added to 3 ppm EE2 in 10 mM phosphate buffer and exposed to low pressure -UV irradiation for 5, 15, and 25 min	Vaddadi *et al.* (2020)
pH-mediated chemical precipitation reaction method	Ag_2O/P-25	Membrane Bioreactor/ nylon membranes	Carbamazepine and Diclofenac (pharmaceutically active compounds)	UV-light	89.10% and 93.5%	180 min of UV irradiation were achieved at catalyst dosage of 0.4 g/L	Gurung *et al.* (2019)
Sol–gel method	Zinc oxide	Three-phase inverse fluidized-bed bioreactor	Textile effluent	–	Maximum COD to 47 ppm (97.24%)	10% inoculum size with low gas velocity 0.0027 m/s and static bed height of 2.43 cm for 280 nm sized ZnO nanoparticles	Sur and Mukhopadhyay (2019)
Sonochemical-combined thermal synthesis method	$CuO/Cu_2O/Cu$	Rotating packed-bed reactor	Safranin O and methylene blue	LED	98.1% and 91.91% for SO and MB, respectively	2.0 g copper acetate monohydrate was dissolved in 50 mL deionized water to which ammonia solution (25% w/w) was added at pH 8 and the resultant blue suspension was irradiated ultrasonic horn	Mosleh *et al.* (2018)
Hydrothermal method	MnO_2	Batch	4-Chlorophenol	UV-LED light	100%	100% degradation achieved in 30 min at pH = 4.0, 1 mM peroxymonosulfate and 0.25 g/L MnO_2	Eslami *et al.* (2018)

Aloe vera leaf extract	Iron oxide	Membrane bioreactor/clay-alumina-based ceramic microfiltration and ceramic ultrafiltration membranes	Domestic and industrial wastewater simulated with atrazine	–	91.5–96%	Synthetic solution of atrazine (0.1–5.0 mg/L) and sludge dose (1–10 g/L). Three bacterial strains were added to each set and incubated at 37°C. The contact time in the bioreactor was optimized time range 6 h–7 days.	Mukherjee *et al.* (2018)
Green algae *Chlamydomonas reinhardtii*, *Scenedesmus* sp. and Cyanobacteria *Anabaena* sp., *Synechococcus* sp.	Iron oxide	Semi-continuous bench-scale photo-bioreactor/Polytetrafluoroethylene membrane filter	Mercury remediation	Fluorescent tube lights	–	Dried algae showed an absorption capacity of 2.2 μg (27.5%) of mercury from a 2 L flask after 5.5 h	Rocha *et al.* (2017)
Ultrasonic treatment	Zinc oxide	Anoxic–aerobic submerged membrane bioreactor/hollow fiber ultrafiltration membrane	Wastewater treatment	–	10 and 50 mg/L ZnO NPs decreased COD removal from 93.1 ± 0.6% to 90.1 ± 0.8% and 86.3 ± 2.3%	Domestic sewage fed from two 70 L glass tanks at 4°C. The mixed liquor suspended solid in the aeration tank was around 3–6 g/L, with an average sludge retention time of 25 d for each MBR. HRT was approximately 10 h and a constant permeate flux of 13–15 L/m^2 h was maintained	Zhang *et al.* (2017)

of the reactor (Figure 3.6a). Finally, if the source light is chosen and the satisfactory compromise is adopted, the scale-up becomes merely a modular operation. Further, the type of photocatalyst, size of the particles and their surface morphology influence internal light scattering, therefore, influencing the efficiency and the reproducibility of the results.

The morphology and dimensions of the nanoparticles are controlled by precursor concentration, reducing agent, stabilizing agent and other key factors include pH, temperature and contact time. Likewise, the proteins, amino acids, enzymes, polysaccharides and organic acids serve as a stabilizing material around the nanoparticles (Khatami *et al.*, 2018; Król *et al.*, 2017; Kuppusamy *et al.*, 2016; Mittal *et al.*, 2013). Quoting about the external parameters, pH plays an important role in the morphology of nanoparticles. Reports suggest that at alkaline pH value hexagonal and triangular gold nanoplates are formed, whereas in case of acidic pHs nanoparticles are not formed (Malarkodi *et al.*, 2014). With the change in temperature, the chemical nature of extracts also changes resulting in homogenous size distribution that can be related to larger size at low temperature and smaller size at high temperature. Reaction time is associated with Ostwald ripening where the synthesis of nanomaterials in the biological entities is controlled by the nucleation reaction leading to the reduction of individual metallic ions (Chung *et al.*, 2016; Hasan *et al.*, 2018).

3.6.2 Photobioreactors

There are various types of photobioreactors (PBRs) (Figure 3.6b) that have been developed so far, including bubble column, airlift reactor, flat-plate, stirred-tank, tubular, conical, torus and seaweed-type PBRs, only a few can be used for mass cultivation of nanoparticles Liu *et al.* (2013). Current progress in these aspects includes light, mixing, mass transfer, pH, temperature, capital cost and operating cost. In the design of an efficient PBR, light penetration and distribution inside the reactor are the dominant factors Huang *et al.* (2008). Moreover, good mixing and mass transfer as well as favorable temperature and pH can significantly improve the growth of the microbial community. Lastly, capital and operating costs are crucial factors for a cost-effective PBR in the long run. The efficiency of this process is determined by the photoconversion efficiency and is limited by the efficiency of photosynthesis and cell anabolism.

A mechanistic model capable of describing the steady-state operation of an inhibitory pollutant degrading algal–bacterial consortium enclosed in a chemostat PBR was developed (Wang *et al.*, 2016). The model was then validated against experimental data on salicylate removal by a *Chlorella sorokiniana/Ralstonia basilensis* consortium cultivated without external O_2 supplied in an enclosed chemostat PBR operated under various conditions of radiation, salicylate inlet concentrations, temperatures and hydraulic retention time (HRT).

3.6.3 Nanobiohybrids

Nanobiohybrids came into the picture where microbial cell factories are integrated with nanomaterials Liu *et al.* (2020a, b). Coupling nanomaterials with microbes opens up a whole spectrum of new metabolic properties where the nanomaterials can capture reducing equivalents in the presence of light sources Dong *et al.* (2018), these equivalents are provided to the cellular metabolism of the microbes (Liu *et al.*, 2015; Zhang *et al.*, 2020). These can be used as catalysts for photocatalytic activity, thereby improving the degradation process. For instance, in the case of *Moorella thermoacetica* – CdS constructs, the Cd precursor was added to the growth medium containing *M. thermoacetica* bacterial cells, where cysteine desulfhydrase plays a significant role in producing sulfide from cysteine to produce CdS (Sakimoto *et al.*, 2016). The construct of the non-sulfur reducing bacterium *Rhodopseudomas palustris* with CdS nanoparticles utilizes equivalents provided by bacteriochlorophyll in the presence of a light source by the Calvin cycle (Wang *et al.*, 2019). The other combinations reported include *Thiobacillus denitrificans*-CdS (Chen *et al.*, 2019a), *M. thermoacetica* – CdS, *Escherichia coli* – CdS hybrids, *Saccharomyces cerevisiae* – InP hybrids, *Cupriavidus necator* – CdS@ZnS, CdSe@ZnS, InP@ZnS, Cu_2ZnSnS_4@ZnS hybrids *Azotobacter vinelandii* – CdS@ZnS, CdSe@ZnS, InP@ZnS, Cu_2ZnSnS_4@ZnS hybrids (Ding *et al.*, 2019).

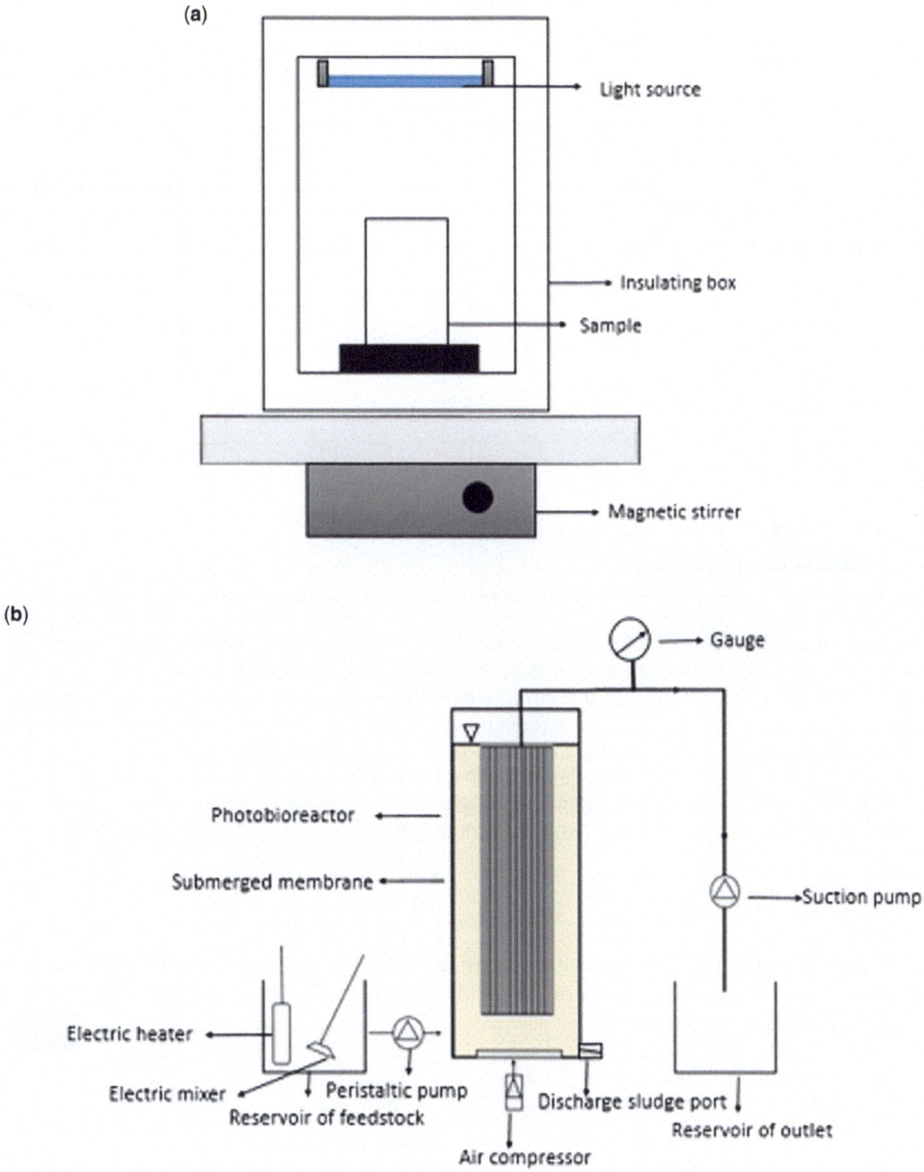

Figure 3.6 Schematic representation of (a) batch and (b) continuous flow PBR.

Microbial co-culturing causes synergistic effects, which help in increasing the tolerance of biomass to the stress conditions, thereby enhancing the efficiency of a nanobiohybrid-based system. Metal organic frameworks wrapped *M. thermoacetica* – CdS compared to the bare nanobiohybrids, as the encapsulated hybrids can continuously undertake CO_2 photoreduction, even under oxidative stress (Figure 3.7) (Cestellos-Blanco *et al.*, 2020). This challenge can be partly addressed by encapsulating the nanobiohybrids in a polymer matrix such as alginate, chitosan or polyethylene glycol. This encapsulation further helps in improving the stability and sensitivity of the nanobiohybrid constructs

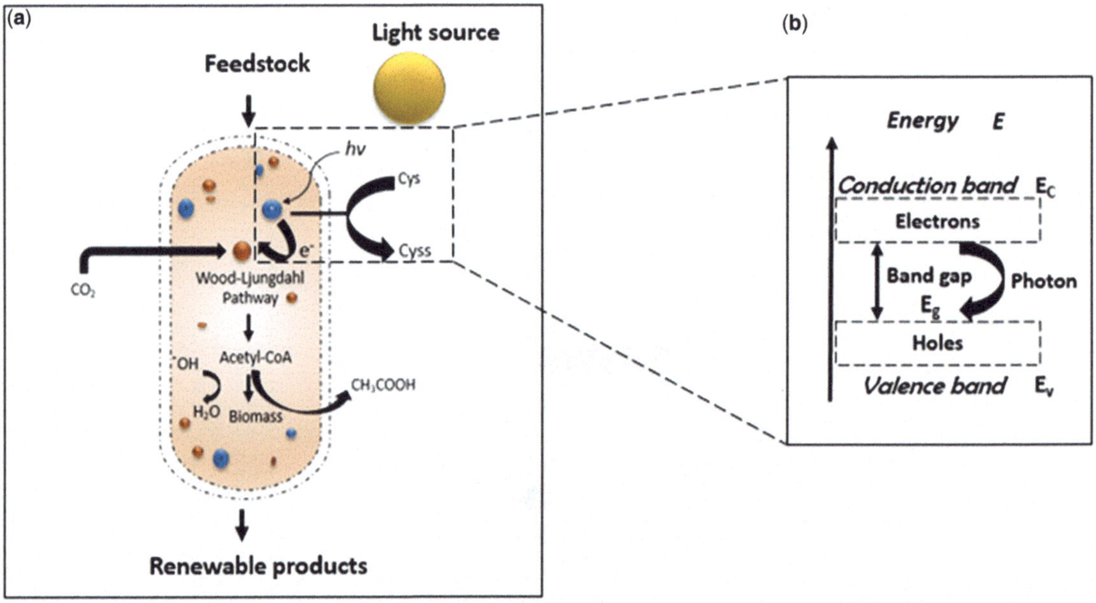

Figure 3.7 Schematic representation of microbe/nanomaterial hybrid systems. (a) When a feedstock (containing CO_2, H_2O, N_2, O_2) is supplied to the nanobiohybrid in the presence of sunlight, the photo-excited electrons from the nanoparticles (blue) are transferred to the enzyme electron acceptors. This leads to an electron–hole in the nanoparticles. The oxidized nanoparticles can be reduced by the red–ox coupled cysteine (*cys*)-cystine (*cyss*), which is the sacrificial electron donor to scavenge the electron–hole. The photogenerated electrons from the intracellular nanoparticles are passed to cytoplasmic mediators (such as ferredoxin, flavoproteins and rubredoxins (brown)) by the Wood–Ljungdhal pathway. (b) In the presence of a light source, the electrons jump from the valence band to the conduction band. *E*: energy, E_c: valence band, E_v: conduction band and E_g: band gap.(Source: Cestellos-Blanco *et al.* 2020).

against external stresses. Further limitations can be avoided by using novel PBR configurations equipped with capsules containing nanobiohybrids that prevent their wash-out and light-emitting diodes for continuous illumination.

3.7 CHALLENGES

3.7.1 Toxicity

The major challenge encountered with the use of nanoparticles is due to their characteristic physicochemical properties when they are released into the environment. Thus, to bridge the gap of knowledge and to exclusively tackle the toxicity issues related to nanoparticles, different strategies aiming to contribute to safe use of NPs are essential. Nanoparticles, due to their unique characteristics including small size, shape, high surface area, charge, chemical properties, solubility and degree of agglomeration, can penetrate into the ecosystem (Kumari *et al.*, 2014). This is the reason why they may pose higher risk than the same mass and material of larger particles. Considering the potential applications of nanoparticles in many fields and to address the knowledge gap, the relevant toxic effects of nanoparticles should be assessed by utilizing in vivo toxicological models, targeting the vital systems. However, in addition designing, adapting and validating such new models in the future for toxicity testing, also the route of exposure, coating material and sterility of nanoparticles and type of cell cultures need to be carefully considered (Tan *et al.*, 2020).

3.7.2 Nanoparticles detection

Particle associations and aggregates pose additional analytical challenges (Das *et al.*, 2012). For instance, the differentiation between a particle of 100 nm and 1–100 nm-sized particles, which would be considered a nanomaterial under all regulations, requires analytical tools capable of assessing the inner structure and type of particle–particle bonding in the material. Only certain imaging methods, such as electron microscopy, can provide such insight. The fact that the morphology of particles can change in response to the surrounding conditions additionally complicates analysis. Other approaches, like fractionation methods, do not distinguish between particles and aggregates of the same size or different particle shapes that may have the same average hydrodynamic diameter. The same holds for inductively coupled plasma mass spectrometry, which only detects total particle composition, with no information on shape/structure. Limitations and problems relate to size and size distribution as well as the challenges associated with detection limits for small sizes (Miernicki *et al.*, 2019).

3.7.3 Light accessibility

Light attenuation in a culture medium occurs due to light absorption and scattering by pigments and water molecules. Attenuation per unit distance is proportional to the irradiance of light, which creates a heterogeneous radiation field inside the culture. The light irradiance is further reduced as the cell grows to high density. Light intensity is one of the most important factors in the cultivation of microbes. In addition, different species respond differently to light wavelengths due to variations in their pigmentation. The absorption bands of these pigments are different, as a result only a certain range of light wavelengths are effective. Due to the dependence of phototropic microbes and algae on light energy, the availability and accessibility of light is crucial. Microbial cultivation systems are designed to maximize surface-to-volume ratio in a manner to encourage light utilization. It can directly affect the photosynthetic kinetics of microbial cells, influence cell growth and the production of metabolites (Cestellos-Blanco *et al.*, 2020; Liu *et al.*, 2020a).

3.8 CONCLUSION

Bioinspired approaches are expected to be effective, a great variety of multifunctional materials with multiscale structures have been prepared by using different synthesis strategies. The design and optimization of nanotechnology enabled water treatment technologies to enhance and improve understanding of (1) nanomaterial structure and activity, (2) relationships of material selection and enhance durability and (3) structural integrity and process reliability. This includes developing green synthesis approaches to scale-up production of functional nanomaterials and cost-effective strategies to regenerate or replenish functionality of nanomaterials. Due to device complexity in nanotechnologies, theoretical modelling is now playing an increasingly important role in device design. Nanotechnology will likely play a critical role by providing a wide range of unprecedented opportunities to enhance some water treatment systems, including adaptable treatment technologies and materials that can be tuned to specific targets. In the near future, nanotechnology-enabled water treatment will likely be implemented mainly in niche applications such as point-of-use devices and in municipal treatment plants.

REFERENCES

Abdelghany T. M., Al-Rajhi A. M. H., Al Abboud M. A., Alawlaqi M. M., Ganash Magdah A., Helmy E. A. M. and Mabrouk A. S. (2018). Recent advances in green synthesis of silver nanoparticles and their applications: about future directions. A review. *Bionanoscience*, **8**, 5–16, https://doi.org/10.1007/s12668-017-0413-3

AbdElhady M. M. (2012). Preparation and characterization of chitosan/zinc oxide nanoparticles for imparting antimicrobial and UV protection to cotton fabric. *International Journal of Carbohydrate Chemistry*, **2012**, 1–6, https://doi.org/10.1155/2012/840591

Abdel-Raouf N., Al-Enazi N. M., Ibraheem I. B. M., Alharbi R. M. and Alkhulaifi M. M. (2019). Biosynthesis of silver nanoparticles by using of the marine brown alga *Padina pavonia* and their characterization. *Saudi Journal of Biological Sciences*, **26**, 1207–1215, https://doi.org/10.1016/j.sjbs.2018.01.007

Abinaya M., Vaseeharan B., Divya M., Sharmili A., Govindarajan M., Alharbi N. S., Kadaikunnan S., Khaled J. M. and Benelli G. (2018). Bacterial exopolysaccharide (EPS)-coated ZnO nanoparticles showed high antibiofilm activity and larvicidal toxicity against malaria and Zika virus vectors. *Journal of Trace Elements in Medicine and Biology*, **45**, 93–103, https://doi.org/10.1016/j.jtemb.2017.10.002

Aboelfetoh E. F., El-Shenody R. A. and Ghobara M. M. (2017). Eco-friendly synthesis of silver nanoparticles using green algae (*Caulerpa serrulata*): reaction optimization, catalytic and antibacterial activities. *Environmental Monitoring and Assessment*, **189**, 1–15, https://doi.org/10.1007/s10661-017-6033-0

Agarwal H., Venkat Kumar S. and Rajeshkumar S. (2017). A review on green synthesis of zinc oxide nanoparticles – an eco-friendly approach. *Resources Technology*, **3**, 406–413, https://doi.org/10.1016/j.reffit.2017.03.002

Ahmed S., Ahmad M., Swami B. L. and Ikram S. (2016a). A review on plants extract mediated synthesis of silver nanoparticles for antimicrobial applications: a green expertise. *Journal of Advanced Research*, **7**, 17–28, https://doi.org/10.1016/j.jare.2015.02.007

Ahmed S., Saifullah Ahmad M., Swami B. L. and Ikram S. (2016b). Green synthesis of silver nanoparticles using *Azadirachta indica* aqueous leaf extract. *Journal of Radiation Research and Applied Sciences*, **9**, 1–7, https://doi.org/10.1016/j.jrras.2015.06.006

Ahmad T., Bustam M. A., Zulfiqar M., Moniruzzaman M., Idris A., Iqbal J., Asghar H. M. A. and Ullah S. (2020). Controllable phytosynthesis of gold nanoparticles and investigation of their size and morphology-dependent photocatalytic activity under visible light. *Journal of Photochemistry and Photobiology A: Chemistry*, **392**, 112429, https://doi.org/10.1016/j.jphotochem.2020.112429

Akhtar M. S., Panwar J. and Yun Y. S. (2013). Biogenic synthesis of metallic nanoparticles by plant extracts. *ACS Sustainable Chemistry & Engineering*, **1**, 591–602, https://doi.org/10.1021/sc300118u

Akturk O. (2020). Colloidal stability and biological activity evaluation of microbial exopolysaccharide levan-capped gold nanoparticles. *Colloids Surfaces B Biointerfaces*, **192**, 111061, https://doi.org/10.1016/j.colsurfb.2020.111061

Al-Mamun R., Kader S. and Islam S. (2021). Environmental nanotechnology, monitoring & management solar-TiO_2 immobilized photocatalytic reactors performance assessment in the degradation of methyl orange dye in aqueous solution. *Environmental Nanotechnology, Monitoring & Management*, **16**, 100514, https://doi.org/10.1016/j.enmm.2021.100514

Alharbi O. M. L., Basheer A. A., Khattab R. A. and Ali I. (2018). Health and environmental effects of persistent organic pollutants. *Journal of Molecular Liquids*, **263**, 442–453, https://doi.org/10.1016/j.molliq.2018.05.029

Ali J., Ali N., Wang L., Waseem H. and Pan G. (2019). Revisiting the mechanistic pathways for bacterial mediated synthesis of noble metal nanoparticles. *Journal of Microbiological Methods*, **159**, 18–25, https://doi.org/10.1016/j.mimet.2019.02.010

Ali S., Sudha K. G., Karunakaran G., Kowsalya M., Kolesnikov E., Gorshenkov M. V. and Rajeshkumar M. P. (2021). Novel *Leea grandifolia* leaves mediated synthesis of ZnO nanorods for photocatalytic and anticancer applications. *Applied Organometallic Chemistry*, **35**, 1–20, https://doi.org/10.1002/aoc.6239

Almalki M. A. and Khalifa A. Y. Z. (2020). Silver nanoparticles synthesis from *Bacillus* sp KFU36 and its anticancer effect in breast cancer MCF-7 cells via induction of apoptotic mechanism. *Journal of Photochemistry and Photobiology B: Biology*, **204**, 111786, https://doi.org/10.1016/j.jphotobiol.2020.111786

AlNadhari S., Al-Enazi N. M., Alshehrei F. and Ameen F. (2021). A review on biogenic synthesis of metal nanoparticles using marine algae and its applications. *Environmental Research*, **194**, 110672, https://doi.org/10.1016/j.envres.2020.110672

Ameen F., AlYahya S., Govarthanan M., ALjahdali N., Al-Enazi N., Alsamhary K., Alshehri W. A., Alwakeel S. S. and Alharbi S. A. (2020). Soil bacteria *Cupriavidus* sp. mediates the extracellular synthesis of antibacterial silver nanoparticles. *Journal of Molecular Structure*, **1202**, 127233, https://doi.org/10.1016/j.molstruc.2019.127233

Ameen F., Dawoud T. and Alnadhari S. (2021). Ecofriendly and low-cost synthesis of ZnO nanoparticles from *Acremonium potronii* for the photocatalytic degradation of azo dyes. *Environmental Research*, **202**, 111700, https://doi.org/10.1016/j.envres.2021.111700

Amoli-Diva M., Irani E. and Pourghazi K. (2020). Photocatalytic filtration reactors equipped with bi-plasmonic nanocomposite/poly acrylic acid-modified polyamide membranes for industrial wastewater treatment. *Separation and Purification Technology*, **236**, 116257, https://doi.org/10.1016/j.seppur.2019.116257

Aritonang H. F., Koleangan H. and Wuntu A. D. (2019). Synthesis of silver nanoparticles using aqueous extract of medicinal plants' (*Impatiens balsamina* and *Lantana camara*) fresh leaves and analysis of antimicrobial activity. *International Journal of Microbiology*, **2019**, 1–8, https://doi.org/10.1155/2019/8642303

Arshad H., Sami M. A., Sadaf S. and Hassan U. (2021). *Salvadora persica* mediated synthesis of silver nanoparticles and their antimicrobial efficacy. *Scientific Reports*, **11**, 1–11, https://doi.org/10.1038/s41598-021-85584-w

Arya A., Gupta K., Chundawat T. S. and Vaya D. (2018). Biogenic synthesis of copper and silver nanoparticles using green alga *Botryococcus braunii* and its antimicrobial activity. *Bioinorganic Chemistry and Applications*, **2018**, 1–9, https://doi.org/10.1155/2018/7879403

Arya A., Mishra V. and Chundawat T. S. (2019). Green synthesis of silver nanoparticles from green algae (*Botryococcus braunii*) and its catalytic behavior for the synthesis of benzimidazoles. *Chemical Data Collections*, **20**, 1–7, https://doi.org/10.1016/j.cdc.2019.100190

Asghar M., Habib S., Zaman W., Hussain S., Ali H. and Saqib S. (2020). Synthesis and characterization of microbial mediated cadmium oxide nanoparticles. *Microscopy Research and Technique*, **83**, 1574–1584, https://doi.org/10.1002/jemt.23553

Aygün A., Özdemir S., Gülcan M., Cellat K. and Şen F. (2020). Synthesis and characterization of Reishi mushroom-mediated green synthesis of silver nanoparticles for the biochemical applications. *Journal of Pharmaceutical and Biomedical Analysis*, **178**, 112970, https://doi.org/10.1016/j.jpba.2019.112970

Bajorowicz B., Kobylański M. P., Gołąbiewska A., Nadolna J., Zaleska-Medynska A. and Malankowska A. (2018). Quantum dot-decorated semiconductor micro- and nanoparticles: a review of their synthesis, characterization and application in photocatalysis. *Advances in Colloid and Interface Science*, **256**, 352–372, https://doi.org/10.1016/j.cis.2018.02.003

Balachandar R., Gurumoorthy P., Karmegam N., Barabadi H., Subbaiya R., Anand K., Boomi P. and Saravanan M. (2019). Plant-mediated synthesis, characterization and bactericidal potential of emerging silver nanoparticles using stem extract of *Phyllanthus pinnatus*: a recent advance in phytonanotechnology. *Journal of Cluster Science*, **30**, 1481–1488, https://doi.org/10.1007/s10876-019-01591-y

Bali R. and Harris A. T. (2010). Biogenic synthesis of Au nanoparticles using vascular plants. *Industrial & Engineering Chemistry Research*, **49**, 12762–12772, https://doi.org/10.1021/ie101600m

Banasiuk R., Krychowiak M., Swigon D., Tomaszewicz W., Michalak A., Chylewska A., Ziabka M., Lapinski M., Koscielska B., Narajczyk M. and Krolicka A. (2020). Carnivorous plants used for green synthesis of silver nanoparticles with broad-spectrum antimicrobial activity. *Arabian Journal of Chemistry*, **13**, 1415–1428, https://doi.org/10.1016/j.arabjc.2017.11.013

Banerjee P., Satapathy M., Mukhopahayay A. and Das P. (2014). Leaf extract mediated green synthesis of silver nanoparticles from widely available Indian plants: synthesis, characterization, antimicrobial property and toxicity analysis. *Bioresources and Bioprocessing*, **1**, 1–10, https://doi.org/10.1186/s40643-014-0003-y

Bari N. K., Kumar G., Bhatt A., Hazra J. P., Garg A., Ali M. E. and Sinha S. (2018). Nanoparticle fabrication on bacterial microcompartment surface for the development of hybrid enzyme-inorganic catalyst. *ACS Catalysis*, **8**, 7742–7748, https://doi.org/10.1021/acscatal.8b02322

Basnet P. and Chatterjee S. (2020). Structure-directing property and growth mechanism induced by capping agents in nanostructured ZnO during hydrothermal synthesis – a systematic review. *Nano-Structures and Nano-Objects*, **22**, 100426, https://doi.org/10.1016/j.nanoso.2020.100426

Bereczk-Tompa É., Vonderviszt F., Horváth B., Szalai I. and Pósfai M. (2017). Biotemplated synthesis of magnetic filaments. *Nanoscale*, **9**, 15062–15069, https://doi.org/10.1039/c7nr04842d

Bharde A., Wani A., Shouche Y., Joy P. A., Prasad B. L. V. and Sastry M. (2005). Bacterial aerobic synthesis of nanocrystalline magnetite. *Journal of the American Chemical Society*, **127**, 9326–9327, https://doi.org/10.1021/ja0508469

Borase H. P., Patil C. D., Suryawanshi R. K., Koli S. H., Mohite B. V., Benelli G. and Patil S. V. (2017). Mechanistic approach for fabrication of gold nanoparticles by *Nitzschia* diatom and their antibacterial activity. *Bioprocess and Biosystems Engineering*, **40**, 1437–1446, https://doi.org/10.1007/s00449-017-1801-3

Bozarth A., Maier U. G. and Zauner S. (2009). Diatoms in biotechnology: modern tools and applications. *Applied Microbiology and Biotechnology*, **82**, 195–201, https://doi.org/10.1007/s00253-008-1804-8

Bruna N., Collao B., Tello A., Caravantes P., Díaz-Silva N., Monrás J. P., Órdenes-Aenishanslins N., Flores M., Espinoza-Gonzalez R., Bravo D. and Pérez-Donoso J. M. (2019). Synthesis of salt-stable fluorescent nanoparticles (quantum dots) by polyextremophile halophilic bacteria. *Scientific Reports*, **9**, 1–13, https://doi.org/10.1038/s41598-018-38330-8

Cabral H., Fonseca V., Sousa T. and Leal M. C. (2019). Synergistic effects of climate change and marine pollution: an overlooked interaction in coastal and estuarine areas. *International Journal of Environmental Research and Public Health*, **16**, 1–17, https://doi.org/10.3390/ijerph16152737

Cestellos-Blanco S., Zhang H., Kim J. M., Shen Y. and Yang P. (2020). Photosynthetic semiconductor biohybrids for solar-driven biocatalysis. *Nature Catalysis*, **3**, 245–255, https://doi.org/10.1038/s41929-020-0428-y

Chandran S. P., Chaudhary M., Pasricha R., Ahmad A. and Sastry M. (2006). Synthesis of gold nanotriangles and silver nanoparticles using *Aloe vera* plant extract. *Biotechnology Progress*, **22**, 577–583, https://doi.org/10.1021/bp0501423

Chandrasekar A., Vasantharaj S., Jagadeesan N. L., Shankar S. N., Pannerselvam B., Bose V. G., Arumugam G. and Shanmugavel M. (2021). Studies on phytomolecules mediated synthesis of copper oxide nanoparticles for biomedical and environmental applications. *Biocatalysis and Agricultural Biotechnology*, **33**, 101994, https://doi.org/10.1016/j.bcab.2021.101994

Chatterjee S., Mahanty S., Das P., Chaudhuri P. and Das S. (2020). Biofabrication of iron oxide nanoparticles using manglicolous fungus *Aspergillus niger* BSC-1 and removal of Cr(VI) from aqueous solution. *Chemical Engineering Journal*, **385**, 123790, https://doi.org/10.1016/j.cej.2019.123790

Chawla P., Kumar N., Bains A., Dhull S. B., Kumar M., Kaushik R. and Punia S. (2020). Gum arabic capped copper nanoparticles: synthesis, characterization, and applications. *International Journal of Biological Macromolecules*, **146**, 232–242, https://doi.org/10.1016/j.ijbiomac.2019.12.260

Chawla P., Najda A., Bains A., Nurzyńska-Wierdak R., Kaushik R. and Tosif M. M. (2021). Potential of gum arabic functionalized iron hydroxide nanoparticles embedded cellulose paper for packaging of paneer. *Nanomaterials*, **11**, 1308, https://doi.org/10.3390/nano11051308

Chen M., Kang H., Gong Y., Guo J., Zhang H. and Liu R. (2015). Bacterial cellulose supported gold nanoparticles with excellent catalytic properties. *ACS Applied Materials & Interfaces*, **7**, 21717–21726, https://doi.org/10.1021/acsami.5b07150

Chen F., Xiao Z., Yue L., Wang J., Feng Y., Zhu X., Wang Z. and Xing B. (2019a). Algae response to engineered nanoparticles: current understanding, mechanisms and implications. *Environmental Science: Nano*, **6**, 1026–1042.

Chen M., Zhou X. F., Yu Y. Q., Liu X., Zeng R. J. X., Zhou S. G. and He Z. (2019b). Light-driven nitrous oxide production via autotrophic denitrification by self-photosensitized *Thiobacillus denitrificans*. *Environment International*, **127**, 353–360, https://doi.org/10.1016/j.envint.2019.03.045

Chetia L., Kalita D. and Ahmed G. A. (2017). Synthesis of Ag nanoparticles using diatom cells for ammonia sensing. *Sensing and Bio-Sensing Research*, **16**, 55–61, https://doi.org/10.1016/j.sbsr.2017.11.004

Chung I. M., Park I., Seung-Hyun K., Thiruvengadam M. and Rajakumar G. (2016). Plant-mediated synthesis of silver nanoparticles: their characteristic properties and therapeutic applications. *Nanoscale Research Letters*, **11**, 1–14, https://doi.org/10.1186/s11671-016-1257-4

Chwalibog A., Sawosz E., Hotowy A., Szeliga J., Mitura S., Mitura K., Grodzik M., Orlowski P. and Sokolowska A. (2010). Visualization of interaction between inorganic nanoparticles and bacteria or fungi. *International Journal of Nanomedicine*, **5**, 1085–1094, https://doi.org/10.2147/IJN.S13532

Crini G. (2005). Recent developments in polysaccharide-based materials used as adsorbents in wastewater treatment. *Progress in Polymer Science*, **30**, 38–70, https://doi.org/10.1016/j.progpolymsci.2004.11.002

Dahoumane S. A., Mechouet M., Wijesekera K., Filipe C. D. M., Sicard C., Bazylinski D. A. and Jeffryes C. (2017). Algae-mediated biosynthesis of inorganic nanomaterials as a promising route in nanobiotechnology – a review. *Green Chemistry*, **19**, 552–587, https://doi.org/10.1039/c6gc02346k

Das S. K., Khan M. M. R., Guha A. K., Das A. R., and Mandal A. B. (2012). Silver-nano biohybride material: synthesis, characterization and application in water purification. *Bioresource technology*, **124**, 495–499, https://doi.org/10.1016/j.biortech.2012.08.071

Das R. K., Pachapur V. L., Lonappan L., Naghdi M., Pulicharla R., Maiti S., Cledon M., Dalila L. M. A., Sarma S. J. and Brar S. K. (2017). Biological synthesis of metallic nanoparticles: plants, animals and microbial aspects. *Nanotechnology for Environmental Engineering*, **2**, 1–21, https://doi.org/10.1007/s41204-017-0029-4

Dauthal P. and Mukhopadhyay M. (2016). Noble metal nanoparticles: plant-mediated synthesis, mechanistic aspects of synthesis, and applications. *Industrial & Engineering Chemistry Research*, **55**, 9557–9577, https://doi.org/10.1021/acs.iecr.6b00861

des Ligneris E., Dumée L. F. and Kong L. (2018). Nanofiber-based materials for persistent organic pollutants in water remediation by adsorption. *Applied Sciences*, **8**(2), 166, https://doi.org/10.3390/app8020166

Devasvaran K. and Lim V. (2021). Green synthesis of metallic nanoparticles using pectin as a reducing agent: a systematic review of the biological activities. *Pharmaceutical Biology*, **59**, 494–503, https://doi.org/10.1080/13880209.2021.1910716

Dhanaraj S., Thirunavukkarasu S., Allen John H., Pandian S., Salmen S. H., Chinnathambi A. and Alharbi S. A. (2020). Novel marine *Nocardiopsis dassonvillei*-DS013 mediated silver nanoparticles characterization and its bactericidal potential against clinical isolates. *Saudi Journal of Biological Sciences*, **27**, 991–995, https://doi.org/10.1016/j.sjbs.2020.01.003

Ding Y., Bertram J. R., Eckert C., Bommareddy R. R., Patel R., Conradie A., Bryan S., Nagpal P., Cestellos-Blanco S., Zhang H. and Yang P. (2019). Nanorg microbial factories: light-driven renewable biochemical synthesis using quantum dot-bacteria nanobiohybrids. *Faraday Discussions*, **215**, 10272–10282, https://doi.org/10.1021/jacs.9b02549

Dinga E., Ekennia A., Ogbonna C.U., Udu D.A., Mthiyane D.M.N., Marume U. and Onwudiwe D.C. (2022). Phyto-mediated synthesis of MgO nanoparticles using Melia azedarach seed extract: Larvicidal and antioxidant activities. *Scientific African*, **17**, 01366. https://doi.org/10.1016/j.sciaf.2022.e01366

Dolat D., Quici N., Kusiak-Nejman E., Morawski A.W. and Puma G.L. (2012). One-step, hydrothermal synthesis of nitrogen, carbon co-doped titanium dioxide (N, $CTiO_2$) photocatalysts. Effect of alcohol degree and chain length as carbon dopant precursors on photocatalytic activity and catalyst deactivation. *Applied Catalysis B: Environmental*, **115**, 81–89. https://doi.org/10.1016/j.apcatb.2011.12.007

Dong H., Sarkes D. A., Rice J. J., Hurley M. M., Fu A. J. and Stratis-Cullum D. N. (2018). Living bacteria-nanoparticle hybrids mediated through surface-displayed peptides. *Langmuir*, **34**, 5837–5848, https://doi.org/10.1021/acs.langmuir.8b00114

Dumur F., Guerlin A., Dumas E., Bertin D., Gigmes D. and Mayer C. R. (2011). Controlled spontaneous generation of gold nanoparticles assisted by dual reducing and capping agents. *Golden Bulletin*, **44**, 119–137, https://doi.org/10.1007/s13404-011-0018-5

Durán N., Marcato P. D., Durán M., Yadav A., Gade A. and Rai M. (2011). Mechanistic aspects in the biogenic synthesis of extracellular metal nanoparticles by peptides, bacteria, fungi, and plants. *Applied Microbiology and Biotechnology*, **90**, 1609–1624, https://doi.org/10.1007/s00253-011-3249-8

Dzimitrowicz A., Jamróz P., diCenzo G. C., Sergiel I., Kozlecki T. and Pohl P. (2019). Preparation and characterization of gold nanoparticles prepared with aqueous extracts of *Lamiaceae* plants and the effect of follow-up treatment with atmospheric pressure glow microdischarge. *Arabian Journal of Chemistry*, **12**, 4118–4130, https://doi.org/10.1016/j.arabjc.2016.04.004

Ebrahiminezhad A., Zare-Hoseinabadi A., Sarmah A. K., Taghizadeh S., Ghasemi Y. and Berenjian A. (2018). Plant-mediated synthesis and applications of iron nanoparticles. *Molecular Biotechnology*, **60**, 154–168, https://doi.org/10.1007/s12033-017-0053-4

El-Saied H. A. A. and Ibrahim A. M. (2020). Effective fabrication and characterization of eco-friendly nano chitosan capped zinc oxide nanoparticles for effective marine fouling inhibition. *Journal of Environmental Chemical Engineering*, **8**, 103949, https://doi.org/10.1016/j.jece.2020.103949

Elamawi R. M., Al-Harbi R. E. and Hendi A. A. (2018). Biosynthesis and characterization of silver nanoparticles using *Trichoderma longibrachiatum* and their effect on phytopathogenic fungi. *Egyptian Journal of Biological Pest Control*, **28**, 1–11, https://doi.org/10.1186/s41938-018-0028-1

Elfick A., Rischitor G., Mouras R., Azfer A., Lungaro L., Uhlarz M., Herrmannsdörfer T., Lucocq J., Gamal W., Bagnaninchi P., Semple S. and Salter D. M. (2017). Biosynthesis of magnetic nanoparticles by human mesenchymal stem cells following transfection with the magnetotactic bacterial gene mms6. *Scientific Reports*, **7**, 1–8, https://doi.org/10.1038/srep39755

Elia P., Zach R., Hazan S., Kolusheva S., Porat Z. and Zeiri Y. (2014). Green synthesis of gold nanoparticles using plant extracts as reducing agents. *International Journal of Nanomedicine*, **9**, 4007–4021, https://doi.org/10.2147/IJN.S57343

Eslami A., Hashemi M. and Ghanbari F. (2018). Degradation of 4-chlorophenol using catalyzed peroxymonosulfate with nano-MnO_2/UV irradiation: toxicity assessment and evaluation for industrial wastewater treatment. *Journal of Cleaner Production*, **195**, 1389–1397, https://doi.org/10.1016/j.jclepro.2018.05.137

Eugenio M., Müller N., Frasés S., Almeida-Paes R., Lima L. M. T. R., Lemgruber L., Farina M., De Souza W. and Sant'Anna C. (2016). Yeast-derived biosynthesis of silver/silver chloride nanoparticles and their antiproliferative activity against bacteria. *RSC Advances*, **6**, 9893–9904, https://doi.org/10.1039/c5ra22727e

Fahimirad S., Ajalloueian F. and Ghorbanpour M. (2019). Synthesis and therapeutic potential of silver nanomaterials derived from plant extracts. *Ecotoxicology and Environmental Safety*, **168**, 260–278, https://doi.org/10.1016/j.ecoenv.2018.10.017

Fang B., Jiang Y., Rotello V. M., Nüsslein K. and Santore M. M. (2014). Easy come easy go: surfaces containing immobilized nanoparticles or isolated polycation chains facilitate removal of captured *Staphylococcus aureus* by retarding bacterial bond maturation. *ACS Nano*, **8**, 1180–1190, https://doi.org/10.1021/nn405845y

Fang X., Wang Y., Wang Z., Jiang Z. and Dong M. (2019). Microorganism assisted synthesized nanoparticles for catalytic applications. *Energies*, **12**, 1–12, https://doi.org/10.3390/en12010190

Fatima R., Priya M., Indurthi L., Radhakrishnan V. and Sudhakaran R. (2020). Biosynthesis of silver nanoparticles using red algae *Portieria hornemannii* and its antibacterial activity against fish pathogens. *Microbial Pathogenesis*, **138**, 103780, https://doi.org/10.1016/j.micpath.2019.103780

Fazlzadeh M., Rahmani K., Zarei A., Abdoallahzadeh H., Nasiri F. and Khosravi R. (2017). A novel green synthesis of zero valent iron nanoparticles (NZVI) using three plant extracts and their efficient application for removal of Cr(VI) from aqueous solutions. *Advanced Powder Technology*, **28**, 122–130, https://doi.org/10.1016/j.apt.2016.09.003

Fernández-García M., Martínez-Arias A., Hanson J. C. and Rodriguez J. A. (2004). Nanostructured oxides in chemistry: characterization and properties. *Chemical Reviews*, **104**, 4063–4104, https://doi.org/10.1021/cr030032f

Fernández-Llamosas H., Castro L., Blázquez M. L., Díaz E. and Carmona M. (2017). Speeding up bioproduction of selenium nanoparticles by using *Vibrio natriegens* as microbial factory. *Scientific Reports*, **7**, 1–9, https://doi.org/10.1038/s41598-017-16252-1

Feroze N., Arshad B., Younas M., Afridi M. I., Saqib S. and Ayaz A. (2020). Fungal mediated synthesis of silver nanoparticles and evaluation of antibacterial activity. *Microscopy Research and Technique*, **83**, 72–80, https://doi.org/10.1002/jemt.23390

Franco-Ulloa S., Tatulli G., Bore S. L., Moglianetti M., Pompa P. P., Cascella M. and De Vivo M. (2020). Dispersion state phase diagram of citrate-coated metallic nanoparticles in saline solutions. *Nature Communications*, **11**, 1–10, https://doi.org/10.1038/s41467-020-19164-3

Gahlawat G. and Choudhury A. R. (2019). A review on the biosynthesis of metal and metal salt nanoparticles by microbes. *RSC Advances*, **9**(23), 12944–12967, https://doi.org/10.1039/c8ra10483b

Ganesan V., Hariram M., Vivekanandhan S. and Muthuramkumar S. (2020). *Periconium* sp. (endophytic fungi) extract mediated sol-gel synthesis of ZnO nanoparticles for antimicrobial and antioxidant applications. *Materials Science in Semiconductor Processing*, **105**, 104739, https://doi.org/10.1016/j.mssp.2019.104739

Gardea-Torresdey J. L., Parsons J. G., Gomez E., Peralta-Videa J., Troiani H. E., Santiago P. and Yacaman M. J. (2002). Formation and growth of Au nanoparticles inside live alfalfa plants. *Nano Letters*, **2**, 397–401, https://doi.org/10.1021/nl015673+

Ghiuta I., Croitoru C., Kost J., Wenkert R. and Munteanu D. (2021). Bacteria-mediated synthesis of silver and silver chloride nanoparticles and their antimicrobial activity. *Applied Sciences*, **11**, 1–11, https://doi.org/10.3390/app11073134

Gingasu D., Mindru I., Culita D. C., Marinescu G., Somacescu S., Ianculescu A., Surdu V. A., Preda S., Oprea O. and Vasile B. S. (2021). *Mentha piperita*-mediated synthesis of cobalt aluminate nanoparticles and their photocatalytic activity. *Journal of Materials Science: Materials in Electronics*, **32**, 11220–11231, https://doi.org/10.1007/s10854-021-05791-z

Golnaraghi-Ghomi A. R., Mohammadi-Khanaposhti M., Sokhansanj A., Saadati Y., Khazraei E., Kobarfard F., Barabadi H. and Golnaraghi A. (2021). Artificial neural network modeling of fungus-mediated extracellular biosynthesis of zirconium nanoparticles using standard *Penicillium* sp. *Journal of Cluster Science*, **7**, 1907–1921, https://doi.org/10.1007/s10876-021-02111-7

Gong C. P., Luo Y. and Pan Y. Y. (2019). Novel synthesized zinc oxide nanoparticles loaded alginate-chitosan biofilm to enhanced wound site activity and anti-septic abilities for the management of complicated abdominal wound dehiscence. *Journal of Photochemistry and Photobiology B: Biology*, **192**, 124–130, https://doi.org/10.1016/j.jphotobiol.2019.01.019

González-Garcinuño Á., Masa R., Hernández M., Domínguez Á., Tabernero A. and Del Valle E. M. (2019). Levan-Capped silver nanoparticles for bactericidal formulations: release and activity modelling. *International Journal of Molecular Sciences*, **20**, 1502, https://doi.org/10.3390/ijms20061502

Grasso G., Zane D. and Dragone R. (2020). Microbial nanotechnology: challenges and prospects for green biocatalytic synthesis of nanoscale materials for sensoristic and biomedical applications. *Nanomaterials*, **10**, 1–11, https://doi.org/10.3390/nano10010011

Griffero L., Alcántara-Durán J., Alonso C., Rodríguez-Gallego L., Moreno-González D., García-Reyes J. F., Molina-Díaz A. and Pérez-Parada A. (2019). Basin-scale monitoring and risk assessment of emerging contaminants in South American Atlantic coastal lagoons. *Science of the Total Environment*, **697**, 134058, https://doi.org/10.1016/j.scitotenv.2019.134058

Gurung K., Ncibi M. C., Thangaraj S. K., Jänis J., Seyedsalehi M. and Sillanpää M. (2019). Removal of pharmaceutically active compounds (PhACs) from real membrane bioreactor (MBR) effluents by photocatalytic degradation using composite Ag_2O/P-25 photocatalyst. *Separation and Purification Technology*, **215**, 317–328, https://doi.org/10.1016/j.seppur.2018.12.069

Hamouda R. A., Hussein M. H., Abo-Elmagd R. A. and Bawazir S. S. (2019). Synthesis and biological characterization of silver nanoparticles derived from the cyanobacterium *Oscillatoria limnetica*. *Scientific Reports*, **9**, 1–17, https://doi.org/10.1038/s41598-019-49444-y

Han D. and Currell M. J. (2017). Persistent organic pollutants in China's surface water systems. *Science of the Total Environment*, **580**, 602–625, https://doi.org/10.1016/j.scitotenv.2016.12.007

Han Z., Zheng X., Hu F., Qu F., Umar A. and Wu X. (2015). Facile synthesis of hollow ZnS nanospheres for environmental remediation. *Materials Letters*, **160**, 271–274, https://doi.org/10.1016/j.matlet.2015.07.152

Hasan M., Ullah I., Zulfiqar H., Naeem K., Iqbal A., Gul H., Ashfaq M. and Mahmood N. (2018). Biological entities as chemical reactors for synthesis of nanomaterials: progress, challenges and future perspective. *Materials Today Chemistry*, **8**, 13–28, https://doi.org/10.1016/j.mtchem.2018.02.003

Hemmati S., Rashtiani A., Zangeneh M. M., Mohammadi P., Zangeneh A. and Veisi H. (2019). Green synthesis and characterization of silver nanoparticles using *Fritillaria* flower extract and their antibacterial activity against some human pathogens. *Polyhedron*, **158**, 8–14, https://doi.org/10.1016/j.poly.2018.10.049

Heng Z. W., Chong W. C., Pang Y. L. and Koo C. H. (2021). An overview of the recent advances of carbon quantum dots/metal oxides in the application of heterogeneous photocatalysis in photodegradation of pollutants towards visible-light and solar energy exploitation. *Journal of Environmental Chemical Engineering*, **9**, 105199, https://doi.org/10.1016/j.jece.2021.105199

Hileuskaya K., Ladutska A., Kulikouskaya V., Kraskouski A., Novik G., Kozerozhets I., Kozlovskiy A. and Agabekov V. (2020). 'Green' approach for obtaining stable pectin-capped silver nanoparticles: physico-chemical characterization and antibacterial activity. *Colloids and Surfaces A: Physicochemical and Engineering Aspects*, **585**, 124141, https://doi.org/10.1016/j.colsurfa.2019.124141

Hodes G. (2007). When small is different: Some recent advances in concepts and applications of nanoscale phenomena. *Advanced Materials*, **19**, 639–655, https://doi.org/10.1002/adma.200601173

Huang J., Lin L., Li Q., Sun D., Wang Y., Lu Y., He N., Yang K., Yang X., Wang H., Wang W. and Lin W. (2008). Continuous-flow biosynthesis of silver nanoparticles by lixivium of sundried *Cinnamomum camphora* leaf in tubular microreactors. *Industrial & Engineering Chemistry Research*, **47**, 6081–6090, https://doi.org/10.1021/ie701698e

Huq M. A. and Akter S. (2021). Bacterial mediated rapid and facile synthesis of silver nanoparticles and their antimicrobial efficacy against pathogenic microorganisms. *Materials (Basel)*, **14**, 1–10, https://doi.org/10.3390/ma14102615

Hussain A., Alajmi M. F., Khan M. A., Pervez S. A., Ahmed F., Amir S., Husain F. M., Khan M. S., Shaik G. M., Hassan I., Khan R. A. and Rehman M. T. (2019). Biosynthesized silver nanoparticle (AgNP) from *Pandanus odorifer* leaf extract exhibits anti-metastasis and anti-biofilm potentials. *Frontiers in Microbiology*, **10**, 1–19, https://doi.org/10.3389/fmicb.2019.00008

Iravani S. and Varma R. S. (2020). Bacteria in heavy metal remediation and nanoparticle biosynthesis. *ACS Sustainable Chemistry & Engineering*, **8**, 5395–5409, https://doi.org/10.1021/acssuschemeng.0c00292

Islam R., Kumar S., Karmoker J., Kamruzzaman M., Rahman M. A., Biswas N., Tran T. K. A. and Rahman M. M. (2018). Bioaccumulation and adverse effects of persistent organic pollutants (POPs) on ecosystems and human exposure: a review study on Bangladesh perspectives. *Environmental Technology & Innovation*, **12**, 115–131, https://doi.org/10.1016/j.eti.2018.08.002

Ismail M., Khan M. I., Khan S. B., Khan M. A., Akhtar K. and Asiri A. M. (2018). Green synthesis of plant supported Cu–Ag and Cu–Ni bimetallic nanoparticles in the reduction of nitrophenols and organic dyes for water treatment. *Journal of Molecular Liquids*, **260**, 78–91, https://doi.org/10.1016/j.molliq.2018.03.058

Jain P., Kumar A., Verma N. and Gupta R. K. (2019). In-situ synthesis of TiO_2 nanoparticles in ACF: photocatalytic degradation under continuous flow. *Solar Energy*, **189**, 35–44, https://doi.org/10.1016/j.solener.2019.07.042

Jamali A.A., Akbari F., Ghorakhlu M.M., de la Guardia M. and Khosroushahi A.Y. (2012). Applications of diatoms as potential microalgae in nanobiotechnology. *BioImpacts: BI*, **2**(2), 83. https://doi.org/10.5681%2Fbi.2012.012

Jones O. G. and McClements D. J. (2011). Recent progress in biopolymer nanoparticle and microparticle formation by heat-treating electrostatic protein-polysaccharide complexes. *Advances in Colloid and Interface Science*, **167**, 49–62, https://doi.org/10.1016/j.cis.2010.10.006

Joye I. J. and McClements D. J. (2014). Biopolymer-based nanoparticles and microparticles: fabrication, characterization, and application. *Current Opinion in Colloid & Interface Science*, **19**, 417–427, https://doi.org/10.1016/j.cocis.2014.07.002

Kang S. H., Bozhilov K. N., Myung N. V., Mulchandani A. and Chen W. (2008). Microbial synthesis of CdS nanocrystals in genetically engineered *E. coli*. *Angewandte Chemie International Edition*, **47**, 5186–5189, https://doi.org/10.1002/anie.200705806

Kannan K., Radhika D., Sadasivuni K. K., Reddy K. R. and Raghu A. V. (2020). Nanostructured metal oxides and its hybrids for photocatalytic and biomedical applications. *Advances in Colloid and Interface Science*, **281**, 102178, https://doi.org/10.1016/j.cis.2020.102178

Kansal S.K., Sood S., Umar A. and Mehta S.K. (2013). Photocatalytic degradation of Eriochrome Black T dye using well-crystalline anatase TiO_2 nanoparticles. *Journal of Alloys and Compounds*, **581**, 392–397. https://doi.org/10.1016/j.jallcom.2013.07.069

Khan A. U., Khan M., Malik N., Cho M. H. and Khan M. M. (2019). Recent progress of algae and blue green algae-assisted synthesis of gold nanoparticles for various applications. *Bioprocess and Biosystems Engineering*, **42**, 1–15, https://doi.org/10.1007/s00449-018-2012-2

Khanna P., Kaur A. and Goyal D. (2019). Algae-based metallic nanoparticles: synthesis, characterization and applications. *Journal of Microbiological Methods*, **163**, 105656, https://doi.org/10.1016/j.mimet.2019.105656

Khatami M., Alijani H. Q., Nejad M. S. and Varma R. S. (2018). Core@shell nanoparticles: greener synthesis using natural plant products. *Applied Sciences*, **8**, 1–17, https://doi.org/10.3390/app8030411

Khatami M., Alijani H. Q., Fakheri B., Mobasseri M. M., Heydarpour M., Farahani Z. K. and Khan A. U. (2019). Super-paramagnetic iron oxide nanoparticles (SPIONs): greener synthesis using *Stevia* plant and evaluation of its antioxidant properties. *Journal of Cleaner Production*, **208**, 1171–1177, https://doi.org/10.1016/j.jclepro.2018.10.182

Koe W. S., Lee J. W., Chong W. C., Pang Y. L. and Sim L. C. (2020). An overview of photocatalytic degradation: photocatalysts, mechanisms, and development of photocatalytic membrane. *Environmental Science and Pollution Research*, **27**, 2522–2565, https://doi.org/10.1007/s11356-019-07193-5

Koole R., Groeneveld E., Vanmaekelbergh D., Meijerink A. and De Mello Donegá C. (2014). Size effects on semiconductor nanoparticles. In: Nanoparticles: Workhorses of Nanoscience. Springer-Verlag, Berlin Heidelberg, C. de Mello Donegá (ed.), pp. 13–51, https://doi.org/10.1007/978-3-662-44823-6_2

Krishna G., Srileka V., Singara Charya M. A., Abu Serea E. S. and Shalan A. E. (2021). Biogenic synthesis and cytotoxic effects of silver nanoparticles mediated by white rot fungi. *Heliyon*, **7**, e06470, https://doi.org/10.1016/j.heliyon.2021.e06470

Król A., Pomastowski P., Rafińska K., Railean-Plugaru V. and Buszewski B. (2017). Zinc oxide nanoparticles: synthesis, antiseptic activity and toxicity mechanism. *Advances in Colloid and Interface Science*, **249**, 37–52, https://doi.org/10.1016/j.cis.2017.07.033

Kulal D. and Shetty Kodialbail V. (2021). Visible light mediated photocatalytic dye degradation using Ag_2O/AgO-TiO_2 nanocomposite synthesized by extracellular bacterial mediated synthesis – an eco-friendly approach for pollution abatement. *Journal of Environmental Chemical Engineering*, **9**, 105389, https://doi.org/10.1016/j.jece.2021.105389

Kumar A., Kumar S., Kiran K., Banerjee S., Pande V. and Dandapat A. (2021). Myco-nanotechnological approach to synthesize silver oxide nanocuboids using endophytic fungus isolated from *Citrus pseudolimon* plant. *Colloids Surfaces B Biointerfaces*, **206**, 111948, https://doi.org/10.1016/j.colsurfb.2021.111948

Kumari K. G. I. D., Moldrup P., Paradelo M., and de Jonge L. W. (2014). Phenanthrene sorption on biochar-amended soils: application rate, aging, and physicochemical properties of soil. *Water, Air, & Soil Pollution*, **225**, 1–3, https://doi.org/10.1007/s11270-014-2105-8

Kuppusamy P., Yusoff M. M., Maniam G. P. and Govindan N. (2016). Biosynthesis of metallic nanoparticles using plant derivatives and their new avenues in pharmacological applications – an updated report. *Saudi Pharmaceutical Journal*, **24**, 473–484, https://doi.org/10.1016/j.jsps.2014.11.013

Lakshmanan G., Sathiyaseelan A., Kalaichelvan P. T. and Murugesan K. (2018). Plant-mediated synthesis of silver nanoparticles using fruit extract of *Cleome viscosa* L.: assessment of their antibacterial and anticancer activity. *Karbala International Journal of Modern Science*, **4**, 61–68, https://doi.org/10.1016/j.kijoms.2017.10.007

Li Y., Wang H., Xie J., Hou J., Song X. and Dionysiou D.D. (2021). Bi_2WO_6-TiO_2/starch composite films with Ag nanoparticle irradiated by γ-ray used for the visible light photocatalytic degradation of ethylene. *Chemical Engineering Journal*, **421**, 129986. https://doi.org/10.1016/j.cej.2021.129986

Liu H., Huang J., Gao Y., Sun D., Li J., Li X., Zhang Z. and Li Q. (2013). Production of silver nanoparticles in a continuous stirred tank reactor based on plant-mediated biosynthesis: flow behaviors and residence time distribution prediction by computational fluid dynamics simulation. *Industrial & Engineering Chemistry Research*, **52**, 2280–2289, https://doi.org/10.1021/ie302145f

Liu C., Gallagher J. J., Sakimoto K. K., Nichols E. M., Chang C. J., Chang M. C. Y. and Yang P. (2015). Nanowire-bacteria hybrids for unassisted solar carbon dioxide fixation to value-added chemicals. *Nano Letters*, **15**, 3634–3639, https://doi.org/10.1021/acs.nanolett.5b01254

Liu Y., Yan S., Li M., Wang K., Zeng D. and Yang H. (2019). A new polyethyleneinmine-gold nanoparticles-protoporphyrin cobalt (II) nanocomposite for electrochemical hydrogen peroxide biosensing. *Colloids and Surfaces A: Physicochemical and Engineering Aspects*, **572**, 345–351, https://doi.org/10.1016/j.colsurfa.2019.04.021

Liu J., Mantell J. and Jones M. R. (2020a). Minding the gap between plant and bacterial photosynthesis within a self-assembling biohybrid photosystem. ACS Nano, https://doi.org/10.1021/acsnano.0c00058

Liu Z., Wang K., Chen Y., Tan T. and Nielsen J. (2020b). Third-generation biorefineries as the means to produce fuels and chemicals from CO_2. *Nature Catalysis*, **3**, 274–288, https://doi.org/10.1038/s41929-019-0421-5

Logeswari P., Silambarasan S. and Abraham J. (2015). Synthesis of silver nanoparticles using plants extract and analysis of their antimicrobial property. *Journal of Saudi Chemical Society*, **19**, 311–317, https://doi.org/10.1016/j.jscs.2012.04.007

López Y. C. and Antuch M. (2020). Morphology control in the plant-mediated synthesis of magnetite nanoparticles. *Current Opinion in Green and Sustainable Chemistry*, **24**, 32–37, https://doi.org/10.1016/j.cogsc.2020.02.001

Mahyoub J. A. (2021). Bioactivity of two marine algae extracts and their synthesized silver nanoparticles as safe controls against Musca domestica housefly. *Entomological Research*, https://doi.org/10.1111/1748-5967.12512

Malarkodi C., Rajeshkumar S., Paulkumar K., Vanaja M., Gnanajobitha G. and Annadurai G. (2014). Biosynthesis and antimicrobial activity of semiconductor nanoparticles against oral pathogens. *Bioinorganic chemistry and applications*, 2014. https://doi.org/10.1155/2014/347167

Marchiol L., Mattiello A., Pošćić F., Giordano C. and Musetti R. (2014). In vivo synthesis of nanomaterials in plants: location of silver nanoparticles and plant metabolism. *Nanoscale Research Letters*, **9**, 1–11, https://doi.org/10.1186/1556-276X-9-101

Mashwani Z.-ur.-R., Khan T., Khan M. A. and Nadhman A. (2015). Synthesis in plants and plant extracts of silver nanoparticles with potent antimicrobial properties: current status and future prospects. *Applied Microbiology and Biotechnology*, **99**, 9923–9934, https://doi.org/10.1007/s00253-015-6987-1

Matussin S., Harunsani M. H., Tan A. L. and Khan M. M. (2020). Plant-extract-mediated SnO_2 nanoparticles: synthesis and applications. *ACS Sustainable Chemistry & Engineering*, **8**, 3040–3054, https://doi.org/10.1021/acssuschemeng.9b06398

Meena M., Swapnil P., Yadav G. and Sonigra P. (2021). Role of fungi in bio-production of nanomaterials at megascale, fungi bio-prospects in sustainable agriculture, environment and nano-technology. **3**, 453–474, INC, https://doi.org/10.1016/b978-0-12-821734-4.00006-x

Miernicki M., Hofmann T., Eisenberger I., von der Kammer F. and Praetorius A. (2019). Legal and practical challenges in classifying nanomaterials according to regulatory definitions. *Nature Nanotechnology*, **14**, 208–216, https://doi.org/10.1038/s41565-019-0396-z

Mirza A. U., Kareem A. and Bhat S. A. (2019). Metal oxide-based photocatalyst for the degradation of organic pollutants in water.

Mitra S., Das A., Sen S. and Mahanty B. (2018). Potential of metabolic engineering in bacterial nanosilver synthesis. *World Journal of Microbiology and Biotechnology*, **34**, 1–10, https://doi.org/10.1007/s11274-018-2522-8

Mittal A. K., Chisti Y. and Banerjee U. C. (2013). Synthesis of metallic nanoparticles using plant extracts. *Biotechnology Advances*, **31**, 346–356, https://doi.org/10.1016/j.biotechadv.2013.01.003

Mofokeng T. P., Moloto M. J., Shumbula P. M., Nyamukamba P., Mubiayi P. K., Takaidza S. and Marais L. (2018). Antimicrobial activity of amino acid-capped zinc and copper sulphide nanoparticles. *Journal of Nanotechnology*, **2018**, 1–9, https://doi.org/10.1155/2018/4902675

Mofokeng T. P., Moloto M. J., Shumbula P. M. and Tetyana P. (2019). Synthesis, characterization and cytotoxicity of alanine-capped CuS nanoparticles using human cervical carcinoma HeLa cells. *Analytical Biochemistry*, **580**, 36–41, https://doi.org/10.1016/j.ab.2019.06.008

Molnár Z., Bódai V., Szakacs G., Erdélyi B., Fogarassy Z., Sáfrán G., Varga T., Kónya Z., Tóth-Szeles E., Szucs R. and Lagzi I. (2018). Green synthesis of gold nanoparticles by thermophilic filamentous fungi. *Scientific Reports*, **8**, 1–12, https://doi.org/10.1038/s41598-018-22112-3

Mosleh S., Rahimi M. R., Ghaedi M., Dashtian K. and Hajati S. (2018). Sonochemical-assisted synthesis of $CuO/Cu_2O/Cu$ nanoparticles as efficient photocatalyst for simultaneous degradation of pollutant dyes in rotating packed bed reactor: LED illumination and central composite design optimization. *Ultrasonics Sonochemistry*, **40**, 601–610, https://doi.org/10.1016/j.ultsonch.2017.08.007

Mukherjee D., Bhattacharya P., Jana A., Bhattacharya S., Sarkar S., Ghosh S., Majumdar S. and Swarnakar S. (2018). Synthesis of ceramic ultrafiltration membrane and application in membrane bioreactor process for pesticide remediation from wastewater. *Process Safety and Environmental Protection*, **116**, 22–33, https://doi.org/10.1016/j.psep.2018.01.010

Mukherjee A., Sarkar D. and Sasmal S. (2021). A review of green synthesis of metal nanoparticles using algae. *Frontiers in Microbiology*, **12**, 693899. https://doi.org/10.3389/fmicb.2021.693899

Murgolo S., De Ceglie C., Di Iaconi C. and Mascolo G. (2021). Novel TiO_2-based catalysts employed in photocatalysis and photoelectrocatalysis for effective degradation of pharmaceuticals (PhACs) in water: a short review. *Current Opinion in Green and Sustainable Chemistry*, **30**, 100473, https://doi.org/10.1016/j.cogsc.2021.100473

Muthukumar H., Palanirajan S. K., Shanmugam M. K. and Gummadi S. N. (2020). Plant extract mediated synthesis enhanced the functional properties of silver ferrite nanoparticles over chemical mediated synthesis. *Biotechnology Reports*, **26**, e00469, https://doi.org/10.1016/j.btre.2020.e00469

Nadeem M., Tungmunnithum D., Hano C., Abbasi B. H., Hashmi S. S., Ahmad W. and Zahir A. (2018). The current trends in the green syntheses of titanium oxide nanoparticles and their applications. *Green Chemistry Letters and Reviews*, **11**, 492–502, https://doi.org/10.1080/17518253.2018.1538430

Neethu S., Midhun S. J., Radhakrishnan E. K. and Jyothis M. (2018). Green synthesized silver nanoparticles by marine endophytic fungus *Penicillium polonicum* and its antibacterial efficacy against biofilm forming, multidrug-resistant *Acinetobacter baumanii*. *Microbial Pathogenesis*, **116**, 263–272, https://doi.org/10.1016/j.micpath.2018.01.033

Nelwamondo S. M. M., Moloto M. J., Krause R. W. M. and Moloto N. (2012). Synthesis and characterization of alanine-capped water soluble copper sulphide quantum dots. *Materials Letters*, **75**, 161–164, https://doi.org/10.1016/j.matlet.2012.01.079

Ogunyemi S. O., Abdallah Y., Zhang M., Fouad H., Hong X., Ibrahim E., Masum M. M. I., Hossain A., Mo J. and Li B. (2019). Green synthesis of zinc oxide nanoparticles using different plant extracts and their antibacterial activity against *Xanthomonas oryzae* pv. *oryzae*. *Artificial Cells, Nanomedicine, and Biotechnology*, **47**, 341–352, https://doi.org/10.1080/21691401.2018.1557671

Olatunji O. S. (2019). Evaluation of selected polychlorinated biphenyls (PCBs) congeners and dichlorodiphenyltrichloroethane (DDT) in fresh root and leafy vegetables using GC-MS. *Scientific Reports*, **9**, 1–10, https://doi.org/10.1038/s41598-018-36996-8

Oliveira C. P. M. de, Viana M. M. and Amaral M. C. S. (2020). Coupling photocatalytic degradation using a green TiO_2 catalyst to membrane bioreactor for petroleum refinery wastewater reclamation. *Journal of Water Process Engineering*, **34**, 101093, https://doi.org/10.1016/j.jwpe.2019.101093

Ong C. B., Mohammad A. W. and Ng L. Y. (2019). Integrated adsorption-solar photocatalytic membrane reactor for degradation of hazardous Congo red using Fe-doped ZnO and Fe-doped ZnO/rGO nanocomposites. *Environmental Science and Pollution Research*, **26**, 33856–33869, https://doi.org/10.1007/s11356-018-2557-2

Osuntokun J., Onwudiwe D. C. and Ebenso E. E. (2019). Green synthesis of ZnO nanoparticles using aqueous *Brassica oleracea* L. var. *italica* and the photocatalytic activity. *Green Chemistry Letters and Reviews*, **12**, 444–457, https://doi.org/10.1080/17518253.2019.1687761

Ovais M., Khalil A. T., Ayaz M., Ahmad I., Nethi S. K. and Mukherjee S. (2018a). Biosynthesis of metal nanoparticles via microbial enzymes: a mechanistic approach. *International Journal of Molecular Sciences*, **19**, 1–20, https://doi.org/10.3390/ijms19124100

Ovais M., Khalil A. T., Islam N. U., Ahmad I., Ayaz M., Saravanan M., Shinwari Z. K. and Mukherjee S. (2018b). Role of plant phytochemicals and microbial enzymes in biosynthesis of metallic nanoparticles. *Applied Microbiology and Biotechnology*, **102**, 6799–6814, https://doi.org/10.1007/s00253-018-9146-7

Patil M. P. and Kim G. D. (2017). Eco-friendly approach for nanoparticles synthesis and mechanism behind antibacterial activity of silver and anticancer activity of gold nanoparticles. *Applied Microbiology and Biotechnology*, **101**, 79–92, https://doi.org/10.1007/s00253-016-8012-8

Pirtarighat S., Ghannadnia M. and Baghshahi S. (2019). Green synthesis of silver nanoparticles using the plant extract of *Salvia spinosa* grown in vitro and their antibacterial activity assessment. *Journal of Nanostructure in Chemistry*, **9**, 1–9, https://doi.org/10.1007/s40097-018-0291-4

Polte J. (2015). Fundamental growth principles of colloidal metal nanoparticles – a new perspective. *CrystEngComm*, **17**, 6809–6830, https://doi.org/10.1039/c5ce01014d

Porras J., Giannakis S., Torres-Palma R. A., Fernandez J. J., Bensimon M. and Pulgarin C. (2018). Fe and Cu in humic acid extracts modify bacterial inactivation pathways during solar disinfection and photo-Fenton processes in water. *Applied Catalysis B: Environmental*, **235**, 75–83, https://doi.org/10.1016/j.apcatb.2018.04.062

Pradeep T. 2007. NANO: The Essentials: Understanding Nanoscience and Nanotechnology. 1st ed. New Delhi McGraw-Hill Offices: Tata McGraw-Hill Publishing Company Limited. https://www.accessengineeringlibrary.com/content/book/9780071548298

Puiatti G. A., de Oliveira Cruz F. S., Nascimento M. A., de Oliveira A. F. and Lopes R. P. (2020). Degradation of the Direct Red 80 dye by chitosan bead-supported Fe/Ni nanoparticles in a fluidized bed reactor. *Chemical Papers*, **74**, 3367–3381, https://doi.org/10.1007/s11696-020-01165-7

Pytlik N., Kaden J., Finger M., Naumann J., Wanke S., Machill S. and Brunner E. (2017). Biological synthesis of gold nanoparticles by the diatom *Stephanopyxis turris* and in vivo SERS analyses. *Algal Research*, **28**, 9–15, https://doi.org/10.1016/j.algal.2017.10.004

Pytlik N., Klemmed B., Machill S., Eychmüller A. and Brunner E. (2019). In vivo uptake of gold nanoparticles by the diatom *Stephanopyxis turris*. *Algal Research*, **39**, 101447, https://doi.org/10.1016/j.algal.2019.101447

Rahmani R., Gharanfoli M., Gholamin M., Darroudi M., Chamani J., Sadri K. and Hashemzadeh A. (2020). Plant-mediated synthesis of superparamagnetic iron oxide nanoparticles (SPIONs) using aloe vera and flaxseed extracts and evaluation of their cellular toxicities. *Ceramics International*, **46**, 3051–3058, https://doi.org/10.1016/j.ceramint.2019.10.005

Rajaram T. and Das A. (2008). Water pollution by industrial effluents in India: discharge scenarios and case for participatory ecosystem specific local regulation. *Futures*, **40**, 56–69, https://doi.org/10.1016/j.futures.2007.06.002

Rajeshkumar S., Nandhini N. T., Manjunath K., Sivaperumal P., Krishna Prasad G., Alotaibi S. S. and Roopan S. M. (2021). Environment friendly synthesis copper oxide nanoparticles and its antioxidant, antibacterial activities using seaweed (*Sargassum longifolium*) extract. *Journal of Molecular Structure*, **1242**, 130724, https://doi.org/10.1016/j.molstruc.2021.130724

Ramos M. M., dos S. Morais E., da S. Sena I., Lima A. L., de Oliveira F. R., de Freitas C. M., Fernandes C. P., de Carvalho J. C. T. and Ferreira I. M. (2020). Silver nanoparticle from whole cells of the fungi *Trichoderma* sp. isolated from Brazilian Amazon. *Biotechnology Letters*, **42**, 833–843, https://doi.org/10.1007/s10529-020-02819-y

Rana A.S., Zubair M., Chen Y., Wang Z., Deng J., Chani M.T.S., Danner A., Teng J. and Mehmood M.Q. (2023). Broadband solar absorption by chromium metasurface for highly efficient solar thermophotovoltaic systems. *Renewable and Sustainable Energy Reviews*, **171**, 113005. https://doi.org/10.1016/j.rser.2022.113005

Rasheed T., Bilal M., Hassan A. A., Nabeel F., Bharagava R. N., Romanholo Ferreira L. F., Tran H. N. and Iqbal H. M. N. (2020). Environmental threatening concern and efficient removal of pharmaceutically active compounds using metal-organic frameworks as adsorbents. *Environmental Research*, **185**, 109436, https://doi.org/10.1016/j.envres.2020.109436

Ravichandran V., Vasanthi S., Shalini S., Shah S. A. A., Tripathy M. and Paliwal N. (2019). Green synthesis, characterization, antibacterial, antioxidant and photocatalytic activity of *Parkia speciosa* leaves extract mediated silver nanoparticles. *Results in Physics*, **15**, 102565, https://doi.org/10.1016/j.rinp.2019.102565

Rocha A. A., Wilde C., Hu Z., Nepotchatykh O., Nazarenko Y. and Ariya P. A. (2017). Development of a hybrid photo-bioreactor and nanoparticle adsorbent system for the removal of CO_2, and selected organic and metal co-pollutants. *Journal of Environmental Sciences*, **57**, 41–53, https://doi.org/10.1016/j.jes.2016.12.010

Saeed S., Iqbal A. and Ashraf M. A. (2020). Bacterial-mediated synthesis of silver nanoparticles and their significant effect against pathogens. *Environmental Science and Pollution Research*, **27**, 37347–37356, https://doi.org/10.1007/s11356-020-07610-0

Saharan P., Chaudhary G. R., Lata S., Mehta S. K. and Mor S. (2015). Ultra fast and effective treatment of dyes from water with the synergistic effect of Ni doped ZnO nanoparticles and ultrasonication. *Ultrasonics Sonochemistry*, **22**, 317–325, https://doi.org/10.1016/j.ultsonch.2014.07.004

Sahoo B.D., Joshi K.D. and Kaushik T.C. (2019). Structural stability of uranium carbide (UC) under high pressure: ab-initio study. *Computational Condensed Matter*, **21**, 00431. https://doi.org/10.1016/j.cocom.2019.e00431

Sakimoto K. K., Wong A. B. and Yang P. (2016). Self-photosensitization of nonphotosynthetic bacteria for solar-to-chemical production. *Science*, **351**, 74–77, https://doi.org/10.1126/science.aad3317

Samanta P., Desai A. V., Let S. and Ghosh S. K. (2019). Advanced porous materials for sensing, capture and detoxification of organic pollutants toward water remediation. *ACS Sustainable Chemistry & Engineering*, **7**, 7456–7478, https://doi.org/10.1021/acssuschemeng.9b00155

Sanaeimehr Z., Javadi I. and Namvar F. (2018). Antiangiogenic and antiapoptotic effects of green-synthesized zinc oxide nanoparticles using *Sargassum muticum* algae extraction. *Cancer Nanotechnology*, **9**, 1–9, https://doi.org/10.1186/s12645-018-0037-5

Santhoshkumar J., Kumar S. V. and Rajeshkumar S. (2017). Synthesis of zinc oxide nanoparticles using plant leaf extract against urinary tract infection pathogen. *Resources Technology*, **3**, 459–465, https://doi.org/10.1016/j.reffit.2017.05.001

Saratale R. G., Saratale G. D., Shin H. S., Jacob J. M., Pugazhendhi A., Bhaisare M. and Kumar G. (2018). New insights on the green synthesis of metallic nanoparticles using plant and waste biomaterials: current knowledge, their agricultural and environmental applications. *Environmental Science and Pollution Research*, **25**, 10164–10183, https://doi.org/10.1007/s11356-017-9912-6

Saravanan C., Rajesh R., Kaviarasan T., Muthukumar K., Kavitake D. and Shetty P. H. (2017). Synthesis of silver nanoparticles using bacterial exopolysaccharide and its application for degradation of azo-dyes. *Biotechnology Reports*, **15**, 33–40, https://doi.org/10.1016/j.btre.2017.02.006

Saravanan M., Barik S. K., MubarakAli D., Prakash P. and Pugazhendhi A. (2018). Synthesis of silver nanoparticles from *Bacillus brevis* (NCIM 2533) and their antibacterial activity against pathogenic bacteria. *Microbial Pathogenesis*, **116**, 221–226, https://doi.org/10.1016/j.micpath.2018.01.038

Sathishkumar R. S., Sundaramanickam A., Srinath R., Ramesh T., Saranya K., Meena M. and Surya P. (2019). Green synthesis of silver nanoparticles by bloom forming marine microalgae *Trichodesmium erythraeum* and its applications in antioxidant, drug-resistant bacteria, and cytotoxicity activity. *Journal of Saudi Chemical Society*, **23**, 1180–1191, https://doi.org/10.1016/j.jscs.2019.07.008

Segets D., Marczak R., Schäfer S., Paula C., Gnichwitz J. F., Hirsch A. and Peukert W. (2011). Experimental and theoretical studies of the colloidal stability of nanoparticles – a general interpretation based on stability maps. *ACS Nano*, **5**, 4658–4669, https://doi.org/10.1021/nn200465b

Shah M., Fawcett D., Sharma S., Tripathy S. K. and Poinern G. E. J. (2015). Green synthesis of metallic nanoparticles via biological entities. *Materials*, **8**, 7278–7308, https://doi.org/10.3390/ma8115377

Shaik M. R., Khan M., Kuniyil M., Al-Warthan A., Alkhathlan H. Z., Siddiqui M. R. H., Shaik J. P., Ahamed A., Mahmood A., Khan M. and Adil S. F. (2018). Plant-extract-assisted green synthesis of silver nanoparticles using *Origanum vulgare* L. extract and their microbicidal activities. *Sustainability*, **10**, 1–14, https://doi.org/10.3390/su10040913

Sharma G., Sharma S., Kumar A., Ala'a H., Naushad M., Ghfar A.A., Mola G.T. and Stadler F.J. (2018). Guar gum and its composites as potential materials for diverse applications: A review. *Carbohydrate polymers*, **199**, 534–545. https://doi.org/10.1016/j.carbpol.2018.07.053

Sharma D., Kanchi S. and Bisetty K. (2019a). Biogenic synthesis of nanoparticles: a review. *Arabian Journal of Chemistry*, **12**, 3576–3600, https://doi.org/10.1016/j.arabjc.2015.11.002

Sharma P., Pant S., Dave V., Tak K., Sadhu V. and Reddy K. R. (2019b). Green synthesis and characterization of copper nanoparticles by *Tinospora cardifolia* to produce nature-friendly copper nano-coated fabric and their antimicrobial evaluation. *Journal of Microbiological Methods*, **160**, 107–116, https://doi.org/10.1016/j.mimet.2019.03.007

Sharma V., Kaushik S., Pandit P., Dhull D., Yadav J. P. and Kaushik S. (2019c). Green synthesis of silver nanoparticles from medicinal plants and evaluation of their antiviral potential against chikungunya virus. *Applied Microbiology and Biotechnology*, **103**, 881–891, https://doi.org/10.1007/s00253-018-9488-1

Shojaei Yeganeh M., Kazemizadeh A. R., Ramazani A., Eskandari P. and Rabbi Angourani H. (2019). Plant-mediated synthesis of $Cu_{0.5}Zn_{0.5}Fe_2O_4$ nanoparticles using *Minidium leavigatum* and their applications as an adsorbent for removal of reactive blue 222 dye. *Materials Research Express*, **6**, 0–12, https://doi.org/10.1088/2053-1591/ab6637

Shu M., He F., Li Z., Zhu X., Ma Y., Zhou Z., Yang Z., Gao F. and Zeng M. (2020). Biosynthesis and antibacterial activity of silver nanoparticles using yeast extract as reducing and capping agents. *Nanoscale Research Letters*, **15**, 1–9, https://doi.org/10.1186/s11671-019-3244-z

Siddiqi K. S. and Husen A. (2016). Fabrication of metal nanoparticles from fungi and metal salts: scope and application. *Nanoscale Research Letters*, **11**, 1–15, https://doi.org/10.1186/s11671-016-1311-2

Singh K. and Arora S. (2011). Removal of synthetic textile dyes from wastewaters: a critical review on present treatment technologies. *Critical Reviews in Environmental Science and Technology*, **41**, 807–878, https://doi.org/10.1080/10643380903218376

Singh P., Kim Y. J., Zhang D. and Yang D. C. (2016). Biological synthesis of nanoparticles from plants and microorganisms. *Trends in Biotechnology*, **34**, 588–599, https://doi.org/10.1016/j.tibtech.2016.02.006

Singh H., Du J., Singh P. and Yi T. H. (2018). Extracellular synthesis of silver nanoparticles by *Pseudomonas* sp. THG-LS1.4 and their antimicrobial application. *Journal of Pharmaceutical Analysis*, **8**, 258–264, https://doi.org/10.1016/j.jpha.2018.04.004

Soleimani M., Ghasemi J. B., Mohammadi Ziarani G., Karimi-Maleh H. and Badiei A. (2021). Photocatalytic degradation of organic pollutants, viral and bacterial pathogens using titania nanoparticles. *Inorganic Chemistry Communications*, **130**, 108688, https://doi.org/10.1016/j.inoche.2021.108688

Sonbol H., Ameen F., AlYahya S., Almansob A. and Alwakeel S. (2021). *Padina boryana* mediated green synthesis of crystalline palladium nanoparticles as potential nanodrug against multidrug resistant bacteria and cancer cells. *Scientific Reports*, **11**, 5444, https://doi.org/10.1038/s41598-021-84794-6

Suba S., Vijayakumar S., Vidhya E., Punitha V. N. and Nilavukkarasi M. (2021). Microbial mediated synthesis of ZnO nanoparticles derived from *Lactobacillus* spp: characterizations, antimicrobial and biocompatibility efficiencies. *Sensors International*, **2**, 100104, https://doi.org/10.1016/j.sintl.2021.100104

Subashini G. and Bhuvaneswari S. (2018). Nanoparticles from fungi (Myconanoparticles). In: Fungi and their Role in Sustainable Development: Current Perspectives, P. Gehlot and J. Singh (eds), Springer, Singapore , pp. 753–779, https://doi.org/10.1007/978-981-13-0393-7_39

Sumanth B., Lakshmeesha T. R., Ansari M. A., Alzohairy M. A., Udayashankar A. C., Shobha B., Niranjana S. R., Srinivas C. and Almatroudi A. (2020). Mycogenic synthesis of extracellular zinc oxide nanoparticles from *Xylaria acuta* and its nanoantibiotic potential. *International Journal of Nanomedicine*, **15**, 8519–8536, https://doi.org/10.2147/IJN.S271743

Sur D. H. and Mukhopadhyay M. (2019). Role of zinc oxide nanoparticles for effluent treatment using *Pseudomonas putida* and *Pseudomonas aureofaciens*. *Bioprocess and Biosystems Engineering*, **42**, 187–198, https://doi.org/10.1007/s00449-018-2024-y

Taherzadeh Soureshjani P., Shadi A. and Mohammadsaleh F. (2021). Algae-mediated route to biogenic cuprous oxide nanoparticles and spindle-like $CaCO_3$: a comparative study, facile synthesis, and biological properties. *RSC Advances*, **11**, 10599–10609, https://doi.org/10.1039/d1ra00187f

Tan D. H. S., Banerjee A., Chen Z. and Meng Y. S. (2020). From nanoscale interface characterization to sustainable energy storage using all-solid-state batteries. *Nature Nanotechnology*, **15**, 170–180, https://doi.org/10.1038/s41565-020-0657-x

Tănase M. A., Marinescu M., Oancea P., Răducan A., Mihaescu C. I., Alexandrescu E., Nistor C. L., Jinga L. I., Diţu L. M., Petcu C. and Cinteza L. O. (2021). Antibacterial and photocatalytic properties of ZnO nanoparticles obtained from chemical versus *Saponaria officinalis* extract-mediated synthesis. *Molecules*, **26**, https://doi.org/10.3390/molecules26072072

Teimouri M., Khosravi-Nejad F., Attar F., Saboury A. A., Kostova I., Benelli G. and Falahati M. (2018). Gold nanoparticles fabrication by plant extracts: synthesis, characterization, degradation of 4-nitrophenol from industrial wastewater, and insecticidal activity – a review. *Journal of Cleaner Production*, **184**, 740–753, https://doi.org/10.1016/j.jclepro.2018.02.268

Thakkar K. N., Mhatre S. S. and Parikh R. Y. (2010). Biological synthesis of metallic nanoparticles. *Nanomedicine: Nanotechnology, Biology and Medicine*, **6**, 257–262, https://doi.org/10.1016/j.nano.2009.07.002

Vaddadi L. P., Avisar D., Vadivel V. K., Menashe O., Kurzbaum E., Cohen-Yaniv V. and Mamane H. (2020). LP-UV-Nano MgO_2 pretreated catalysis followed by small bioreactor platform capsules treatment for superior kinetic degradation performance of 17α-ethynylestradiol. *Materials (Basel)*, **13**, 1–13, https://doi.org/10.3390/ma13010083

Vanti G. L., Kurjogi M., Basavesha K. N., Teradal N. L., Masaphy S. and Nargund V. B. (2020). Synthesis and antibacterial activity of *Solanum torvum* mediated silver nanoparticle against *Xanthomonas axonopodis pv. punicae* and *Ralstonia solanacearum*. *Journal of Biotechnology*, **309**, 20–28, https://doi.org/10.1016/j.jbiotec.2019.12.009

Verma A. and Bharadvaja N. (2021). Plant-mediated synthesis and characterization of silver and copper oxide nanoparticles: antibacterial and heavy metal removal activity. *Journal of Cluster Science*, **8**, https://doi.org/10.1007/s10876-021-02091-8

Vetchinkina E., Loshchinina E., Kupryashina M., Burov A. and Nikitina V. (2019). Shape and size diversity of gold, silver, selenium, and silica nanoparticles prepared by green synthesis using fungi and bacteria. *Industrial & Engineering Chemistry Research*, **58**, 17207–17218, https://doi.org/10.1021/acs.iecr.9b03345

Wang Q., Wang Z., Awasthi M.K., Jiang Y., Li R., Ren X., Zhao J., Shen F., Wang M. and Zhang Z. (2016). Evaluation of medical stone amendment for the reduction of nitrogen loss and bioavailability of heavy metals during pig manure composting. *Bioresource technology*, **220**, 297–304. https://doi.org/10.1016/j.biortech.2016.08.081

Wang B., Jiang Z., Yu J. C., Wang J. and Wong P. K. (2019). Enhanced CO_2 reduction and valuable C_{2+} chemical production by a CdS-photosynthetic hybrid system. *Nanoscale*, **11**, 9296–9301, https://doi.org/10.1039/c9nr02896j

Wang Q., Wu X., Jiang L., Fang C., Wang H. and Chen L. (2020). Effective degradation of Di-n-butyl phthalate by reusable, magnetic Fe_3O_4 nanoparticle-immobilized *Pseudomonas* sp. W1 and its application in simulation. *Chemosphere*, **250**, 126339, https://doi.org/10.1016/j.chemosphere.2020.126339

Xiong L., Zhang X., Huang Y. X., Liu W. J., Chen Y. L., Yu S. S., Hu X., Cheng L., Liu D. F. and Yu H. Q. (2018). Biogenic synthesis of Pd-based nanoparticles with enhanced catalytic activity. *ACS Applied Nano Materials*, **1**, 1467–1475, https://doi.org/10.1021/acsanm.7b00322

Xu G., Zhang Z., Li T., Du M., Guan Y. and Guo C. (2020). Continuous high-efficient degradation of organic pollutants based on sea urchin-like Fe_3O_4/ZnO/ZnSe heterostructures in photocatalytic magnetically fixed bed reactor. *Colloids and Surfaces A: Physicochemical and Engineering Aspects*, **603**, 125198, https://doi.org/10.1016/j.colsurfa.2020.125198

Xue X., Ji W., Mao Z., Mao H., Wang Y., Wang X., Ruan W., Zhao B. and Lombardi J. R. (2012). Raman investigation of nanosized TiO_2: effect of crystallite size and quantum confinement. *The Journal of Physical Chemistry C*, **116**, 8792–8797, https://doi.org/10.1021/jp2122196

Yanhong L., Dejun W., Qidong Z., Min Y. and Qinglin Z. (2004). A study of quantum confinement properties of photogenerated charges in ZnO nanoparticles by surface photovoltage spectroscopy. *The Journal of Physical Chemistry B*, **108**, 3202–3206, https://doi.org/10.1021/jp037201k

Yao X., Xu Q., Tian D., Wang N., Fang Y., Deng Z., Phillips G. O. and Lu J. (2013). Physical and chemical stability of gum arabic-stabilized conjugated linoleic acid oil-in-water emulsions. *Journal of Agricultural and Food Chemistry*, **61**, 4639–4645, https://doi.org/10.1021/jf400439d

Yılmaz Öztürk B., Yenice Gürsu B. and Dağ İ. (2020). Antibiofilm and antimicrobial activities of green synthesized silver nanoparticles using marine red algae *Gelidium corneum*. *Process Biochemistry*, **89**, 208–219, https://doi.org/10.1016/j.procbio.2019.10.027

Yu Y., Li C., Huang S., Hu Z., Chen Z. and Gao H. (2018). BiOBr hybrids for organic pollutant removal by the combined treatments of adsorption and photocatalysis. *RSC Advances*, **8**, 32368–32376, https://doi.org/10.1039/c8ra03673j

Yu Y., Zhang Q., Chi M., Jiang H., Liu X., Wang S. and Min D. (2021). Porous wood decorated with gold nanoparticles as flow-through membrane reactor for catalytic hydrogenation of methylene blue and 4-nitrophenol. *Cellulose*, **28**, 7283–7294, https://doi.org/10.1007/s10570-021-03934-3

Zafar R., Zia K. M., Tabasum S., Jabeen F., Noreen A. and Zuber M. (2016). Polysaccharide based bionanocomposites, properties and applications: a review. *International Journal of Biological Macromolecules*, **92**, 1012–1024, https://doi.org/10.1016/j.ijbiomac.2016.07.102

Zeng D., Zhao J., Luk K. H., Cheung S. T., Wong K. H. and Chen T. (2019). Potentiation of in vivo anticancer efficacy of selenium nanoparticles by mushroom polysaccharides surface decoration. *Journal of Agricultural and Food Chemistry*, **67**, 2865–2876, https://doi.org/10.1021/acs.jafc.9b00193

Zhang D. Q., Eng C. Y., Stuckey D. C. and Zhou Y. (2017). Effects of ZnO nanoparticle exposure on wastewater treatment and soluble microbial products (SMPs) in an anoxic-aerobic membrane bioreactor. *Chemosphere*, **171**, 446–459, https://doi.org/10.1016/j.chemosphere.2016.12.053

Zhang Y., Ni Q., Xu C., Wan B., Geng Y., Zheng G., Yang Z., Tao J., Zhao Y., Wen J., Zhang J., Wang S., Tang Y., Li Y., Zhang Q., Liu L., Teng Z. and Lu G. (2019). Smart bacterial magnetic nanoparticles for tumor-targeting magnetic resonance imaging of HER2-positive breast cancers. *ACS Applied Materials & Interfaces*, **11**, 3654–3665, https://doi.org/10.1021/acsami.8b15838

Zhang R., He Y., Yi J., Zhang L., Shen C., Liu S., Liu L., Liu B. and Qiao L. (2020). Proteomic and metabolic elucidation of solar-powered biomanufacturing by bio-abiotic hybrid system. *Chem*, **6**, 234–249, https://doi.org/10.1016/j.chempr.2019.11.002

doi: 10.2166/9781789063592_0061

Chapter 4

Energy storage devices: batteries and supercapacitors

Taniya Rose Abraham[1*], *Kevin Sunil*[2*], *Muhammed Shah*[2*], *Neelesh Ashok*[2**] and *Sabu Thomas*[3*]

[1],[2*],[3*]School of Energy Materials, Mahatma Gandhi University, Kottayam, Kerala, India*
*[2**]Amrita School of Artificial Intelligence, Amrita Vishwa Vidyapeetham, Coimbatore, Tamil Nadu, India*
Email: taniyakizhakkel@gmail.com

ABSTRACT

Energy storage devices play a crucial role in our modern society, enabling the efficient utilization of renewable energy sources, powering portable electronics, and facilitating the transition to electric transportation. The chapter begins by exploring the need of energy storage in the modern world, the fundamental principles and working mechanisms of batteries, including the different types such as lithium-ion, nickel-metal hydride and lead-acid battery. The design, manufacturing and performance characteristics of each battery type are discussed, along with their advantages and limitations. Additionally, the chapter discusses the latest advancements in battery technology, including the development of solid-state batteries, high-energy-density batteries, and next-generation battery materials. Moreover, the chapter examines supercapacitors, an emerging energy storage technology with unique advantages such as high-power density, fast charging, and greater cycle life. The underlying principles and operating principles of supercapacitors are explained, with a focus on different types such as electrochemical double-layer capacitors and pseudocapacitors. Also, the chapter explores the various electrode materials, electrolytes, and device architectures employed in supercapacitors, highlighting recent advancements and future prospects.

Keywords: batteries, electric transportation, energy storage devices, renewable energy, supercapacitors.

4.1 INTRODUCTION

Currently, the demand for reliable and sustainable energy solutions has become increasingly significant. With the rising demand of renewable energy sources, such as from the wind and sun, the intermittent nature of these sources poses challenges in continuous energy supply with demand. The transportation electrification and portable electronics require efficient and long-lasting energy storage solutions. Therefore, there is a critical need for effective energy storage devices that can store surplus energy available at the time of low demand and can use it on high demand.

Energy storage methods play a crucial role in addressing the intermittency and variability of renewable energy sources, stabilizing power grids, and optimizing the utilization of energy resources (Aneke & Wang, 2016). Currently, there exists a wide range of energy storage methods, each with its unique characteristics, advantages, and disadvantages. They can be broadly categorized into mechanical, electrochemical, thermal, and chemical storage systems.

Mechanical energy storage technologies, like the pumped storage hydropower and the compressed air energy storage, utilize the potential energy of water or compressed air to store and release energy. These systems offer large-scale and grid-level storage solutions with high energy storage capacity and relatively long discharge durations. Thermal energy storage systems store excess heat or cold for later use, commonly employed in district heating and cooling applications, concentrating solar power plants, and industrial processes (Dinçer & Rosen, 2011). Chemical energy storage, including batteries and fuel cells, involves the conversion of chemical energy into electrical energy for various applications (Larcher & Tarascon, 2014). Among the different energy storage technologies, batteries and supercapacitors have emerged as key players due to their versatility, efficiency, and continuous advancements. Batteries are widely utilized in applications ranging from small-scale portable electronics to large-scale grid storage. They store electrical energy through reversible electrochemical reactions, making them a reliable and portable source of power.

4.2 BATTERIES: PRINCIPLES AND OPERATION

Batteries are electrochemical energy storage devices that transfer chemical energy into electrical energy through a series of electrochemical reactions. They consist of one or more electrochemical cells, which comprise positive and negative electrodes, an electrolyte, and a separator. When a load is connected to the battery, an electrochemical reaction occurs within the cell, leading to the flow of electrons and the generation of electrical current (Winter & Brodd, 2004).

4.2.1 Battery basics

4.2.1.1 Structure and components

Understanding the structure and components of batteries is fundamental to comprehending their functionality. They consist of several components working together to facilitate the flow of electrons. The schematic representation of the general structure and components of a battery is shown in Figure 4.1.

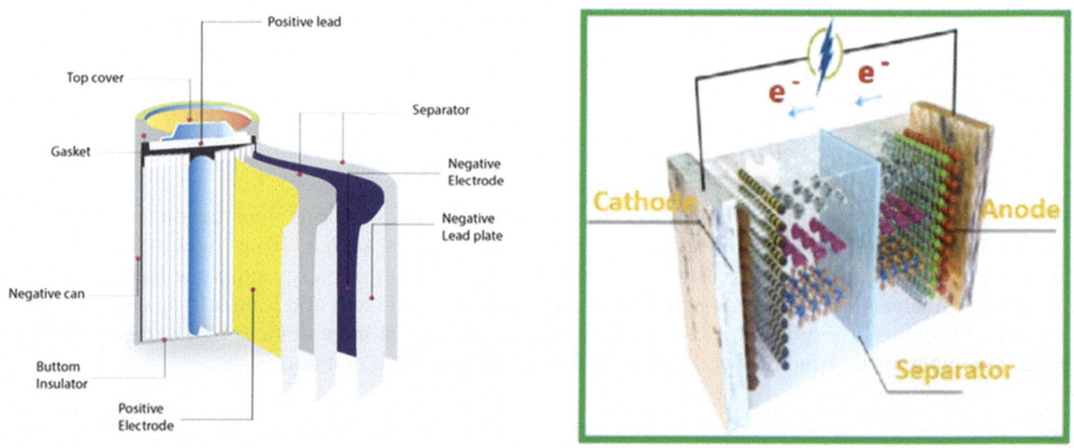

Figure 4.1 General battery structure and components.

At the core of a battery is an electrochemical cell, which comprises two electrodes immersed in an electrolyte. The electrodes are typically made of different materials and are known as the cathode and the anode. Here, the cathode is the positive electrode, while the anode is the negative electrode. These electrodes facilitate the redox (reduction–oxidation) reactions that occur during operation. The electrolyte acts as a medium for ion transport which separates the electrodes and also allows the ions to flow between them. It can be a liquid, gel, or solid, depending on the type of battery. The electrolyte carries the ions necessary for the redox reactions and completes the electrical circuit. In addition to the electrodes and electrolyte, batteries often include other components to enhance performance and safety. These may include separators, which prevent the electrodes from coming into direct contact and causing short circuits. Separators are typically porous membranes that allow the flow of ions while preventing physical contact between the electrodes. Furthermore, batteries may have current collectors, which are conductive materials that collect and distribute the generated electrical current. These collectors are usually metallic foils, such as copper or aluminum, attached to the electrodes.

4.2.1.2 Electrochemical reactions in batteries

Electrochemical reactions play a significant role in the battery operations, which are essential for the energy storage devices used in various applications.

The operation of a battery relies on the transfer of electrons within two chemical reactions, that is an oxidation and a reduction reaction. The physical separation between these reactions helps to set out from the other reactions like rusting. This separation allows for the insertion of a load between them. The voltage of the battery is the one which drives the load and which is determined by the electrochemical potential difference between the two reactions that occur in each cathode and anode. These electrons will pass through the external circuit which corresponds to the current passing through the load. The batter components, depicted in Figure 4.2, include electrodes and electrolytes for both the reduction and oxidation reactions.

The fundamental characteristics of a battery are largely determined by the type of materials utilized for the electrodes and electrolytes in both reactions. The electrode serves as a central aspect of the redox reaction where the electron transfer takes place. In certain battery systems like lead-acid and alkaline batteries, the electrode not only facilitates the electron transfer but participates actively in the chemical reaction that may consume or produce the electrons. However, in other battery systems like

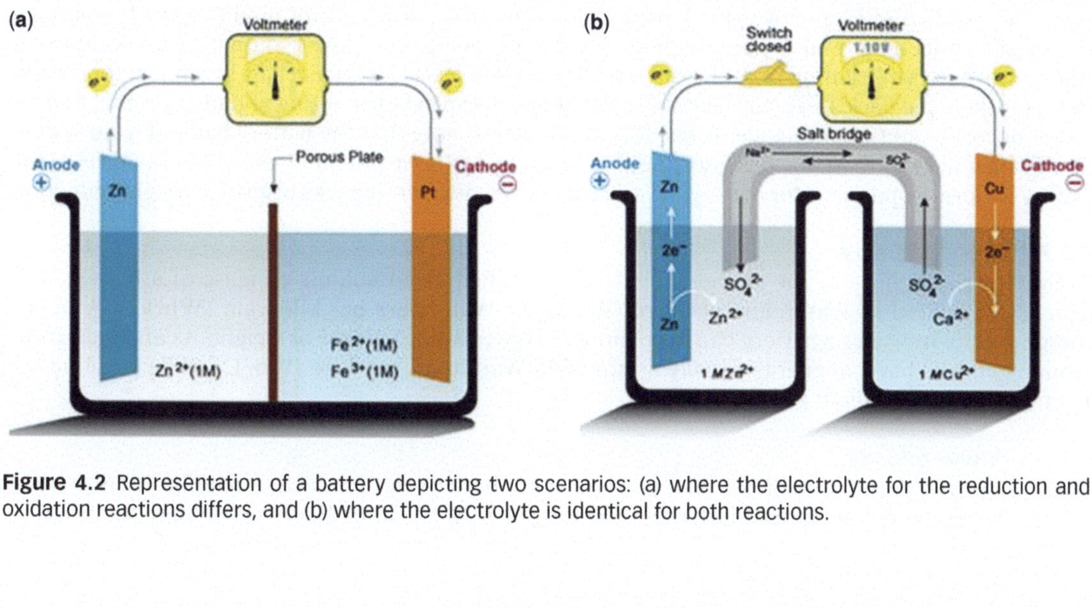

Figure 4.2 Representation of a battery depicting two scenarios: (a) where the electrolyte for the reduction and oxidation reactions differs, and (b) where the electrolyte is identical for both reactions.

fuel cells, the electrode material itself is inert and solely serves as a place for the electron movement between reactants.

During battery discharge, the electrode responsible for the oxidation reaction is known as the anode, characterized by a positive voltage according to its definition. Conversely, the electrode responsible for the reduction reaction is called the cathode and possesses a negative voltage. These distinctions are vital in understanding the flow of electrons within a discharging battery.

The flow of current within a battery results from electron transfer between the electrodes. During the discharge process, the electrons are generated from the anode during oxidation, while the electrons are consumed at the cathode during reduction. This shows the movement of electrons from the anode to the cathode (Figure 4.2a). To facilitate their transfer, a conventional electrical connection, like a wire, links the anode and cathode, enabling easy movement of electrons between them. However, in a battery circuit, electrons are not the sole carriers of charge. While electrons travel from the anode, they do not return from the cathode simultaneously. To maintain electrical neutrality, ion movement occurs within the electrolyte. With the help of a salt bridge, the electrolytes on each anode and cathode side are connected for the redox reaction (Figure 4.2b). This bridge ensures that ions move in a way that prevents charge accumulation at either positive or negative electrodes.

In many practical battery systems, the same electrolyte is employed for both the anode and the cathode, enabling ion transport through the electrolyte itself. In this case, a separator is incorporated between the anode and the cathode to remove the physical contact. The separator is necessary because the electrodes are usually in close proximity, and direct contact would cause a short circuit, bypassing the external circuit and load and allowing electrons to transfer directly. The redox reactions involved in a given battery system play a crucial role in determining various fundamental characteristics of the battery. Additionally, the design of the battery, such as the number of materials used and the electrode configuration, also influences important battery properties, including capacity, charging/discharging performance, and other practical considerations (Döge *et al.*, 2016).

4.2.2 Battery performance metrics

When assessing battery performance, several crucial metrics and considerations should be considered.

4.2.2.1 Cell, module, and pack level

When evaluating energy and power densities, it is crucial to consider whether the data pertain to a particular cell or a complete battery pack. Cells consistently possess the highest energy and power capacities relative to size or weight. This is due to the extra structure of modules, cooling system, electrical connections and weight without increasing energy or power output. The decrease in efficiency when transitioning from the cell level to the pack level is influenced by several factors. These factors include the pack's size, the cell's heat tolerance, the speed of charging and discharging, and the extent of costly lightweighting measures. It is important to note that the battery bank of a car will be significantly heavier than one designed for an aircraft, despite the potential use of the same cells. Car battery packs are optimized for battery life and cost, whereas aircraft packs prioritize weight reduction.

4.2.2.2 Energy density

Energy density defines as the energy stored in a specified given volume or mass of a battery. It is typically measured in Watt-hours per litre (Wh/L) or Watt-hours per kilogram (Wh/kg). A higher energy density indicates a battery can keep more energy in a tiny volume or weight. As an illustration, lithium-ion cells have an energy density around 690 Watt-hours per litre (Wh/L), while gasoline has an energy density of about 9500 Wh/L.

4.2.2.3 Power density

The term 'power density' describes how much power a battery can produce in relation to its mass or volume. It is generally measured in terms of Watts per litre (W/L) or Watts per kilogram (W/kg). A higher power density indicates that the battery can deliver more power relative to its size or weight. The charging speed (C-Rate) is closely linked to the power density because the power transfer into or out of a battery is influenced by internal resistance and the capacity to manage heat dissipation or resist heat.

4.2.2.4 Specific energy (or gravimetric energy density)

The amount of energy stored in a battery per unit mass. For example, the energy density in commercial electric vehicle (EV) cells is typically around 260 Wh/kg. However, this value decreases to approximately 150 Wh/kg at the pack level. In high-performance aerospace packs, the energy density can reach up to 220 Wh/kg.

4.2.2.5 Specific power (or gravimetric power density)

The specific power refers to the measure of power output per unit mass. It is commonly measured in terms of Watts per kilogram (W/kg). Similar to power density, the specific power of a battery is influenced by various factors. For instance, commercial lithium-ion EV cells typically achieve a specific power of around 340 W/kg. In contrast, state-of-the-art aluminum-ion batteries, such as aluminum graphene, have illustrated the laboratory's impressive specific power of 7000 W/kg.

4.2.2.6 Cycle life

It refers to the total number of charge–discharge cycles a battery can go through before its performance degrades to a point where it is no longer useful. However, defining cycle life as a single numerical value is a simplification because battery degradation occurs gradually over time.

4.2.2.7 Charge–discharge efficiency

It simply denotes as the battery's energy efficiency. Charge–discharge efficiency measures how effectively the battery converts the energy input during charging into energy output.

4.2.2.8 Self-discharge rate

It refers to the rate at which a fully charged battery loses energy without being used. When the battery is charged or stored at elevated temperatures, the self-discharge rate will be higher. Generally, lithium-ion batteries self-discharge at a rate of approximately 2–3% per month.

4.2.2.9 Operating temperature

A cell can function optimally in a range of temperatures. Cells that operate at higher temperatures need less cooling, leading to lighter battery packs and extended cycle life. Additionally, operating at low temperatures can eliminate the need for cold-start heating.

4.2.2.10 Impedance

Impedance refers to the resistance encountered within a cell when an electrical current passes through it. Higher impedance levels indicate a weakness within the battery, resulting in the transition of stored energy into heat instead of a usable current when the battery is utilized.

4.2.2.11 Round-trip efficiency

The energy given to a battery throughout charging cannot be fully retrieved during discharge due to some internal losses and its material degradation. The round-trip efficiency is identified by the quantity of energy that can be taken out from the battery during discharge relative to the energy given during charging. This efficiency is influenced by the charging and discharging currents, with higher currents leading to increased thermal losses and decreased efficiency (Döge *et al.*, 2016, Boehnstedt & Whear, 2009).

4.3 TYPES OF BATTERIES

There are various kinds of batteries, each with its unique characteristics and applications. Figure 4.3 gives some familiar types of batteries. Batteries can be classified based on chemical composition, form factor, size and specific applications. There are two main kinds of batteries: (1) primary batteries that cannot be recharged once depleted and (2) secondary batteries that are composed of electrochemical cells that allow their chemical reactions to be reversed by applying a specific voltage to the battery in the opposite direction. Unlike primary cells, secondary cells can be recharged once the energy stored in the battery has been depleted (Boehnstedt & Whear, 2009). The various types of rechargeable batteries are described below.

4.3.1 Nickel–cadmium batteries

The nickel–cadmium battery or NiCd or NiCad battery (Figure 4.4a) is a rechargeable battery that utilizes its electrodes as nickel oxide hydroxide and metallic cadmium. In these cells, cadmium undergoes oxidation to form cadmium hydroxide, while nickel undergoes reduction. As a result, nickel–cadmium cells produce an output voltage of 1.46 V.

$$Cd + NiO_2 + 2H_2O \rightarrow Cd(OH)_2 + Ni(OH)_2 \tag{4.1}$$

NiCd batteries are known for retaining voltage and holding a charge even when not in operation. Even so, one drawback of NiCd batteries is their susceptibility to the 'memory' effect, which occurs when a moderately charged battery is recharged, leading to a decrease in the future capacity of the battery (Yoshino, 2012).

The nickel-metal hydride (NiMH) battery (Figure 4.4b) is a rechargeable secondary cell that closely resembles the nickel–cadmium cell, with the main difference being the use of a hydrogen-absorbing alloy instead of cadmium. As compared to nickel–cadmium cells, NiMH batteries offer approximately 2–3 times the capacity; they can store and deliver more energy. A detailed comparison with battery performance parameters for Ni–Cd and Ni-MH batteries is given in Table 4.1.

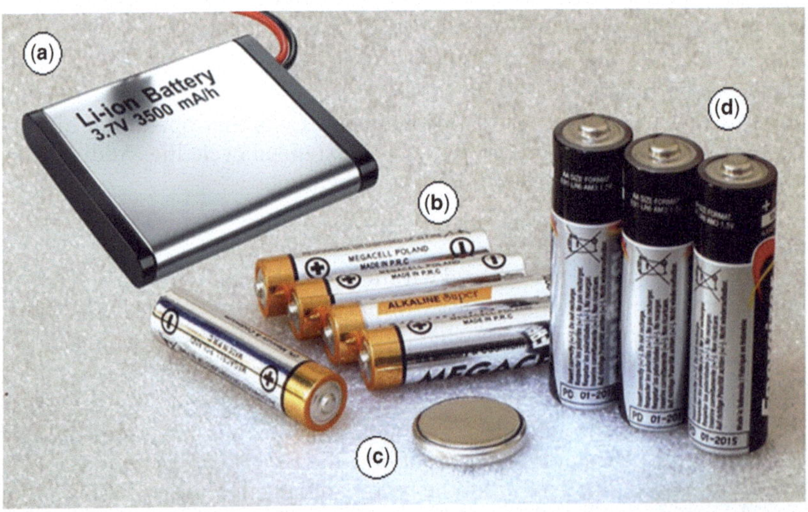

Figure 4.3 Different types of batteries: (a) lithium-ion batteries, (b) alkaline batteries, (c) button cell and (d) Ni-zinc batteries.

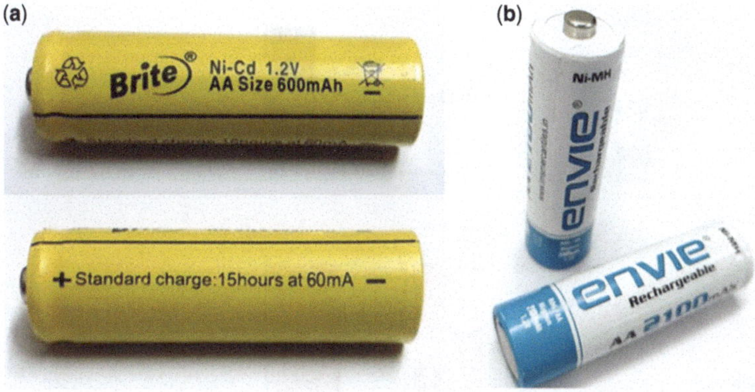

Figure 4.4 Photographs of (a) nickel-cadmium batteries and (b) nickel metal hydride (Ni-MH) batteries.

Table 4.1 Comparison of the performance parameters of Ni-Cd and Ni-MH batteries.

Performance Parameters	Ni-Cd Batteries	Nickel metal hydride (Ni-MH) Battery
Specific energy	40–60 W-h/kg	60–120 h/kg
Energy density	50–150 W-h/L	140–300 Wh/L
Specific power	150 W/kg	250–1000 W/kg
Charge/discharge efficiency	70–90%	66%–92%
Self-discharge rate	10%/month	1.3–2.9%/month at 20°C
Cycle durability/life	2000 cycles	180–2000 cycles

4.3.2 Lead–acid batteries

4.3.2.1 Lead–acid battery composition

The lead–acid battery, which is commonly used in automobiles and trucks, is made up of six electrochemical cells connected in series. Each cell in a lead–acid battery produces 2 V of power. This battery's electrodes are made of lead and are immersed in sulfuric acid. Positive electrodes are lead infused with lead oxide, while negative electrodes are spongy lead metal. Metallic lead at the negative electrodes is oxidized to lead sulfate during the discharge process, while lead oxide at the positive electrodes is reduced to lead sulfate:

$$Pb + PbO_2 + 2H_2SO_4 \rightarrow 2PbSO_4 + 2H_2O$$

When a lead-acid battery is recharged using an alternator, electrons are forced in the opposite direction, reversing the reaction (Nishi, 2001; Yoshino, 2012). Lead-acid batteries generally exhibit coulombic efficiencies around 85% and energy efficiencies in the order of 70% (Figure 4.5).

4.3.2.2 Working principle of lead acid battery

When sulfuric acid is dissolved, its molecules separate into negatively charged sulfate ions (SO_4^{2-}) and positively charged hydrogen ions ($^2H^+$). The hydrogen ions are attracted to the electrodes connected to the supply's negative terminal when two electrodes are placed in the solution and connected to a direct current (DC) power supply. The sulfate ions move toward the electrodes connected to the positive terminal of the power supply, also known as the anode. At the cathode, each hydrogen ion

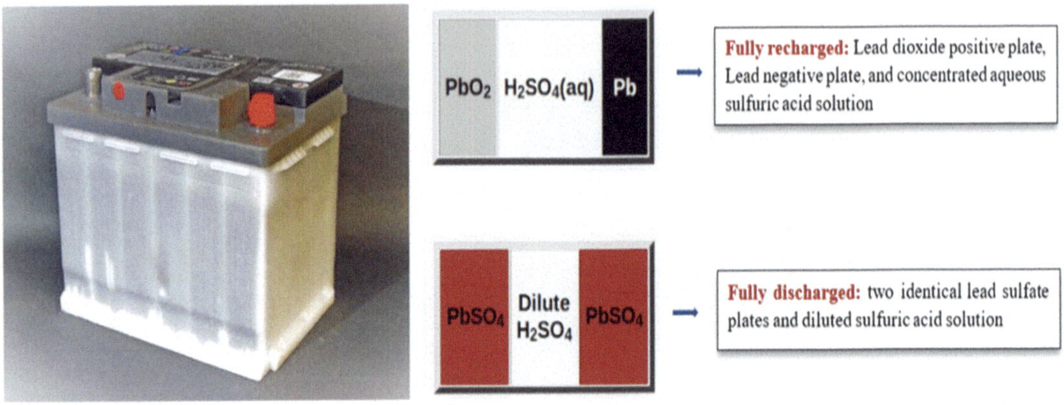

Figure 4.5 Lead-acid battery and its electrochemistry.

accepts one electron. In contrast, each sulfate ion takes the two negatively charged ions at the anode, reacting with water to form sulfuric acid, as shown in Figure 4.6c.

The oxygen produced by the subsequent reaction then reacts with lead oxide to form lead peroxide (PbO_2). As a result, during the charging process of lead-acid batteries, the lead cathode remains as lead, while the lead anode is transformed into brown-colored lead peroxide (Figure 4.6c).

When the DC power supply is turned off, and a voltmeter is connected between the electrodes, the potential difference between them is displayed. If a wire is connected between the electrodes, an electric current will flow from the positive to the negative plate via the external circuit. This means the cell can generate electrical energy. When fully discharged, the anode is made of lead peroxide (PbO_2) and the cathode is made of metallic sponge lead (Pb). When the electrodes are connected via a resistor, the cell discharges and electrons flow in the opposite direction as during charging.

The hydrogen ions migrate toward the anode, where they receive an electron from the anode, transforming into hydrogen atoms. These hydrogen atoms come into contact with PbO_2 and react, forming lead sulfate ($PbSO_4$), which is whitish in color and water as described by the chemical equation is shown in Figure 4.6c. Each sulfate ion (SO_4^{2-}) moves toward the cathode, it releases two electrons, becoming a sulfate radical ($SO_4^{\cdot}$). The sulfate radical then reacts with the metallic lead cathode, forming lead sulfate, also whitish in color.

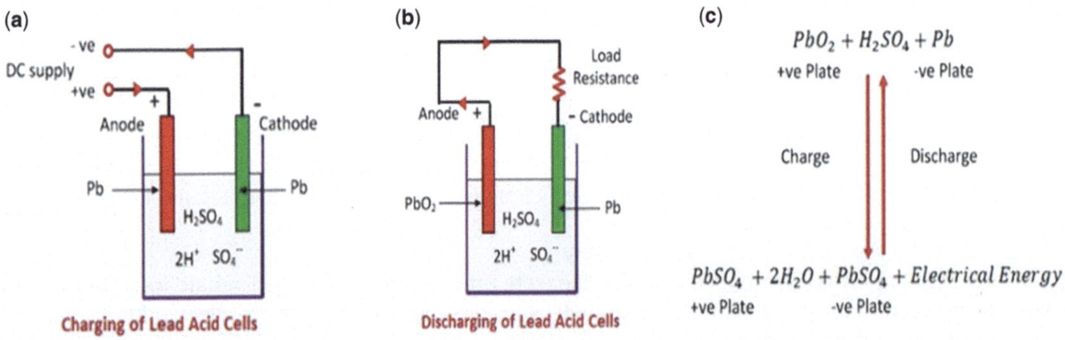

Figure 4.6 Single reversible equation represents the charging and discharging of lead-acid battery. (a) Charging, (b) discharging and (c) reaction scheme.

4.3.2.3 *Market perspective*

The demand for cost-effective, portable energy storage solutions is expected to grow as it has over the past century. Despite advancements in battery technology, lead-acid batteries will remain crucial, particularly when facing new challenges and seeking innovative solutions (Dong *et al.*, 2023). This includes ongoing efforts to enhance separators used in lead-acid batteries (Boehnstedt & Whear, 2009).

4.3.3 Lithium-ion batteries

4.3.3.1 *Lithium-ion battery composition*

Li-ion is a common terminology used to cover various types of batteries, as shown in Table 4.2 and several combinations for specific applications.

4.3.3.2 *Working principle of lithium-ion battery*

A lithium-ion battery, also known as a Li-ion battery, is a modern battery technology that uses lithium ions as a key component in its chemical process (Yoshino, 2012). When a battery discharges, the anode's lithium atoms lose electrons and become ionized. These lithium ions then migrate to the cathode via the electrolyte. They combine with electrons at the cathode and are electrically neutralized. The movement of lithium ions is facilitated by a microporous separator positioned between the anode and cathode as shown in Figure 4.7a. Due to the compact size of lithium ions (which ranks third after hydrogen and helium), Li-ion batteries possess a high voltage and the ability to store a significant amount of charge density (Nishi, 2001).

Li-ion batteries' electrodes can be made of a variety of materials. The typical configuration includes a cathode of lithium cobalt oxide and an anode of graphite. This combination is commonly found in mobile electronic devices such as cell phones and laptop computers. Lithium manganese oxide and lithium iron phosphate are two alternative cathode materials. Li-ion batteries typically use ethers as the electrolyte, which are organic compounds (Kraytsberg & Ein-Eli, 2011).

Table 4.2 Characteristics of different Li-ion batteries.

	LCO $LiCoO_2$	NCA $LiNiCoAlO_2$	NMC $LiNiMnCoO_2$	LMO $LiMn_2O_4$	LFP $LiFePO_4$	LTO* $Li_4Ti_5O_1$	Si/C*
Cathode	**Lithium Cobalt Oxide**	**Lithium Nickel Cobalt Aluminum Oxide**	**Lithium Nickel Manganese Cobalt Oxide**	**Lithium Manganese Spinel**	**Lithium Iron Phosphate**	**Lithium Titanate**	**Silicon Carbon Composite**
Cell voltage (100% SOC)	4.2 V	4.0 V	4.2 V	4.2 V	3.6 V	2.8 V	4.2 V
Energy	++	+++	+++	+	++	–	+++
Power	++	+++	++	+++	++	+	++
Calendar life	+	+++	+	–	++	–	–
Cycle life	+	++	++	++	++	+++	--
Safety	+	+	+	++	+++	+++	+
Cost	–	+	++	++	+	–	++

SOC = state of charge; *LTO and Si/C are anodes, which can be combined with any cathode.

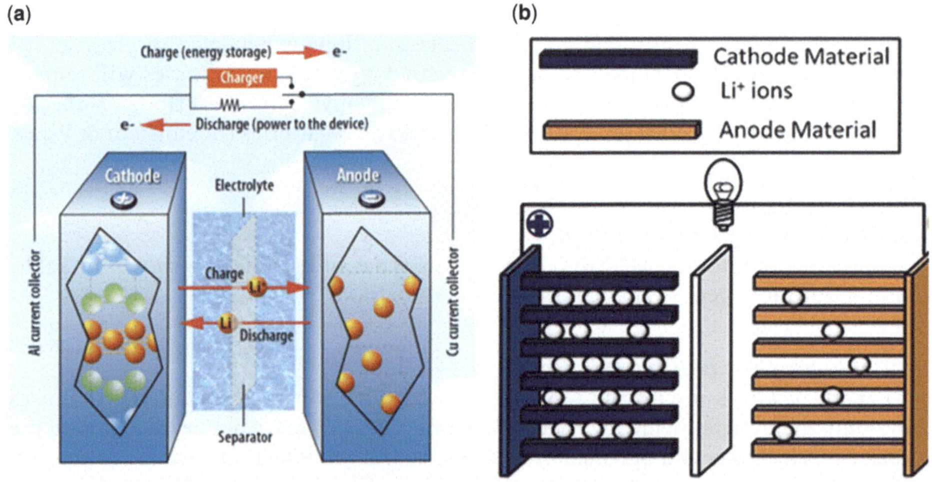

Figure 4.7 Schematic of a lithium-ion battery: (a) schematic representation and (b) working principle.

Li-ion batteries possess a substantial energy density, exhibit minimal memory effect, and experience low self-discharge. They are widely utilized in household and portable electronic devices such as laptops. Various examples of Li-ion batteries are given in **Figure 4.8**. Moreover, their popularity is rapidly increasing for applications in battery electric vehicles and aerospace applications. In many cases, Li-ion batteries are now replacing traditional lead acid batteries in various fields.

4.3.3.3 Market perspective

Li-ion batteries will dominate the rechargeable battery market for at least the next decade, owing to their numerous benefits. One significant advantage is their design flexibility, as they can be manufactured in various shapes and sizes to best fit the available space within the devices. Unlike Ni–Cd batteries, Li-ion batteries are not affected by the memory effect. Furthermore, Li-ion batteries have significantly higher voltages, nearly three times those of standard Ni-based batteries. Because of the high single-cell voltage, fewer cells are required in a battery module or pack to achieve a specific output voltage.

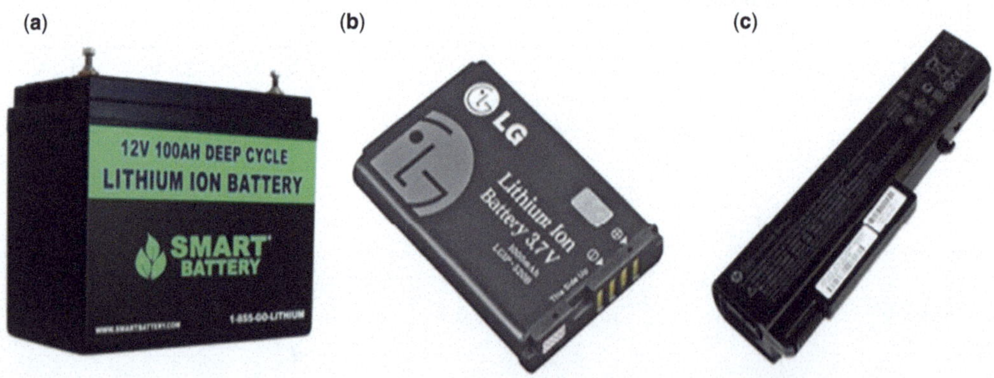

Figure 4.8 Examples of lithium-ion batteries: (a) deep cycle battery, (b) mobile phone battery and (c) laptop battery.

Consequently, it reduces the need for additional hardware, enhancing reliability, weight savings, and reduced part count in the battery module or pack. Another notable advantage is Li-ion batteries' remarkably low self-discharge rate, typically less than 5% monthly. This is a substantial improvement compared to the 20–30% self-discharge rate of Ni-based batteries, further contributing to the favorable characteristics of Li-ion batteries (Dong *et al.*, 2023). Figure 4.9 compares the specific and volumetric energy densities of various battery types.

4.3.4 Sodium-ion batteries

A sodium-ion (Na-ion) battery system is an energy storage system that charges and discharges using electrochemical reactions. These reactions take place between a positive electrode (cathode) containing layered materials composed of sodium and a negative electrode (anode) containing hard carbons or intercalation compounds. A porous material separates the electrodes, allowing ions to flow between them (Ellis & Nazar, 2012). They are submerged in an electrolyte, which can either be an aqueous medium like Na_2SO_4 solution or a non-aqueous solution like salts dissolved in propylene carbonate. Sodium (Na) atoms within the cathode release electrons to the external circuit during the charging process, transforming into ions as shown in Figure 4.10. These transformed ions then move through the electrolyte, making their way toward the anode. At the anode, they join with electrons from the external circuit and interact with the layered material of the anode. This entire process is reversed when the battery is discharged.

Due to the cost effectiveness, minimum environmental pollution and many cycles, sodium batteries are expected to be favored by energy storage and low-speed vehicles. The general applications of this type of batteries include solar energy storage systems, home energy storage systems, low-speed vehicle and electric boats.

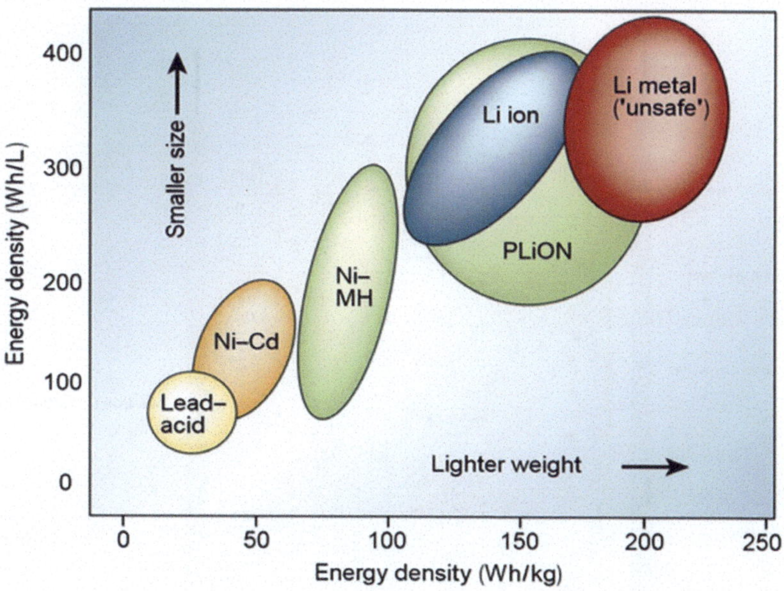

Figure 4.9 Comparison of energy densities and specific energy of different rechargeable batteries. (Source: Dong *et al.* 2023).

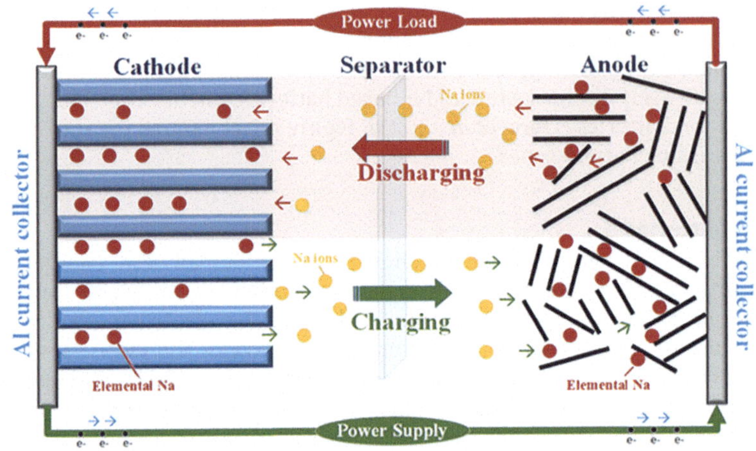

Figure 4.10 Illustration of working principle of Na-ion battery.

4.3.5 Zinc–air batteries

Zinc–air batteries (primary battery) and zinc–air fuel cells (mechanically rechargeable) are metal–air batteries that oxidize zinc by absorbing oxygen from the atmosphere. These batteries have a high energy density and can be produced relatively cheaply. They come in various sizes, including tiny button cells designed for hearing aids and larger batteries that were formerly powered by mercury in film cameras, and even large-scale batteries used in electric vehicles and for storing energy on a grid level (Caramia & Bozzini, 2014).

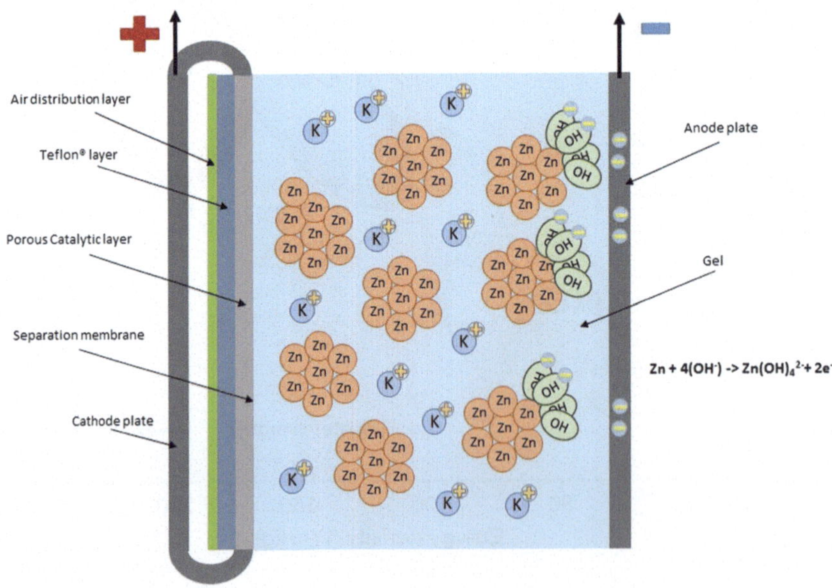

Figure 4.11 Schematic representation of a zinc-air cell.

During the discharge process of these batteries, a set of zinc particles forms a porous anode which becomes infused with an electrolyte. At the cathode, oxygen from the surrounding air reacts, generating hydroxyl ions. These ions then migrate into the zinc paste and transform into zincate ($Zn(OH)_4^{2-}$), releasing electrons that travel toward the cathode. The zincate slowly degrades into zinc oxide and water before returning to the electrolyte. Importantly, the anode's water and hydroxyl are recycled at the cathode, so no water is used in this process. The chemical reactions theoretically produce a voltage of 1.65 V, but in practice, it is usually reduced to a range of 1.35–1.4 V. Zinc–air batteries have several applications due to their high energy density and cost-effectiveness (Shang *et al.*, 2022) (Figure 4.11).

4.4 SUPERCAPACITORS

4.4.1 Principles and operations

Supercapacitors, also known as ultracapacitors or double-layer capacitors, are electrochemical energy storage devices with a very high-power density but a much lower energy density than batteries. This prepares them for ideal applications requiring short bursts of high power, such as hybrid vehicles, electric vehicles or electronic devices (Funabashi, 2016).

Supercapacitors work by storing energy in the following two ways:

4.4.1.1 Electric double-layer capacitance

This is the dominant mechanism in most supercapacitors. It occurs when the positive and negative charges of the electrolyte are attracted to the positive and negative electrodes, respectively. This creates a control layer on each electrode, separated by a thin electrolyte layer. The thickness of this layer is only a few nanometers, which is why supercapacitors have such a high surface area-to-volume ratio (Jalal *et al.*, 2021).

4.4.1.2 Faradaic capacitance

This mechanism is less common but can occur in supercapacitors that use active redox materials. In this case, the ions in the electrolyte can be oxidized or reduced at the electrodes, which stores energy in the form of chemical bonds (Conway, 1999). The charging and discharging of a supercapacitor can be done very quickly, which makes them ideal for applications that require high power but not a lot of energy (Shiraishi, 2003).

Supercapacitors have several advantages, including high power density, fast charging and discharging rates, and a long-life cycle. Supercapacitors have a high-power density, which means they can deliver a large amount of power in a short period of time. This makes them ideal for high-power applications such as hybrid vehicles and electronic devices (Shiraishi, 2003; Soneda, 2013). Furthermore, supercapacitors can be charged and discharged very quickly, making them ideal for applications that require quick response times. They have a very long cycle life, which means that they can be used for an extended period of time without losing its ability to store energy (Shiraishi, 2003).

But supercapacitors also have some drawbacks such as low energy density, high cost, and limited temperature range. The supercapacitors can only store a small amount of energy, that is they have a low energy density. This limits their use in applications requiring high amount of energy such as electric vehicles (Dong *et al.*, 2023). Supercapacitors are currently more expensive than batteries. This is due to the high cost of the materials used to make them. But they are not as tolerant to high temperatures as batteries. This limits their use in applications with high temperatures (Kumaravel *et al.*, 2021).

Supercapacitors are widely used in many electrical devices and appliances, such as hybrid and electric vehicles, laptops, smart phones, and some energy-storing devices. In hybrid and electric vehicles, supercapacitors help to improve the fuel efficiency of the vehicles, reduce emissions, and extend the battery life and its range (Libich *et al.*, 2018). Supercapacitors are also used in renewable energy sources such as wind and solar power generation. This can help to make these sources of energy more reliable and affordable. Supercapacitors are a promising technology with a wide range of potential applications. As the technology continues to develop, supercapacitors are likely to become more widely used in a variety of applications (Yadlapalli *et al.*, 2022).

4.4.2 Supercapacitor electrode materials

The electrode materials have a significant impact on the performance of supercapacitors. To achieve outstanding supercapacitor performance, the material's capacitance must be high. The capacitance of SC is determined by the effective surface area of the electrode material. However, because not all of the effective area is accessible to the electrolyte electrode, the capacitance of the electrode materials is not proportional to their effective surface area (Lobato *et al.*, 2017). It is possible to designate the electrochemically accessible area as an active electrochemical surface area. Additionally, the pore size of the conducting materials determines the electrochemically active surface area and can be easily modified using nanostructures. According to the investigation, a maximum capacitance of 0.7 nm was achieved (Sharma & Kumar, 2020).

When the pore size of electrode materials is increased, the distance between the pores also increases, leading to a decrease in capacitance. At some point, the pore size, the cross-sectional area, and the capacitance all become interdependent. Therefore, optimizing the pore size is crucial to achieve high capacitance while maintaining an appropriate distance between the pores (Heimböckel *et al.*, 2019).

Moreover, the purity of the electrode material is critical for achieving good cycle life and minimizing leakage current in supercapacitors. Contaminants in the electrolyte or electrode material can facilitate undesirable leakage current and subsequent self-discharge of the SC. While self-discharge may not considerably decrease output for simple-to-construct devices, it can still be a significant issue for more complex systems or applications that require long-term stability (Forouzandeh *et al.*, 2020).

The electrodes for supercapacitors are classified based on electrodes taken for each type of supercapacitor (see section 4.5). They are classified as EDLC electrode materials, pseudocapacitor electrode materials and hybrid supercapacitors electrode materials (Jalal *et al.*, 2021)

4.4.2.1 Electrode materials for EDLC

The practical electrode materials for electric double-layer capacitors (EDLCs) are porous carbons, such as activated carbons (Funabashi, 2016). Although various electrode materials such as carbon materials, metal oxides, and conductive polymers are synthesized for EDLCs, activated carbon materials remain the most commonly used electrode materials for EDLCs due to their low cost and numerous industrial applications (Hung *et al.*, 2022; Yang *et al.*, 2020). The main EDLC electrode materials are activated carbon, carbon aerogels, graphene, and carbon nanotubes (CNT) (Forouzandeh *et al.*, 2020).

4.4.2.2 Electrode materials for pseudocapacitor

Normally, the transition metal oxides and electrically conductive polymers are used as pseudocapacitors' electrode materials. The main pseudocapacitor electrode materials are metal oxides (Co_3O_4, NiO, RuO_2, MnO_2 and V_2O_5) and conducting polymers (like the polyaniline, polypyrrol and polyethylene dioxythiophene) (Jalal *et al.*, 2021; Yi *et al.*, 2018).

The redox processes are used on the surfaces of the electro-active materials. At the same time, the redox reactions change by discharging and charging and are reliant on the electrode voltage. This method has a higher energy density and capacitance than EDLCs made entirely of carbon. But its life cycles are often undesirable and cancel out its advantages. Cycling problems have reduced the performance of pseudocapacitors, while EDLCs can reach upto 500,000 cycles (Yi *et al.*, 2018). The performance of pseudo-capacitive devices decreases as cycling goes on, due to repeated chemical reaction cycles that damage the materials and change their shape unintentionally. To deal with this drawback, carbon supports were often used.

4.4.2.3 Electrode materials for hybrid supercapacitor

Various materials have been used in hybrid SCs: the components of pseudocapacitors and EDLC materials that provide a chemical and physical charge storage process in one electrode are incorporated into composite materials, mixtures of conducting polymers, or metal oxides with carbon materials (Gupta *et al.*, 2020).

4.5 TYPES OF SUPERCAPACITORS

Supercapacitors are classified based on the energy-storing technology used in it. They are divided into three groups (Figure 4.12).

(1) Electrochemical double-layer capacitors (EDLC)
(2) Hybrid capacitors
(3) Pseudocapacitors

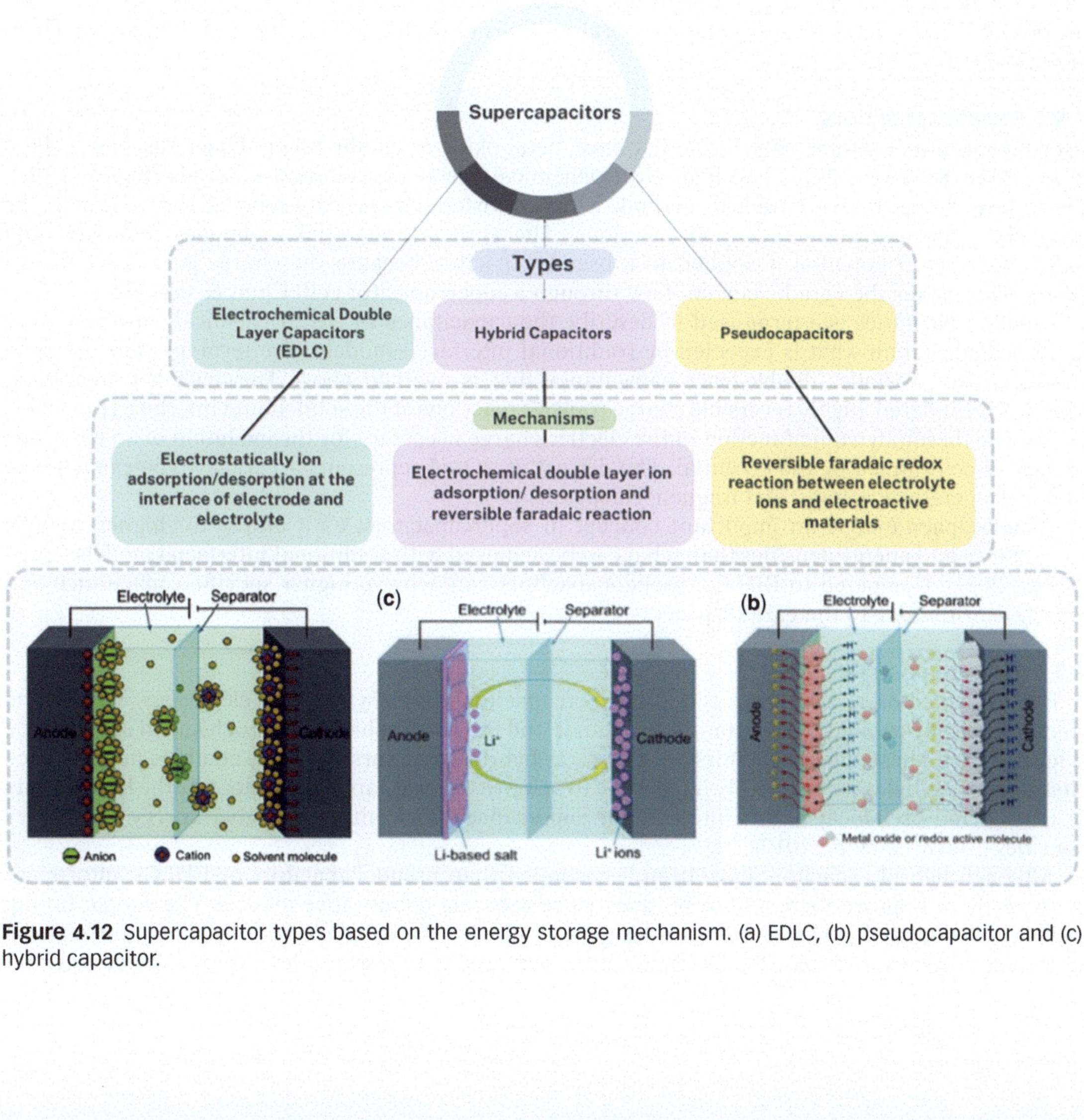

Figure 4.12 Supercapacitor types based on the energy storage mechanism. (a) EDLC, (b) pseudocapacitor and (c) hybrid capacitor.

4.5.1 Electrochemical double-layer capacitors

The EDLC works with an electrochemical double layer mechanism, employing conducting polymers as electrodes. The supercapacitors have a higher power density of about 10 kW/kg and will be fully charged in a very short time, typically about 30–60s (Funabashi, 2016). EDLCs have two carbonaceous porous electrodes with high surface area and are separated by a very thin dielectric separator medium to reduce the distance between the electrodes (Figure 4.12a). The opposite static electricity is stored on each electrode and the dielectric separator restricts the electrons to transfer between the electrodes (Jalal *et al.*, 2021; Shiraishi, 2003). The EDLCs charge in a short period and have long cycle life and shelf life.

In an EDLC, the charge stored in the porous electrodes makes a potential difference between the electrodes, resulting in the ions in the electrolyte diffusing toward the surface of the opposite-charged electrodes and the separator. An electrical double layer is created between the electrodes and the dielectric medium, preventing ions' recombination in the electrodes. Although the increased surface area and reduced gap between the electrodes promote the energy density of the EDLCs, the electrical double layer provides its high energy density (Jalal *et al.*, 2021).

Even though its name shows an electrochemical process, the EDLCs are not working with any electrochemical reactions: it is a non-Faradic phase process. The lack of electrochemical reactions inside the EDLCs helps it to withstand several millions of cycles of charging and discharging (Jalal *et al.*, 2021).

4.5.2 Pseudocapacitors

Pseudocapacitors are supercapacitors that have metal oxide or conducting polymer electrodes (Jalal *et al.*, 2021; Kurzweil, 2015) and high electrochemical pseudocapacitance materials (Figure 4.12c). They store charge by the Faradaic method, which transfers charge between the electrode and the electrolyte, like oxidation–reduction reactions. The electrode material undergoes reduction and oxidation when a potential is applied to a pseudocapacitor, because the charge goes through two layers. This makes the Faradic current flow through a supercapacitor cell (Chen & Lee, 2021).

Pseudocapacitance is a term used to describe the capacitance in electrochemical capacitors that is much higher than what is expected by traditional interfacial models. The term 'pseudo' refers to the ideal, 'electrostatic', double-layer capacitance that is overlaid above Faraday's law-compliant, kinetically hindered, highly reversible charge transfer reactions at the solid–liquid interface (Kurzweil, 2015). The thermodynamic function of the electric charge necessary for the evolution of an electrode process is determined by the potential. The derivative dQ/d4 represents a Faraday-type capacitance that is temperature, voltage, and frequency dependent.

Pseudocapacitance is an important concept in supercapacitors, as it allows for higher specific capacitance and energy densities than what can be achieved with traditional interfacial models (Chen & Lee, 2021). Compared to EDLCs, pseudocapacitors can achieve higher specific capacitance and energy densities due to the Faradic process.

4.5.3 Hybrid capacitor

A hybrid capacitor is a specific type of supercapacitor with asymmetric electrodes, where one electrode mainly exhibits electrostatic capacitance and the other exhibits electrochemical capacitance (Figure 4.12c). These supercapacitors are known as hybrid capacitors, and they combine performance characteristics that were previously impossible to achieve. Hybrid capacitors integrate the best aspects of EDLCs and pseudocapacitors into a single supercapacitor, resulting in higher energy and power densities (Muzaffar *et al.*, 2019).

Although hybrid capacitors have been less studied than pseudocapacitors or EDLCs, efforts are being made to improve them and to produce more accurate quantitative models. The simple tuning of hybrid capacitors' performance and design has allowed them to become the dominant class of supercapacitors, overtaking EDLCs. There is an increased emphasis on developing high energy and high cycle life supercapacitors, and hybrid capacitors are well-positioned to meet these demands. In addition to the increased emphasis on developing high energy and high cycle life, the simple tuning of hybrid capacitors' performance and design allowed them to overtake EDLCs as the dominant SCs class (Chatterjee & Nandi, 2021). Combining the two types mentioned above yields the mechanism for storing charge. Asymmetric, composite, and battery-type hybrid capacitors were categorized into these three groups, which can be distinguished by their electrode design (Jalal *et al.*, 2021). Based on their electrode design, hybrid capacitors are classified into three types: asymmetric, composite, and battery-type. By combining a pseudocapacitor electrode with EDLCs, asymmetric hybrid capacitors combine Faradic and non-Faradic processes. As a result, the carbon material is a negative electrode, while a conducting polymer or metal oxide is a positive electrode (Kumar *et al.*, 2019).

Carbon-based materials with polymer conduction or metal oxides are combined in a single electrode, resulting in an electrode with two types of energy storage mechanisms: chemical and physical. Composites can be classified as binary or ternary depending on whether the electrode is made of two or three materials (Jalal *et al.*, 2021).

Battery-type hybrid capacitors are a one-of-a-kind combination of a battery electrode and a SC electrode that are intended to provide both battery and SC characteristics in a single cell. By combining SC recharging times and battery properties, this design meets the demand for higher-power batteries and high-energy capacitors (Jalal *et al.*, 2021; Zuo *et al.*, 2017).

4.6 APPLICATIONS OF BATTERIES AND SUPERCAPACITORS

Batteries and supercapacitors have extensive applications in various industries and serve multiple purposes. These energy storage devices are mainly used in computers, tablets, and mobile phones. They also play a vital role in powering hybrid and electric vehicles. In medicine, batteries are used in cardiac pacemakers. In aerospace, they provide the energy needed for satellites and spacecraft. Batteries also store energy produced by solar panels and wind turbines in renewable energy systems.

Unlike traditional batteries that cannot sustain high power density demands or support rapid energy release rates, ultracapacitors (or supercapacitors) excel in these traits. It makes them an essential innovation in various applications. The ability to charge and discharge energy much faster is critical when powering machinery that demands quick bursts of energy, such as regenerative braking systems found in electric vehicles.

Additionally, supercapacitors are indispensable components of renewable-based electricity systems, requiring quick storage solutions for better energy management and power load balancing when facing rising fluctuations. These devices also offer reliable backup power sources necessary for the long-term survival of emergency equipment or uninterruptible power supplies (UPS), even when running at extreme temperatures.

4.6.1 Portable electronics and consumer applications

Portable electronic devices (PEDs) serve wide-ranging purposes, such as efficient data gathering or dissemination, which makes them essential components across industries. Batteries and supercapacitors are crucial in portable electronics, powering various consumer applications. PEDs have been using electrochemical energy storage systems, especially rechargeable batteries, for decades and have enjoyed a thriving growth in popularity (Li *et al.*, 2018a). Significant advancements in the electrochemical performances of rechargeable batteries have been made to meet the PEDs' constantly increasing requirements (Liu *et al.*, 2018; Li *et al.*, 2018b; Xie *et al.*, 2017). Nickel–cadmium, lead-acid, nickel-metal hydride, lithium-ion, and other types of rechargeable batteries have all been used in PEDs (Figure 4.13). As time goes on, their unique energy and power are expected to further improve dramatically.

The rapid expansion of PEDs means that existing battery technology needs to be able to keep up (Tarascon & Armand, 2001). Modern rechargeable battery technology for PEDs has revealed several

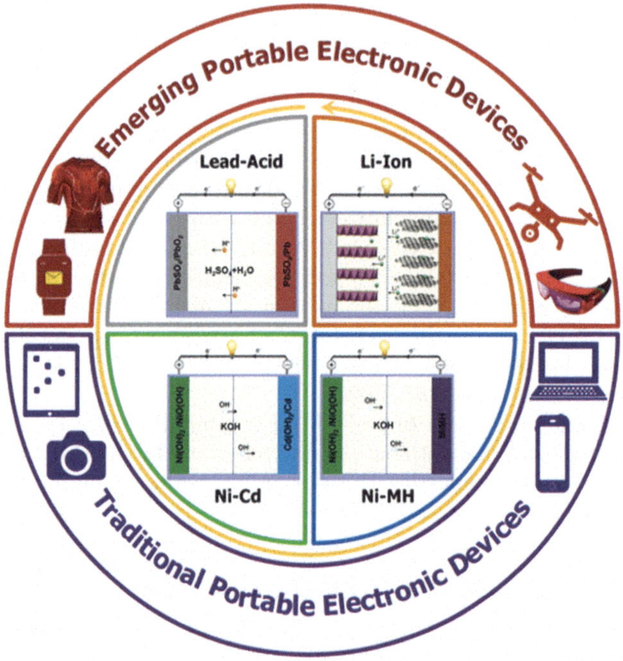

Figure 4.13 Applications of rechargeable batteries in traditional and emerging portable electronic devices. Source: Liang *et al.* (2019).

problems, including low energy storage capacity, limited cycle life, and high self-discharge. These have become a limiting barrier to PED development (Cui *et al.*, 2018; Chen & Fan, 2018). For example, multifunctional PEDs' high-power consumption necessitates energy storage systems with increased energy, greater efficiency, smaller space, lighter weight, and longer operational duration. However, the increasing demands of new electrical and electronic equipment put current batteries to the test. As a result, the practical design and fabrication of innovative batteries has been a relentlessly pursued goal for future PEDs.

Lithium batteries dominate the energy storage sector due to their superior energy density. The consumer electronics industry has scaled production, optimized the supply chain, and reduced prices. However, lithium battery capabilities are being challenged as portable devices demand higher power density (Liang *et al.*, 2019). The research aims to develop new high-energy density technologies, while alternative storage options like supercapacitors emerge as strong contenders to address lithium battery limitations (Riaz *et al.*, 2021).

Although less frequent in consumer electronics, supercapacitors are becoming more popular for specialized uses like portable speakers and camera flashes (Behera, 2022). Due to their quick energy storage and release, they offer short bursts of power for bright flashes or high-quality music. Additionally, their extended cycle life makes them appropriate for applications requiring frequent charging and discharging, providing customers with long-lasting performance.

4.6.2 Mobility of the future

4.6.2.1 Electric vehicles and hybrid vehicles

Because of their ability to store large amounts of energy in a compact and lightweight form while providing sufficient power for various applications, batteries are commonly used as vehicles' primary electrical energy storage devices. Despite having a limited shelf life and cycle life, batteries have remained the preferred choice due to the lack of a viable alternative. However, rising power demands in many applications have outstripped the capabilities of standard batteries, necessitating the development of specialized high-power pulse batteries. Power output is prioritized in these pulse batteries, but energy density and cycle life are frequently sacrificed. Supercapacitors have emerged as a viable alternative to pulse batteries, providing significantly higher power capabilities and significantly longer shelf and cycle life than batteries (Rajagopal *et al.*, 2022).

Figure 4.14 shows a simplified schematic illustration of an emergent all-electric architecture. Both supercapacitors and batteries can store electrical energy. The discharge of electric energy storage via electrical systems supplying power to the wheels is depicted by red arrows. Green arrows indicate the reverse when power from the wheels charges the electric energy storage while braking. Blue arrows show power sharing between device types. The motor receives current as needed, which drives the wheels to accelerate the vehicle. The current from the supercapacitor bank flows through a bidirectional DC/DC converter to regulate the voltage so that it can be matched to the battery bank as needed, managing the current flow between them. Both electric energy storage systems are linked to inverters. These transform DC into the AC needed to power the motor.

4.6.2.2 Aerospace applications

Battery and supercapacitor technologies are widely used in aerospace applications to provide reliable and efficient energy storage solutions. Batteries are critical in aircraft applications for powering numerous onboard systems, such as avionics, lights, communication equipment, and emergency backup systems (Sahoo *et al.*, 2020). For example, lithium-ion batteries have a greater energy density and a long cycle life, making them ideal for powering electric propulsion systems in aircraft and spacecraft. They can store significant amounts of energy for longer periods, allowing for long-distance flights. Furthermore, batteries are essential in electric ground support equipment, lowering emissions and enhancing airport operating efficiency.

Supercapacitors are also extensively used in aerospace. Due to their high-power density and rapid charge–discharge characteristics, they are useful for applications needing brief bursts of energy, such as powering actuators, launching systems, and emergency backup systems. Supercapacitors can produce large power outputs in short bursts, improving critical system performance and responsiveness. Because of their capacity to survive extreme weather conditions, vast temperature ranges, and vibrations, they are trusted energy storage devices in aerospace, where durability and robustness are critical (Mohsan *et al.*, 2023).

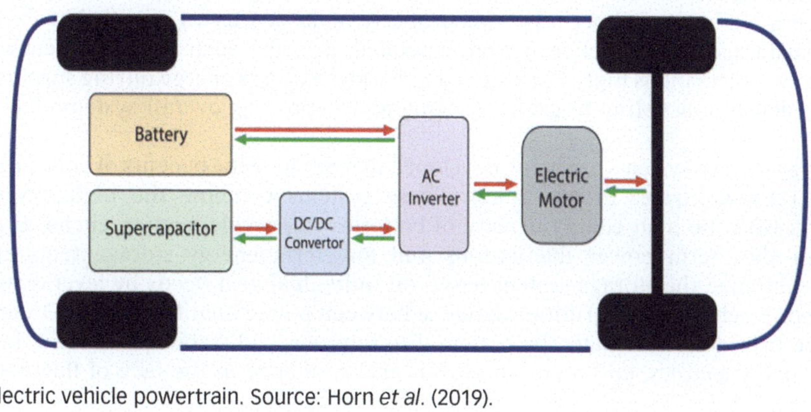

Figure 4.14 Electric vehicle powertrain. Source: Horn *et al.* (2019).

4.6.3 New energy technologies

4.6.3.1 Renewable energy integration

In recent years, renewable energy sources have gained significant attention because of their potential to mitigate climate change and reduce dependence on fossil fuels (Gielen *et al.*, 2019). Due to their intermittent nature, these energy sources pose a challenge for energy storage and management. Battery and supercapacitor technology plays a crucial role here (Tan *et al.*, 2021). When supercapacitors are integrated with renewable energy systems, such as solar panels or wind turbines, excess energy generated during peak production periods can be stored in supercapacitors and released when demand exceeds supply (Yadlapalli *et al.*, 2022).

In contrast, batteries are widely used in renewable energy systems to store long-term energy (Dowling *et al.*, 2020). They can store more energy for extended periods than supercapacitors because they have higher energy densities. Batteries, particularly lithium-ion batteries, can deliver a steady power supply even when renewable energy production is low (Chen *et al.*, 2020). During off-peak hours, they can be charged with excess renewable energy and discharged during periods of high energy demand. Integrating batteries with renewable energy systems ensures a smooth transition from fossil fuel-based energy sources to renewable energy sources.

Supercapacitors and batteries work well together to integrate renewable energy sources. Supercapacitors perform very well in high-power applications like storing energy from jolting wind gusts or intermittent solar panel output surges. They are the best at coping with changes in energy production and demand since they can charge and discharge quickly. On the other hand, batteries offer long-term energy storage and can meet lower power demands over lengthy periods. Supercapacitors and batteries are combined to create a hybrid energy storage system to ensure a dependable and effective integration of renewable energy sources into the grid (Díaz-González *et al.*, 2022).

4.6.3.2 Grid-scale energy storage

Grid-scale energy storage is essential for balancing renewable energy sources' intermittent nature while maintaining a consistent and predictable power supply (Al Essa, 2020). Supercapacitors and batteries are two significant technologies used in this area. When used for grid-scale energy storage, supercapacitors have various advantages. They have a high-power density, allowing them to easily manage rapid charge and discharge cycles. Because of this property, supercapacitors are well-suited for dealing with short-term changes in electricity demand, offering quick response times, and smoothing out power oscillations in the grid. Furthermore, because supercapacitors have a long cycle life, they can withstand repeated charge–discharge cycles without significant degradation, making them durable and reliable grid-scale energy storage components (Kumar *et al.*, 2022).

Batteries, on the other hand, are a popular grid-scale energy storage solution (Posada *et al.*, 2017). Battery systems, such as lithium-ion batteries, have a high energy density, allowing for storing large amounts of energy for extended periods. This property makes batteries appropriate for applications that require sustained power delivery over extended periods, such as when renewable energy generation is low or demand is high. Batteries can also store excess energy during surplus and release it during peak demand, assisting in grid balancing and improving overall system efficiency (Mitali *et al.*, 2022).

Hybrid energy storage systems are being developed to maximize the benefits of both supercapacitors and batteries (Díaz-González *et al.*, 2022). These systems combine the high-power density of supercapacitors with the high energy density of batteries to provide a comprehensive solution that addresses both short-term power fluctuations and long-term energy storage requirements. Grid operators can optimize the storage system based on individual grid needs by leveraging the unique properties of both technologies, striking a balance between power quality, response time, and energy capacity. These hybrid systems have the potential to improve grid stability, increase the integration of renewable energy sources, and increase overall grid resilience in the face of fluctuating demand (Worku, 2022).

4.6.4 Defence application

Batteries and supercapacitors are vital components in modern defence applications, ensuring maximum performance, mobility, and dependability in a wide range of military equipment (Zhang *et al.*, 2022). As defence technologies advance, the synergy between these energy storage systems continues to be essential for ensuring national security, promoting innovation, and ensuring the success of military operations.

In defence, batteries are essential energy sources across various uses, including communication tools, night vision gear, and missile guidance systems. Notably, lithium-ion batteries provide substantial energy density and extended operational longevity, fueling essential electronics over prolonged durations. This capability facilitates effective soldier connectivity and navigation in demanding terrains, thus amplifying situational awareness and mission accomplishments. Furthermore, innovations like solid-state batteries are driving the development of safer and more compact energy storage options, particularly crucial for wearables and unmanned aerial vehicles (Liang *et al.*, 2023).

Conversely, supercapacitors excel at rapidly delivering bursts of power, making them useful in defence systems that demand rapid energy discharge. Supercapacitors provide essential high-power output in applications such as electromagnetic railguns or laser weapons, requiring enormous energy in short bursts (Burke & Zhao, 2021). Their capacity to withstand many charge and discharge cycles without substantial deterioration extends the lifespan of military equipment. Supercapacitors are also used in hybrid propulsion systems for military vehicles, where they optimize energy utilization and reduce fuel consumption, contributing to increased operating efficiency and reduced environmental impact.

4.7 CONCLUSION

Batteries and supercapacitors are the prominent devices to store energy using electrochemical methods. Here we discussed about the different types of electrical energy storage devices and their working methods and electrochemistry. Batteries are working with the conversion of electric charge into chemical charge and vice versa. Batteries have higher energy density and lower power density, whereas supercapacitors have higher power density. It delivers power to the external circuit but stores very little amount of energy. The combinations of both batteries and supercapacitors are used in many electronic devices and have a variety of applications in the mobility, aerospace and renewable energy sector as well as in daily electrical appliances.

REFERENCES

Al Essa M. J. M. (2020). Power management of grid-integrated energy storage batteries with intermittent renewables. *Journal of Energy Storage*, Elsevier, **31**, 101762, https://doi.org/10.1016/j.est.2020.101762

Aneke M. and Wang M. (2016). Energy storage technologies and real life applications – a state of the art review. *Applied Energy*, Elsevier Ltd, **179**, 350–377, https://doi.org/10.1016/j.apenergy.2016.06.097

Behera A. (2022). Advanced Semiconductor/Conductor Materials. In: Advanced Materials, Springer, Cham, https://doi.org/10.1007/978-3-030-80359-9_16

Boehnstedt W. and Whear J. K. (2009). Secondary Batteries – Lead– Acid Systems | Separators, https://doi.org/10.1016/B978-044452745-5.00134-9

Burke A. F. and Zhao J. (2021). Past, present and future of electrochemical capacitors: technologies, performance and applications. *Journal of Energy Storage*, Elsevier, **35**, 102310, https://doi.org/10.1016/j.est.2021.102310

Caramia V. and Bozzini B. (2014). Materials science aspects of zinc-air batteries: a review. *Materials for Renewable and Sustainable Energy*, SpringerOpen, **3**(2), 1. Available online at: https://link.springer.com/article/10.1007/s40243-014-0028-3 (accessed 18 June 2023), https://doi.org/10.1007/s40243-014-0028-3

Chatterjee D. P. and Nandi A. K. (2021). A review on the recent advances in hybrid supercapacitors. *Journal of Materials Chemistry A*, **9**(29), 15880–15918, https://doi.org/10.1039/D1TA02505H

Chen L. and Fan L. Z. (2018). Dendrite-free Li metal deposition in all-solid-state lithium sulfur batteries with polymer-in-salt polysiloxane electrolyte. *Energy Storage Materials*, Elsevier, **15**, 37–45, https://doi.org/10.1016/j.ensm.2018.03.015

Chen J. and Lee P. S. (2021). Electrochemical supercapacitors: from mechanism understanding to multifunctional applications. *Advanced Energy Materials*, **11**(6), 2003311. Available online at: https://onlinelibrary.wiley.com/doi/full/10.1002/aenm.202003311 (accessed 17 June 2023).

Chen T., Jin Y., Lv H., Yang A., Liu M., Chen B., Xie Y. and Chen Q. (2020). Applications of Lithium-Ion Batteries in Grid-Scale Energy Storage Systems. *Trans. Tianjin Univ.*, **26**, 208–217, https://doi.org/10.1007/s12209-020-00236-w

Conway B. E. (1999). Electrochemical Supercapacitors. Springer US, Boston, MA.

Cui C., Wei Z., Xu J., Zhang Y., Liu S., Liu H., Mao M., Wang S., Ma J. and Dou S. (2018). Three-dimensional carbon frameworks enabling MoS2 as anode for dual ion batteries with superior sodium storage properties. *Energy Storage Materials*, Elsevier, **15**, 22–30, https://doi.org/10.1016/j.ensm.2018.03.011

Díaz-González F., Chillón-Antón C., Llonch-Masachs M., Galceran-Arellano S., Rull-Duran J., Bergas-Jané J. and Bullich-Massagué E. (2022). A hybrid energy storage solution based on supercapacitors and batteries for the grid integration of utility scale photovoltaic plants. *Journal of Energy Storage*, Elsevier, **51**, 104446, https://doi.org/10.1016/j.est.2022.104446

Dincer I. and Rosen M. A. (2011). Thermal Energy Storage: Systems and Applications, 2. ed., Wiley, Chichester, ISBN 978-0-470-97073-7.

Döge V., Kurtulus C., Hennige V. and Wegmann R. (2016). Scalable energy storage systems for effective electrified mobility concepts. *Transportation Research Procedia*, Elsevier, **14**, 3621–3630, https://doi.org/10.1016/j.trpro.2016.05.430

Dong W., Xie M., Zhao S., Qin Q. and Huang F. (2023). Materials design and preparation for high energy density and high power density electrochemical supercapacitors. *Materials Science and Engineering: R: Reports*, **152**, 100713, https://doi.org/10.1016/j.mser.2022.100713

Dowling J. A., Rinaldi K. Z., Ruggles T. H., Davis S. J., Yuan M., Tong F., Lewis N. S. and Caldeira K. (2020). Role of long-duration energy storage in variable renewable electricity systems. *Joule*, Cell Press, **4**(9), 1907–1928. Available online at: http://www.cell.com/article/S2542435120303251/fulltext (accessed 21 June 2023).

Ellis B. L. and Nazar L. F. (2012). Sodium and sodium-ion energy storage batteries. *Current Opinion in Solid State and Materials Science*, Elsevier, **16**(4), 168–177, https://doi.org/10.1016/j.cossms.2012.04.002

Forouzandeh P., Kumaravel V. and Pillai S. C. (2020). Electrode materials for supercapacitors: a review of recent advances. *Catalysts*, **10**(9), 969, https://doi.org/10.3390/catal10090969

Funabashi T. (2016). Introduction. In: Integration of Distributed Energy Resources in Power Systems. Elsevier, pp. 1–14. https://doi.org/10.1016/B978-0-12-803212-1.00001-5

Gielen D., Boshell F., Saygin D., Bazilian M. D., Wagner N. and Gorini R. (2019). The role of renewable energy in the global energy transformation. *Energy Strategy Reviews*, Elsevier, **24**, 38–50, https://doi.org/10.1016/j.esr.2019.01.006

Gupta A., Sardana S., Dalal J., Lather S., Maan A. S., Tripathi R., Punia R., Singh K. and Ohlan A. (2020). Nanostructured polyaniline/graphene/Fe_2O_3 composites hydrogel as a high-performance flexible supercapacitor electrode material. *ACS Applied Energy Materials*, **3**(7), 6434–6446, https://doi.org/10.1021/acsaem.0c00684

Heimböckel R., Hoffmann F. and Fröba M. (2019). Insights into the influence of the pore size and surface area of activated carbons on the energy storage of electric double layer capacitors with a new potentially universally applicable capacitor model. *Physical Chemistry Chemical Physics*, **21**(6), 3122–3133, https://doi.org/10.1039/C8CP06443A

Horn M., MacLeod J., Liu M., Webb J. and Motta N. (2019) Supercapacitors: a new source of power for electric cars? *Economic Analysis and Policy*, Elsevier, **61**, 93–103, https://doi.org/10.1016/j.eap.2018.08.003

Hung P.-Y., Zhang H., Lin H., Guo Q., Lau K. and Jia B. (2022). Specializing liquid electrolytes and carbon-based materials in EDLCs for low-temperature applications. *Journal of Energy Chemistry*, Elsevier, **68**, 580–602, https://doi. org/10.1016/j.jechem.2021.12.012

Jalal N. I., Ibrahim R. I. and Oudah M. K. (2021). A review on supercapacitors: types and components. *Journal of Physics: Conference Series*, **1973**(1), 012015, https://doi.org/10.1088/1742-6596/1973/1/012015

Kraytsberg A. and Ein-Eli Y. (2011). Review on Li–air batteries – opportunities, limitations and perspective. *Journal of Power Sources*, Elsevier, **196**(3), 886–893, https://doi.org/10.1016/j.jpowsour.2010.09.031

Kumar Y., Rawal S., Joshi B. and Hashmi S. A. (2019). Background, fundamental understanding and progress in electrochemical capacitors. *Journal of Solid State Electrochemistry*, **23**(3), 667–692, https://doi.org/10.1007/s10008-018-4160-3

Kumar N., Kim S.-B., Lee S.-Y. and Park S.-J. (2022). Recent Advanced Supercapacitor: A Review of Storage Mechanisms, Electrode Materials, Modification, and Perspectives. *Nanomaterials*, MDPI AG, **12**(20), 3708, http://dx.doi.org/10.3390/nano12203708

Kumaravel V., Bartlett J. and Pillai S. C. (2021). Solid electrolytes for high-temperature stable batteries and supercapacitors. *Advanced Energy Materials*, **11**(3), 2002869, https://doi.org/10.1002/aenm.202002869

Kurzweil P. (2015). Electrochemical Double-layer Capacitors. In: *Electrochemical Energy Storage for Renewable Sources and Grid Balancing*, Elsevier, pp. 345–407, https://doi.org/10.1016/B978-0-444-62616-5.00019-X

Larcher D. and Tarascon J. M. (2014). Towards greener and more sustainable batteries for electrical energy storage. *Nature Chemistry* 2014 7:1, Nature Publishing Group, **7**(1), 19–29. Available online at: https://www.nature.com/articles/nchem.2085 (accessed 18 June 2023)

Li F., Kaiser M. R., Ma J., Guo Z., Liu H. and Wang J. (2018). Free-standing sulfur-polypyrrole cathode in conjunction with polypyrrole-coated separator for flexible Li-S batteries. *Energy Storage Materials*, **13**, 312–322, https://doi.org/10.1016/J.ENSM.2018.02.007

Li Y., Xu H., Chien P. H., Wu N., Xin S., Xue L., Park K., Hu Y. Y. and Goodenough J. B. (2018). A Perovskite Electrolyte That Is Stable in Moist Air for Lithium-Ion Batteries. *Angewandte Chemie (International Ed. in English)*, **57**(28), 8587–8591, https://doi.org/10.1002/ANIE.201804114

Liang Y., Zhao C. Z., Yuan H., Chen Y., Zhang W., Huang J. Q., Yu D., Liu Y., Titirici M. M., Chueh Y. L., Yu H. and Zhang Q. (2019). A review of rechargeable batteries for portable electronic devices. *InfoMat*, **1**(1), 6–32, https://doi.org/10.1002/INF2.12000

Liang Z., He J., Hu C., Pu X., Khani H., Dai L., Fan D. (Emma), Manthiram A. and Wang Z.-L. (2023). Next-Generation Energy Harvesting and Storage Technologies for Robots Across All Scales. *Advanced Intelligent Systems*, **5**(4), 2200045, https://doi.org/10.1002/AISY.202200045

Libich J., Máca J., Vondrák J., Čech O. and Sedlaříková M. (2018). Supercapacitors: properties and applications. *Journal of Energy Storage*, **17**, 224–227, https://doi.org/10.1016/j.est.2018.03.012

Liu Z., Yang S., Sun B., Chang X., Zheng J. and Li X. (2018). A Peapod-like CoP@C Nanostructure from Phosphorization in a Low-Temperature Molten Salt for High-Performance Lithium-Ion Batteries. *Angewandte Chemie International Edition*, **57**(32), 10187–10191, https://doi.org/10.1002/ANIE.201805468

Lobato B., Suárez L., Guardia L. and Centeno T. A. (2017). Capacitance and surface of carbons in supercapacitors. *Carbon*, Pergamon, **122**, 434–445, https://doi.org/10.1016/j.carbon.2017.06.083

Mitali J., Dhinakaran S. and Mohamad A. A. (2022). Energy storage systems: a review. *Energy Storage and Saving*, Elsevier, **1**(3), 166–216, https://doi.org/10.1016/j.enss.2022.07.002

Mohsan S. A. H., Othman N. Q. H., Li Y., Alsharif M. H. and Khan M. A. (2023). Unmanned aerial vehicles (UAVs): practical aspects, applications, open challenges, security issues, and future trends. *Intelligent Service Robotics* 2023 16:1, Springer, **16**(1), 109–137. Available online at: https://link.springer.com/article/10.1007/s11370-022-00452-4 (accessed 21 June 2023).

Muzaffar A., Ahamed M. B., Deshmukh K. and Thirumalai J. (2019). A review on recent advances in hybrid supercapacitors: design, fabrication and applications. *Renewable and Sustainable Energy Reviews*, **101**, 123–145, https://doi.org/10.1016/j.rser.2018.10.026

Nishi Y. (2001). The development of lithium ion secondary batteries. *The Chemical Record*, John Wiley & Sons, Ltd, **1**(5), 406–413. Available online at: https://onlinelibrary.wiley.com/doi/full/10.1002/tcr.1024 (accessed 18 June 2023).

Posada J. O. G., Rennie A. J. R., Villar S. P., Martins V. L., Marinaccio J., Barnes A., Glover C. F., Worsley D. A. and Hall P. J. (2017). Aqueous batteries as grid scale energy storage solutions. *Renewable and Sustainable Energy Reviews*, **68**, 1174–1182, https://doi.org/10.1016/J.RSER.2016.02.024

Rajagopal S., Pulapparambil Vallikkattil R., Mohamed Ibrahim M. and Velev D. G. (2022). Electrode materials for supercapacitors in hybrid electric vehicles: challenges and current progress. *Condensed Matter* 2022, Vol. 7, Page 6, Multidisciplinary Digital Publishing Institute, **7**(1), 6. Available online at: https://www.mdpi.com/2410-3896/7/1/6/htm (accessed 21 June 2023).

Riaz A., Sarker M. R., Saad M. H. M. and Mohamed R. (2021). Review on comparison of different energy storage technologies used in micro-energy harvesting, WSNs, low-cost microelectronic devices: challenges and recommendations. *Sensors* 2021, Vol. 21, Page 5041, Multidisciplinary Digital Publishing Institute, **21**(15), 5041. Available online at: https://www.mdpi.com/1424-8220/21/15/5041/htm (accessed 21 June 2023).

Sahoo S., Zhao X. and Kyprianidis K. (2020). A review of concepts, benefits, and challenges for future electrical propulsion-based aircraft. *Aerospace* 2020, Vol. 7, Page 44, Multidisciplinary Digital Publishing Institute, **7**(4), 44. Available online at: https://www.mdpi.com/2226-4310/7/4/44/htm (accessed 21 June 2023).

Shang W., Wang H., Yu W., Li R., Wu Z. and Tan P. (2022). *A zinc-air battery capable of working in anaerobic conditions and fast environmental energy harvesting*, https://doi.org/10.1016/j.xcrp.2022.100904

Sharma P. and Kumar V. (2020). Current technology of supercapacitors: a review. *Journal of Electronic Materials*, **49**(6), 3520–3532, https://doi.org/10.1007/s11664-020-07992-4

Shiraishi S. (2003). Electric Double Layer Capacitors. In: *Carbon Alloys*, Elsevier, pp. 447–457, https://doi.org/10.1016/B978-008044163-4/50027-9

Soneda Y. (2013). Carbons for Supercapacitors. In: *Handbook of Advanced Ceramics*, Elsevier, pp. 211–222, https://doi.org/10.1016/B978-0-12-385469-8.00013-7

Tan K. M., Babu T. S., Ramachandaramurthy V. K., Kasinathan P., Solanki S. G. and Raveendran S. K. (2021). Empowering smart grid: A comprehensive review of energy storage technology and application with renewable energy integration. *Journal of Energy Storage*, **39**, 102591, https://doi.org/10.1016/J.EST.2021.102591

Tarascon J. M. and Armand M. (2001). Issues and challenges facing rechargeable lithium batteries. *Nature 2001 414:6861*, Nature Publishing Group, **414**(6861), 359–367. Available online at: https://www.nature.com/articles/35104644 (accessed 19 June 2023).

Winter M. and Brodd R. J. (2004). What are batteries, fuel cells, and supercapacitors? *Chemical Reviews*, American Chemical Society, **104**(10), 4245–4270. Available online at: https://pubs.acs.org/doi/full/10.1021/cr020730k (accessed 18 June 2023).

Worku M. Y. (2022). Recent advances in energy storage systems for renewable source grid integration: a comprehensive review. *Sustainability* 2022, Vol. 14, Page 5985, Multidisciplinary Digital Publishing Institute, **14**(10), 5985. Available online at: https://www.mdpi.com/2071-1050/14/10/5985/htm (accessed 21 June 2023).

Xie J., Peng H. J., Huang J. Q., Xu W. T., Chen X. and Zhang Q. (2017). A Supramolecular Capsule for Reversible Polysulfide Storage/Delivery in Lithium-Sulfur Batteries. *Angewandte Chemie International Edition*, **56**(51), 16223–16227, https://doi.org/10.1002/ANIE.201710025

Yadlapalli R. T., Alla R. K. R., Kandipati R. and Kotapati A. (2022). Super capacitors for energy storage: progress, applications and challenges. *Journal of Energy Storage*, Elsevier, **49**, 104194, https://doi.org/10.1016/j.est.2022.104194

Yang B., Ruan D., Zhang Y., Wang C. and Qiao Z. (2020). Structural changes of activated carbon electrodes for EDLCs in the manufacturing process. *Transactions of Tianjin University*, Tianjin University, **26**(5), 391–398. Available online at: https://link.springer.com/article/10.1007/s12209-020-00268-2 (accessed 18 June 2023).

Yi C., Zou J., Yang H. and Leng X. (2018). Recent advances in pseudocapacitor electrode materials: transition metal oxides and nitrides. *Transactions of Nonferrous Metals Society of China*, **28**(10), 1980–2001, https://doi.org/10.1016/S1003-6326(18)64843-5

Yoshino A. (2012). The birth of the lithium-ion battery. *Angewandte Chemie International Edition*, John Wiley & Sons, Ltd, **51**(24), 5798–5800. Available online at: https://onlinelibrary.wiley.com/doi/full/10.1002/anie.201105006 (accessed 18 June 2023).

Zhang H., Zhang J., Gao X., Wen L., Li W. and Zhao D. (2022). Advances in materials and structures of supercapacitors. *Ionics*, **28**(2), 515–531, https://doi.org/10.1007/S11581-021-04359-5/FIGURES/6

Zuo W., Li R., Zhou C., Li Y., Xia J. and Liu J. (2017). Battery-Supercapacitor Hybrid Devices: Recent Progress and Future Prospects. *Advanced Science*, **4**(7), 1600539, https://doi.org/10.1002/advs.201600539

Part 2

Utility of Organic, Inorganic and Magnetic Nanoparticles

doi: 10.2166/9781789063592_0087

Chapter 5

Nanobiohybrids using organic nanoparticles for applications in water and wastewater treatment

Upasana Priyadarshini and Remya Neelancherry

Indian Institute of Technology Bhubaneswar, Argul, Odisha 752050, India
Corresponding Author: Remya Neelancherry, Indian Institute of Technology Bhubaneswar, Odisha 752050, India
E-mail: remya@iitbbs.ac.in

ABSTRACT

To deal with emergencies like water shortages and water pollution, scientists have worked on making materials and technologies that function well enough. The current technologies are not '100%' efficient at providing potable water. Therefore, it is essential to come up with new and better ways to purify water. Nanotechnology offers inventive water treatment techniques. Nanomaterials are fabricated with useful properties for adsorption, catalysis, sensing, and optoelectronics applications, such as reactivity, high-aspect ratio, controlled pore volume as well as electrostatic, hydrophilic, and hydrophobic interactions. Anions, metal ions, microorganisms, and organic compounds can be eliminated using nanoparticles (NPs). Minimal NP concentrations are required for water treatment, making their application cost-effective. Biopolymeric materials have a number of advantages over their synthetic counterparts due to their easy biodegradability and low impact on the environment. Because of their minimum toxicity, improved adsorption capability, excellent biodegradability, significant photocatalytic potential and biocompatibility, nanobiohybrids have a significant role in the field of water and wastewater treatment. This chapter summarizes various production methods of diversified nanobiohybrids synthesized in recent years, with a particular focus on nanobiocomposites.

Keywords: adsorbent, nanobiohybrid, nanobiocatalyst, nanomaterials, nanotechnology, pollutant, wastewater treatment

5.1 INTRODUCTION

Water is the most valuable resource of human existence, and potable water provision is a necessity for all humans. There is widespread agreement that the water crisis and pollution are major issues that require urgent attention. Large volumes of wastewater from human habitations and industrial processes are the most common cause of water pollution. About 80% of all municipal and industrial effluents are discarded without much pretreatment, which either directly or through indirect means makes the water polluted (Joshi *et al.*, 2021). Organic, inorganic, and biological pollutants all exist in water.

Numerous environmental pollutants are known to be poisonous and carcinogenic, with consequently disastrous consequences for human health and natural environments. Heavy metals are well-known contaminants because of their extreme toxicity. Among the compounds that may be present in wastewater and difficult to remove are persistent organic pollutants (POPs). Organic compounds that are resistant to being broken down by chemical, photolytic, or microbiological degradation are known as POPs. They are distinguished by their toxicity, semi-volatility, hydrophobicity, bioaccumulation and prolonged persistence in the environment. Emerging pollutants include solvents such as colouring agents, insecticides, and industrial chemical by-products with a variety of elemental compositions, such as dioxins, polycyclic aromatic hydrocarbons (PAHs) and perfluorinated compounds (Trojanowicz, 2020). It is essential to safeguard natural water systems, wildlife, humans, and vegetation from harmful contaminants. Conventional effluent treatment technologies, such as adsorption, membrane filtration, electrocoagulation, flocculation, and coagulation have been vastly enhanced over the last several decades to overcome their inefficiencies.

In recent years, a new variety of diversified materials known as nanomaterials (NMs) has quickly seized market share in several key sectors. The electronics business, the manufacturing sector, and the environmental protection sector have all reacted positively to NMs. There has been extensive worldwide use of nanoelectronic components, nano-pharmaceutical goods, and nanoadsorbents. Nanohybrids have been developed as advantageous alternatives to vanquish the constraints of many engineered materials in the present day. Nanohybrids are classified according to their scattered matrices and phase materials (Rizwan *et al.*, 2022). Recent progress in nanotechnology has shown that a wide variety of techniques, including mechanical, chemical, combustion, microwave, sonochemical, and sol–gel can be utilized to create nanomaterials.

The term nanobiocomposites was introduced to express a new class of nanobiohybrid materials formed by the nanoscale reinforcement of biopolymers like nucleic acids, proteins (gelatin, gluten, casein, and soy), lipids (cutin), and polysaccharides (cellulose, chitosan, glycogen, pectin, and alginate). When compared to other synthetic materials, biopolymeric materials excel due to their greater biodegradability and environmental friendliness. Physical, mechanical, and chemical properties of these natural substances make them suitable for extensive usage in food, medical, pharmaceutical, and environmental applications. The development of biopolymeric components for commercialization in several domains, such as the food industry, cosmetics industry, pharmaceuticals industry, and petroleum industry are receiving a significant amount of focus from scientists. In this chapter, emphasis is given on the preparation method of various nanobiohybrids along with their application in the field of water and wastewater treatment as adsorbent and biocatalyst.

5.2 PRODUCTION OF NANOBIOHYBRIDS

Nanobiotechnology is being applied in the area of environmental engineering, specifically as it pertains to the combination of biological components with nanomaterials for the synthesis of novel nanobiohybrid materials with a high capacity for water and wastewater purification. Biomolecular species and inorganic nanocomponents are assembled to create nanobiohybrid materials. The diverse chemical compositions, structures, and textures of inorganic solids assembled with biological species identify the characteristics of the nanobiohybrids that are developed. The inorganic parts of nanobiohybrids are typically single elements such as transition metals and carbon particles, metal oxides and hydroxides, silica, silicates, carbonates, and phosphates. The interaction mechanism that governs the production processes is the bonding between the inorganic and organic equivalents, which determines the stability of the ensuing nanobiohybrids. Hydrogen bonding and metal complexing are the major mechanisms responsible for the coating of micro- or nano-particulated materials with biopolymers (Ruiz-Hitzky *et al.*, 2008). Figure 5.1 shows various biocomponents of nanobiohybrids along with their sources and the governing mechanisms.

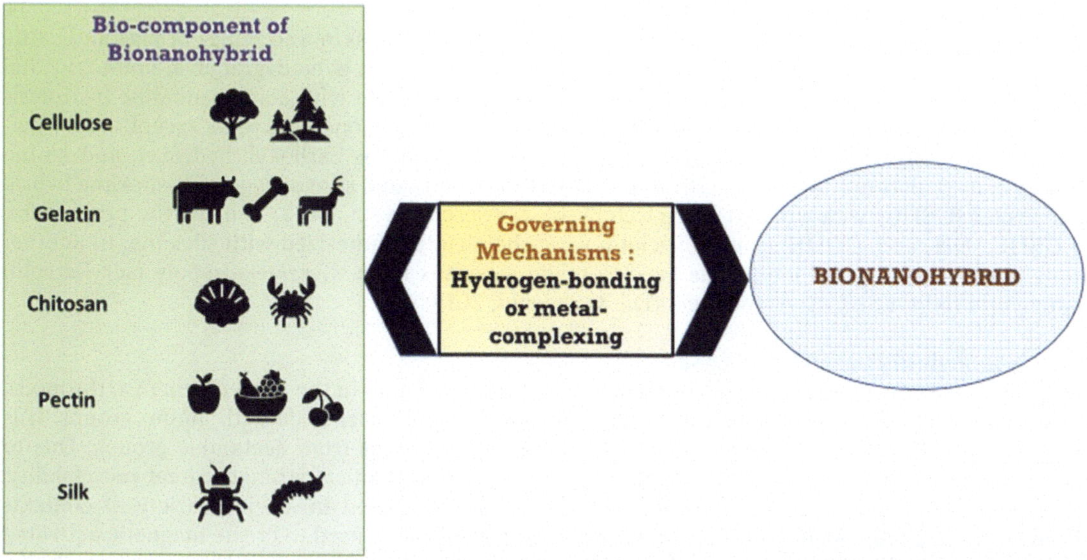

Figure 5.1 Biopolymeric component of nanobiohybrids, their sources and major mechanisms involved in nanobiohybrid formation.

5.2.1 Nanohybrids based on cellulose

Cellulose is very common, mostly non-toxic, and the most easily broken-down polymer made from carbohydrates. Cellulose is hydrophilic, but it does not dissolve in water, and its unique structure and properties make it applicable in various fields. Furthermore, cellulose imparts strength and rigidity to the cell structure. Several nanocellulose modifications are reported for nanohybrid synthesis, which is widely used for effluent treatment, and these nanohybrids exhibit exceptional adsorption of various contaminants (Daochalermwong *et al.*, 2020). The maleic acid cellulose-based nanohybrid (MA-CNC) is synthesized when the cellulose nanohybrid is combined with maleic anhydride (Kara *et al.*, 2021). Through H_2SO_4 hydrolysis, cellulose nanocrystalline (CNC) and maleic anhydride-functionalized cellulose nanocrystalline (MA-CNC) were obtained using the weed *Eichhornia crassipes*. A probable mechanism can be depicted in equation (5.1) (Rizwan *et al.*, 2022):

$$Cd^{2+}(aq) + R(OH,CO,COOH) \rightarrow Cd\left(R(OH,CO,COOH)\right)^{2+} \tag{5.1}$$

The absorption process was aided by physicochemical interlinkages between -OH and carbonyl moieties on the surface of CNC, and due to the Cd(II) ions' electrostatic interaction with these molecules.

A hydroxyethyl methacrylate modified cellulose nanohybrid (HEMA-CNC) was produced when hydroxyethyl methacrylate was combined with a cellulose nanohybrid using a microwave-assisted grafting copolymerization technique (Gouda & Aljaafari, 2021). Another study shows fabrication of the kaolinite-cellulose and cobalt oxide-based nanohybrid (Kao-Cel/CO_3O_4) (Hussain *et al.*, 2022). *Berberis lycium* leaf extract was used as a reducing and capping agent for cobalt oxide precipitation in ethanol over kaolinite–cellulose clay (Kao–Cel). In this study, ion-exchange, chelation, and electrostatic interactions phenomena were collectively responsible for the adsorption of contaminants onto the nanohybrid surface.

5.2.2 Nanohybrids based on gelatin

Gelatin is a polypeptide extracted from animal materials like the skin and bones of pigs and cattle with a translucent appearance and has a high molecular weight. It is biodegradable, non-toxic, and gelatinous when exposed to moisture. Partially denaturing collagen with acid or alkaline hydrolysis allows for its extraction. Gelatin can be utilized as polymeric backbones due to its exceptional water-retention capacity, film formation, and emulsification. Gelatin has carboxyl, hydroxyl, and amino groups in its structure (Chen *et al.*, 2014). By combining gelatin and siloxane, gelatin–siloxane hybrid composites (GSHC) were created (Wojciechowska & Cierpiszewski, 2022). During the preparation of GSHC, -NH_2 and -COOH groups of gelatin were chemically synthesized with siloxane. In another study, a cross-linking technique was used to make a porous nano-silver/gelatin-poly (acrylic acid) nanohybrid (NPGESNC-AcA) hydrogel (Dil & Sadeghi, 2018).

5.2.3 Nanohybrids based on chitosan

Chitin is the second most common biopolymer after cellulose found in the exoskeleton of arthropods, crustacean shells, and the cell wall of fungi. Chitosan is a polysaccharide with amino groups. It is made by deacetylating chitin, which means making amine groups from acetamide groups. Due to its unique properties, such as biodegradability, exceptional mechanical strength, great recyclability, biocompatibility, and non-toxic behaviour, chitosan has been used in a wide variety of contexts (Rizwan *et al.*, 2022). Spent coffee grounds and shrimp shells were used to create magnetic activated carbon/chitosan (MACCS) by combining chitosan and magnetic activated carbon (Le *et al.*, 2020). Recently, the β-cyclodextrin-ethylenediaminetetraacetic acid-chitosan (β-CD-EDTA-CS) polymer was combined with a multifunctional nanohybrid using cross-linking (amindation reaction) (Verma *et al.*, 2022). In another study, nanohybrid beads were produced when graphene oxide (GO) and chitosan (Ch) were modified with iron oxide (Fe_2O_3) nanoparticles (NPs) (Fe_3O_4-GO/Ch) via a simple chemical modification (Mishra & Tripathi, 2022). A hydrothermal method is used to synthesize Fe_3O_4 NPs and with Hummer's method graphene oxide was produced using ferric chloride as the iron precursor. At the onset of polymerization, Fe_3O_4-GO was introduced to the chitosan solution, and glutaraldehyde was used as a cross-linking agent to create Fe_3O_4-GO/Ch NPs.

5.2.4 Nanohybrids based on pectin

Pectin is found in the cell membranes of plants and is a high-molecular-weight anionic linear polysaccharide that is biodegradable and non-toxic. Chains of d-galacturonate molecules linked by α-(1–4) glycosidic links give rise to pectin. These links spread out -COOH/-OH groups along the backbone of the molecule (Kayan & Kayan, 2021). In a study, nanobiohybrid gel beads of chitosan–pectin were synthesized by employing a simplistic cross-linking approach (Shao *et al.*, 2021). In another investigation, the nanobiohybrid of pectin was combined with graphene oxide to produce a pectin–graphene oxide nanocomposite (Pc/GO) by applying the sol–gel method (Kaushal *et al.*, 2020). Using an environmental-friendly synthesis, a nanocomposite gel was obtained when a copolymer gel based on pectin was incorporated with silver NPs. When pectin in aqueous solution is combined with silver nitrate and accompanied by microwave irradiation, this yields silver NPs. The nanocomposite hydrogel was developed by polymerizing 2-acrylamido-2-methyl-1-propane sulphonic acid (AMPS) and acrylamide (AAm) in a pectin solution that included silver particles while microwave heating (Kodoth & Badalamoole, 2020).

5.2.5 Nanohybrid based on silk protein

Naturally occurring polymer silk (*Bombyx mori*) is a material that mainly contains fibroin and sericin (a protein having adhesive properties). Various organisms, such as silkworms, spiders, flies, and silverfish are good sources for sericin. The heavy H-chain (390 kDa) and the light L-chain (26 kDa) of silk fibroin are linked by a disulphide bond (Wahab *et al.*, 2019). In a study, using the sol–gel technique, silica–gelatin aerogels were prepared that is both inexpensive and environmental friendly (Herman *et al.*, 2020). The aim was to synthesize gelatin containing aerogels along with open mesoporous structures. In another study, a simple chemical cross-linking technique was used to create a functional hydrogel of silk fibroin/poly-ethylenimine (PEI) which is also a low cost and sustainable nanobiohybrid (Godiya *et al.*, 2019).

5.3 NANOBIOHYBRID APPLICATIONS IN WATER AND WASTEWATER TREATMENT

5.3.1 Nanobiohybrids as adsorbent

Adsorption is a surface phenomenon that involves the shift of contaminants onto a solid surface. In general, physical forces are responsible for adsorption, but this can sometimes be attributed to feeble chemical bonds. Particle size, contact time, temperature, initial metal concentration, adsorbent dose and pH are some of the variables influencing the process of adsorption. It is possible that the surface area and lack of selectivity of conventional adsorbents limit their effectiveness (Goyal *et al.*, 2011). In order to address the limitations of conventional adsorbents, researchers developed nanoadsorbents for wastewater treatment. Nanoadsorbents have a wide range of physicochemical qualities. These include conductivity, catalytic activity, a large surface area, high chemical reactivity, selectivity, magnetic and optical properties (Sharma *et al.*, 2022). Figure 5.2 shows the physical and chemical interaction for adsorption of various contaminants by nanobiohybrids.

In the past decade, a variety of wastewater treatment methods, including precipitation, advanced oxidation, ozonation, membrane filtration, ion exchange, flocculation–coagulation, and oxidation have been investigated but they have numerous disadvantages, including high energy costs, material regeneration, by-product production, and disposal requirements. Nanoadsorbents are NPs derived from organic or inorganic sources with a high adsorption affinity. They are extremely efficient at removing contaminants of all various molecular sizes because of their enormous specific surface area, active surface, and porosity. Regeneration is also a crucial characteristic of nanoadsorbents. A range of nanoadsorbents are utilized for wastewater treatment. They have been found to be exceptionally effective at removing pollutants from effluents (Sharma *et al.*, 2022). It is therefore impossible to conduct a direct comparison of capacity because they are reliant on several variables, including the shape and size of the NPs as well as working variables such as pH, temperature, reaction time, and experimental design. Titania nanotubes, ordered mesoporous carbon (OMC), carbon nanofibres, graphene, and carbon nanotubes are a few of the well-known nanomaterials used in wastewater

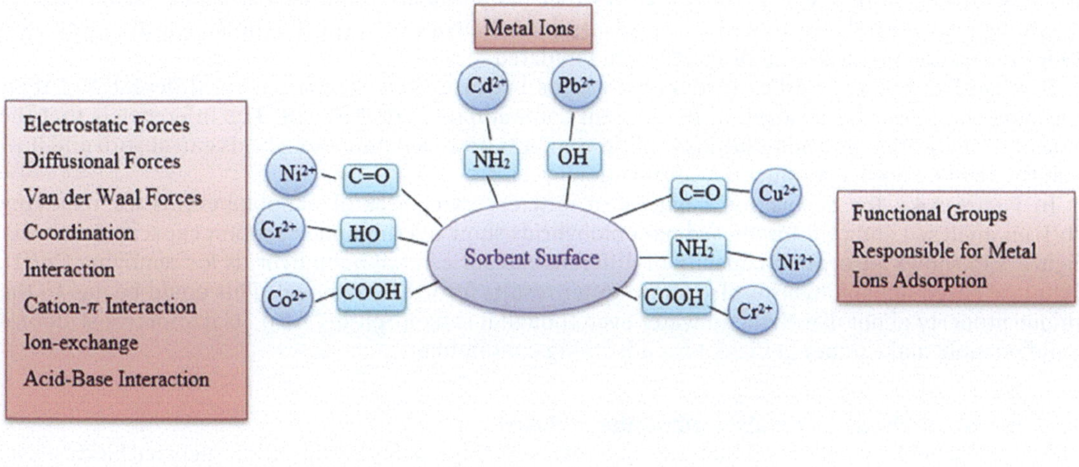

Figure 5.2 Physical and chemical interface for adsorption of various contaminants.

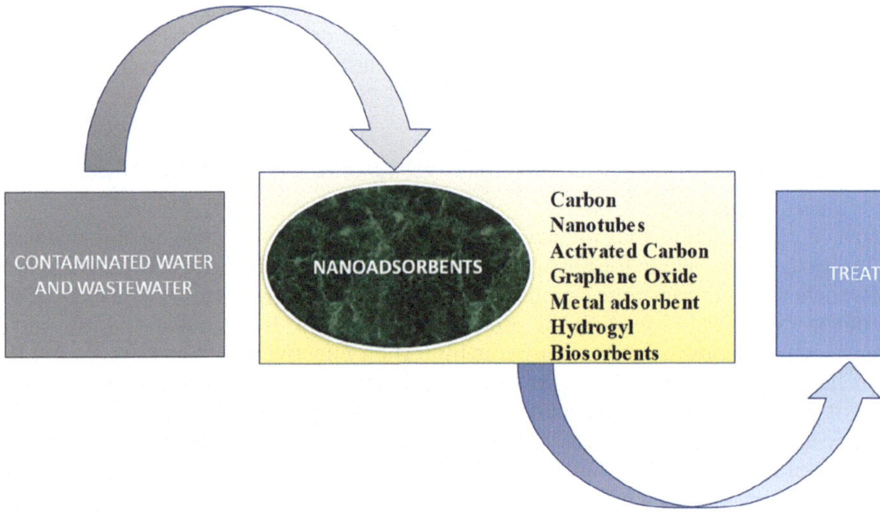

Figure 5.3 Water and wastewater effluent treatment using different nanoabsorbents.

treatment (Sharma *et al.*, 2022). A higher concentration of surface defects results from the large surface area, which promotes interaction with pollutants. NPs are occasionally coated with extra coating agents to enhance their adsorption properties. In Figure 5.3, the treatment of water and wastewater using various nanoadsorbents is given.

The development of hybrids of biopolymers with other nanomaterials generates biopolymeric nanohybrids or nanobiohybrids with exceptional morphological, structural features and physiochemical behaviour. In recent years, the application of nanobiohybrids in wastewater treatment has evolved significantly. Heavy metals are the major target contaminant that is removed by nanobiohybrids. Nanobiohybrids have a substantial surface area and reliably eliminate metal ions in soluble state. In addition, nanobiohybrids have the ability to form films; activated functionalities, and high catalytic and adsorptive capacity are the other advantages. Biopolymers do not have much elasticity and are not very strong, so they can be mixed with other materials and modifiers to make them better at absorbing heavy metals. In Table 5.1, the adsorption efficiency of various nanobiohybrids along with their preparation methods and properties are tabulated.

It is hard to compare different nanobiohybrids because their structures are different and their optimum conditions for adsorption of different contaminants are different. The inference is that the type of configuration and adsorption conditions affects how well nanobiohybrids can absorb and how well they can be used again in a new adsorption cycle.

In Figure 5.4a, the adsorption capacity of different categories of nanobiohybrids for removing Pb(II) is analysed and cellulose-based nanobiohybrids show maximum adsorption capacity for Pb(II). Figure 5.4b shows adsorption capacity of different categories of nanobiohybrids for removing Cd(II). Cellulose-based nanobiohybrids also show better results for Cd(II) removal. This could be due to the unique property of not dissolving in water even though it is hydrophilic. Also, its distinctive structure could possibly make it more efficient for adsorbing contaminants.

5.3.2 Nanobiohybrids as catalyst (nanobiocatalysis)

It has become evident that a combined approach of producing carrier-based nanostructured biocatalysts can serve as a nanobiocatalysis interface. A novel technique referred to as nanobiocatalysis has the

Table 5.1 Synthesis method and adsorption capacity of various nanobiohybrids for effective removal of target pollutant.

Nanobiohybrid	Preparation Method	Surface Area (m^2/g)	Target Pollutant	Adsorption Capacity (mg/g)	References
Kaolinite–cellulose/cobalt oxide (Kao-Cel/Co_3O_4)	Facile chemical method	85.92	Pb(II) Cd(II)	293.67 267.84	Hussain *et al.* (2022)
Gelatine–siloxane hybrid monoliths	Facile chemical method	_	Cu(II) Cd(II) Pb(II)	1.76 1.50 3.75	Wojciechowska and Cierpiszewski (2022)
β-Cyclodextrin–ethylenediaminetetraacetic acid-chitosan (β-CD-EDTA-CS)	Cross-linking method (amindation reaction)	2.39	Hg(II) Cd(II)	346.30 202.90	Verma *et al.* (2022)
Magnetic iron NPs/ graphene oxide-chitosan (Fe_3O_4-GO/Ch)	Polymerization and cross linking, Hummer's method, Hydrothermal treatment	39.80	Pb(II)	187.00	Mishra and Tripathi (2022)
Maleic anhydride Nanocellulose (MA-CNC)	Hydrolysis using sulphuric acid	123.90	Cd(II)	215.52	Kara *et al.* (2021)
Hydroxyethyl Methacrylate-modified Cellulose nanofibres (HEMA-CNC)	Grafting copolymerization assisted by microwave-assisted technique	_	Cd(II) Pb(II)	840.00 926.00	Gouda and Aljaafari (2021)
Chitosan–pectin gel beads	Facile cross linking	_	Cu(II) Cd(II) Hg(II) Pb(II)	169.40 177.60 208.50 _	Shao *et al.* (2021)
Magnetic activated carbon/chitosan (MACCS)	Chemical method	105.74	Ni(II)	108.70	Le *et al.* (2020)
Dual-responsive pectin/ graphene oxide (Pc/GO)	Sol–gel technique	_	Cr(III)	3333.00 mol/g	Kaushal *et al.* (2020)
Silver NP embedded pectin-based hydrogel	Microwave-assisted polymerization	_	Cu(II) Pb(II)	111.00 130.00	Kodoth and Badalamoole (2020)
Silica-gelatin aerogels	Sol–gel technique	910±60	Hg(II)	209.00	Herman *et al.* (2020)
Silk Fibroin/ polyethylenimine functional hydrogel	Cross-linking technique	_	Cu(II) Pb(II) Cd(II) Zn(II) Ni(II) Ag(II)	163.90 185.20 169.50 125.00 140.00 200.00	Godiya *et al.* (2019)
Nanosilver/gelatine-poly (acrylic acid) nanohybrid (NPGESNC-AcA)	Cross-linking technique	_	Cu(II)	147.10	Dil and Sadeghi (2018)

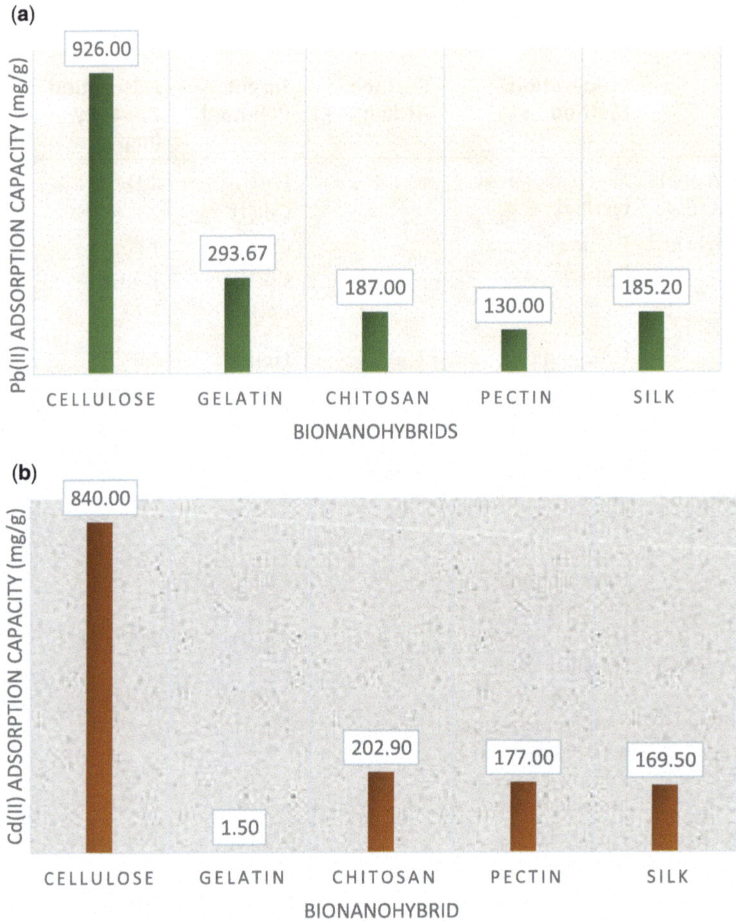

Figure 5.4 Adsorption capacity of various nanobiohybrids for (a) Pb(II) and (b) Cd(II).

ability to enhance the structure and manufacturing of nanoscale catalysts with enhanced catalytic properties for use in multipurpose bioprocessing. In addition, nanocomposites with distinct physical, functional, chemical, and structural properties have contributed significantly to imitating the microenvironment of enzymes. By adjusting the enzyme's microenvironment with nanobiocatalysis, greater selectivity and specificity can be attained. Nanobiocatalysis is achieving popularity as a potent biocatalysis benchmark with many strictly controlled properties, such as stability, selectivity, induced activity, specificity, resistivity, reaction efficiency, optimal yield, multi-usability, high catalytic turnover, enhanced mass transfer efficiency, ease of recovery and cost-effectiveness (Bilal *et al.*, 2021a).

Enzymes are highly effective and powerful catalysts which are widely applied throughout research and industry. The extraordinary selectivity, activity and sensitivity of enzymes makes them potential biocatalysts for a vast range of fascinating applications, including biocatalysis and in the field of medicine (Ayub *et al.*, 2022). Enzymes are not merely well-developed catalysts but are a part of the regulatory system of the cell and hence can detect matter other than their specified substrate as well as the product. Biocatalysis is the enzymatic transformation of substrates which are biochemicals in nature into the target product. The significant barriers of nanobiocatalysts may be eliminated through

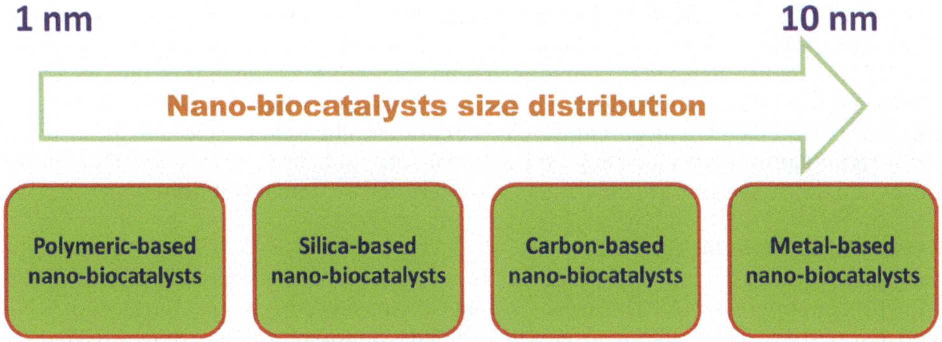

Figure 5.5 Different types of materials used in nanobiocatalysis.

the use of enzyme immobilization, a method to lower the overall production expenses of a process by streamlining and improving repetitive batch procedures and subsequent processing. Based on enzyme immobilization, nanobiocatalysts can be classified into four categories which are shown in Figure 5.5.

5.3.2.1 Polymeric nanobiocatalyst

In recent times, researchers encased individual enzymes in particular enzyme NPs. Within a NP, the enzyme was rendered stable by a thin shell, either undisturbed from the surface of the enzyme or by forming a polymer around it beforehand (Chapman & Stenzel, 2019). When compared to the use of other support materials, immobilizing enzymes using nanosupports based on polymers has numerous benefits. Nanobiocatalysts with improved stability and multiple functions are now available, owing to large-scale manufacturing and an abundance of activated functional groups (Ayub *et al.*, 2022). Due to their changing physical characteristics and unpredictability of molecular size, polymeric materials are inappropriate for biocatalytic immobility. In contrast, nanostructured polymers have several distinctive characteristics that render them suitable for multifunctionality, recycling, and enzyme stabilization. Laccase molecules have been blended into bacterial nanocellulose networks employing soaking and adsorption processes. Utilizing bacterial nanocellulose immobilized laccase, dye effluent from textile industry was decoloured. A horizontal rotating reactor enabled improved reusability of immobilized enzymes, enhanced detoxification, and decrease of COD and carbon, indicating that it has great potential for dye discharge decolorization (Yuan *et al.*, 2020).

5.3.2.2 Silica-based nanobiocatalysts

The adaptability of mesoporous silica materials in terms of their structural and textural properties is gaining increasing interest. These characteristics include surface area, volume, particle, and pore size. An innovative laccase enzyme carrier pathway with 83% immobilization yield is an excellent nanobiocatalyst for wastewater management and dye removal (Wehaidy *et al.*, 2019).

5.3.2.3 Carbon-based nanobiocatalysts

Carbon-based nanobiocatalysis has been utilized extensively for enzyme immobilization due to its inertness, biocompatibility, and thermal stability. The development of a novel multi-walled carbon nanotube (MWCNT)/cordierite composite support was studied (Li *et al.*, 2017). MWCNTs were effectively loaded onto a cordierite matrix using a noncovalent self-assembly technique involving the amination reaction of N-aminoethylaminopropyltrimethoxysilane. Both the surface and interior cavities of the cordierite matrix were loaded with MWCNTs. Physical adsorption was effective in immobilizing horseradish peroxidase (HRP) upon this composite support. In another study, by adding chitosan to halloysite, a naturally occurring nanotubular aluminosilicate, chitosan–halloysite hybrid nanotubes (CTS-HNT) were developed (Zhai *et al.*, 2013). The horseradish peroxidase enzyme was immobilized using CTS-HNT as it is a natural, plentiful, inexpensive, and environmentally favourable resource. The horseradish peroxidase immobilized on CTS-HNT was able to retain its activity over a wide range and exhibited exceptional storage stability. While horseradish peroxidase that was immobilized on CTS-HNT has been employed to eliminate phenolic compounds from wastewater, it showed a significantly high efficiency and removal rate, indicating that CTS-HNT immobilized horseradish peroxidase systems are efficient as a wastewater treatment solution.

5.3.2.4 Metal-based nanobiocatalysts

As an approach of constructing biocatalytic systems, metal-based nanoparticles (MNPs) have attracted significant interest. MNPs have exceptional enzyme recycling ability. Surfaces of MNPs were formulated and synthesized by adding groups like thiolate, amino phosphate, or carboxylate to create strong interface bonds with biocatalysts and enhance the immobilization yield (Bilal *et al.*, 2021b). In a study, tyrosinase-MNPs were utilized as a magnetic nanobiocatalyst to remediate wastewater containing phenolic compounds. The effect of temperature, pH, catalyst dosage, reusability, and initial phenol concentration on the phenolic compound removal efficiency of the fabricated nanobiocatalyst was analysed. The results demonstrated that the specific concentration of nanobiocatalyst could degrade phenolic compounds over a broad pH and temperature range (Abdollahi *et al.*, 2018). More than 70% of phenol was degraded by the immobilized tyrosinase when a very high concentration of the substrate (2500 mg/L) was applied for phenol removal (Abdollahi *et al.*, 2018).

5.4 CONCLUSION

At this point in time, we are in need of water purification methods that can supply high-quality drinkable water, eliminate micropollutants, and intensify the processes involved in industrial processes. This is made possible by nanotechnology; the one-of-a-kind characteristics of NPs make them excellent candidates for the development of a technology for rapid water treatment. NPs have the ability to eliminate anions and metal ions, as well as organic contaminants and microorganisms. Due to the low doses of NPs that are necessary for the water treatment process, their application is comparatively economical. One of the main problems with traditional biosorbents is their lack of selectivity for a certain metal or a certain chemical species of an element. There is thus a drift towards making novel biomaterials like nanobiohybrids that can fix these problems. This chapter elaborated different bio-components have been used with various nanomaterials like metal-based NPs and carbon-based NPs to remove contaminants from wastewater as an adsorbent. It has been concluded that cellulose-based nanobiohybrids are the best nanobioadsorbent. Recent advances in nanobiotechnology have enabled the incorporation of eco-friendly biocatalysts into a variety of nanostructures with distinct properties. This gave rise to novel biocatalysts that are nanobiocatalysts. This chapter showed the application of nanobiohybrids as biocatalyst is a novel approach and an efficient way to treat water and wastewater. Hence, nanobiohybrids are very promising that can be a remedy to water pollution.

REFERENCES

Abdollahi K., Yazdani F., Panahi R. and Mokhtarani, B. (2018). Biotransformation of phenol in synthetic wastewater using the functionalized magnetic nano-biocatalyst particles carrying tyrosinase. *3 Biotech*, **8**(10), 419, https://doi.org/10.1007/s13205-018-1445-2

Ayub J., Saeed M. U., Hussain N., Zulfiqar I., Mehmood T., Iqbal H. M. N. and Bilal M. (2022). Designing robust nano-biocatalysts using nanomaterials as multifunctional carriers – expanding the application scope of bio-enzymes. *Topics in Catalysis*, **66**, 625–648, https://doi.org/10.1007/s11244-022-01657-8

Bilal M., Hussain N., Américo-Pinheiro J. H. P., Almulaiky Y. Q. and Iqbal H. M. N. (2021a). Multi-enzyme co-immobilized nano-assemblies: bringing enzymes together for expanding bio-catalysis scope to meet biotechnological challenges. *International Journal of Biological Macromolecules*, **186**, 735–749, https://doi.org/10.1016/j.ijbiomac.2021.07.064

Bilal M., Qamar S. A., Ashraf S. S., Rodríguez-Couto S. and Iqbal H. M. N. (2021b). Robust nanocarriers to engineer nanobiocatalysts for bioprocessing applications. *Advances in Colloid and Interface Science*, **293**, 102438, https://doi.org/10.1016/j.cis.2021.102438

Chapman R. and Stenzel M. H. (2019). All wrapped up: stabilization of enzymes within single enzyme nanoparticles. *Journal of the American Chemical Society*, **141**(7), 2754–2769, https://doi.org/10.1021/jacs.8b10338

Chen G., Qiao C., Wang Y. and Yao J. (2014). Synthesis of magnetic gelatin and its adsorption property for Cr (VI). *Industrial & Engineering Chemistry Research*, **53**, 15576–15581, https://doi.org/10.1021/ie502709u

Daochalermwong A., Chanka N., Songsrirote K., Dittanet P., Niamnuy C. and Seubsai A. (2020). Removal of heavy metal ions using modified celluloses prepared from pineapple leaf fiber. *ACS Omega*, **5**(10), 5285–5296, https://doi.org/10.1021/acsomega.9b04326

Dil N. N. and Sadeghi M. (2018). Free radical synthesis of nanosilver/gelatin-poly (acrylic acid) nanocomposite hydrogels employed for antibacterial activity and removal of Cu(II) metal ions. *Journal of Hazardous Materials*, **351**, 38–53, https://doi.org/10.1016/j.jhazmat.2018.02.017

Godiya C. B., Cheng X., Deng G., Li D. and Lu X. (2019). Silk fibroin/polyethylenimine functional hydrogel for metal ion adsorption and upcycling utilization. *Journal of Environmental Chemical Engineering*, **7**(1), 102806, https://doi.org/10.1016/j.jece.2018.11.050

Gouda M. and Aljaafari A. (2021). Removal of heavy metal ions from wastewater using hydroxyethyl methacrylate-modified cellulose nanofibers: kinetic, equilibrium, and thermodynamic analysis. *International Journal of Environmental Research and Public Health*, **18**(12), 6581, https://doi.org/10.3390/ijerph18126581

Goyal A., Johal E. and Rath G. (2011). Nanotechnology for water treatment. *Current Nanoscience*, **7**, 640–654, https://doi.org/10.2174/157341311796196772

Herman P., Fábián I. and Kalmár J. (2020). Mesoporous silica-gelatin aerogels for the selective adsorption of aqueous Hg(II). *ACS Applied Nano Materials*, **3**(1), 195–206, https://doi.org/10.1021/acsanm.9b01903

Hussain D., Khan S. A., Alharthi S. S. and Khan T. A. (2022). Insight into the performance of novel kaolinite-cellulose/cobalt oxide nanocomposite as green adsorbent for liquid phase abatement of heavy metal ions: modelling and mechanism. *Arabian Journal of Chemistry*, **15**(7), 103925, https://doi.org/10.1016/j.arabjc.2022.103925

Joshi N., Bhardwaj U., Khan S. and Iqbal H. M. N. (2021). Synergistic role of bacterial consortium to biodegrade toxic dyes containing wastewater and its simultaneous reuse as an added value. *Chemosphere*, **284**, 131273, https://doi.org/10.1016/j.chemosphere.2021.131273

Kara H. T., Anshebo S. T. and Sabir F. K. (2021). Adsorptive removal of Cd(II) ions from wastewater using maleic anhydride nanocellulose. Journal of Nanotechnology, **2021**, 1–15, https://doi.org/10.1155/2021/9966811

Kaushal S., Kaur N., Kaur M. and Singh P. P. (2020). Dual-responsive pectin/graphene oxide (Pc/GO) nano-composite as an efficient adsorbent for Cr (III) ions and photocatalyst for degradation of organic dyes in waste water. *Journal of Photochemistry and Photobiology A: Chemistry*, **403**(6), 112841, https://doi.org/10.1016/j.jphotochem

Kayan G. Ö. and Kayan A. (2021). Composite of natural polymers and their adsorbent properties on the dyes and heavy metal ions. *Journal of Polymers and the Environment*, **29**(11), 3477–3496, https://doi.org/10.1007/s10924-021-02154-x

Kodoth A. K. and Badalamoole V. (2020). Silver nanoparticle-embedded pectin-based hydrogel for adsorptive removal of dyes and metal ions. *Polymer Bulletin*, **77**(2), 541–564, https://doi.org/10.1007/s00289-019-02757-4

Le V. T., Dao M. U., Le H. S., Tran D. L., Doan V. D. and Nguyen H. T. (2020). Adsorption of Ni(II) ions by magnetic activated carbon/chitosan beads prepared from spent coffee grounds, shrimp shells and green tea extract. *Environmental Technology*, **41**(21), 2817–2832, https://doi.org/10.1080/09593330.2019.1584250

Li Z. L., Cheng L., Zhang L. W., Liu W., Ma W. Q. and Liu L. (2017). Preparation of a novel multi-walled-carbon-nanotube/cordierite composite support and its immobilization effect on horseradish peroxidase. *Process Safety and Environmental Protection*, **107**, 463–467, https://doi.org/10.1016/j.psep.2017.02.021

Mishra S. and Tripathi A. (2022). Adsorptive removal of Pb(II) via green route using magnetic iron nanoparticle sprinkled graphene oxide-chitosan beads in aqueous medium. *Environmental Nanotechnology, Monitoring and Management*, **17**(5), 100632, https://doi.org/10.1016/j.enmm.2021.100632

Rizwan K., Babar Z. B., Munir S., Arshad A. and Rauf A. (2022). Recent advancements in engineered biopolymeric-nanohybrids: a greener approach for adsorptive-remediation of noxious metals from aqueous matrices. *Environmental Research*, **215**(3), 114398, https://doi.org/10.1016/j.envres.2022.114398

Ruiz-Hitzky E., Darder M. and Aranda P. (2008). An introduction to bio-nanohybrid materials. In: Bio-Inorganic Hybrid Nanomaterials: Strategies, Syntheses, Characterization and Applications, E. Ruiz-Hitzky, K. Ariga and M. Lvov (eds.), Wiley-VCH Verlag GmbH & Co. KGaA, https://doi.org/10.1002/9783527621446.ch1

Shao Z., Lu J., Ding J., Fan F., Sun X., Li P., Fang Y. and Hu Q. (2021). Novel green chitosan-pectin gel beads for the removal of Cu(II), Cd(II), Hg(II) and Pb(II) from aqueous solution. *International Journal of Biological Macromolecules*, **176**, 217–225, https://doi.org/10.1016/j.ijbiomac.2021.02.037

Sharma P., Nanda K., Yadav M., Shukla A., Srivastava S. K., Kumar S. and Singh S. P. (2022). Remediation of noxious wastewater using nanohybrid adsorbent for preventing water pollution. *Chemosphere*, **292**(3), 133380, https://doi.org/10.1016/j.chemosphere.2021.133380

Trojanowicz M. (2020). Removal of persistent organic pollutants (POPs) from waters and wastewaters by the use of ionizing radiation. The Science of the total environment, **718**, 134425, https://doi.org/10.1016/j.scitotenv.2019.134425

Verma M., Lee I., Hong Y., Kumar V. and Kim H. (2022). Multifunctional β-cyclodextrin-EDTA-chitosan polymer adsorbent synthesis for simultaneous removal of heavy metals and organic dyes from wastewater. *Environmental Pollution*, **292**(2890), 118447, https://doi.org/10.1016/j.envpol.2021.118447

Wahab N., Saeed M., Ibrahim M., Munir A., Saleem M., Zahra M. and Waseem A. (2019). Synthesis, characterization, and applications of silk/bentonite clay composite for heavy metal removal from aqueous solution. *Frontiers in Chemistry*, **7**, 1–12, https://doi.org/10.3389/fchem.2019.00654

Wehaidy H. R., Abdel-Naby M. A., El-Hennawi H. M. and Youssef H. F. (2019). Nanoporous zeolite-X as a new carrier for laccase immobilization and its application in dyes decolorization. *Biocatalysis and Agricultural Biotechnology*, **19**, 101135, https://doi.org/10.1016/j.bcab.2019.101135

Wojciechowska P. and Cierpiszewski R. (2022). Gelatin – siloxane hybrid monoliths as novel heavy metal adsorbents applied sciences gelatin – siloxane hybrid monoliths as novel heavy metal adsorbents. *Applied Sciences*, **12**(3), 1258, https://doi.org/10.3390/app12031258

Yuan H., Chen L., Cao Z. and Hong F. F. (2020). Enhanced decolourization efficiency of textile dye Reactive Blue 19 in a horizontal rotating reactor using strips of BNC-immobilized laccase: optimization of conditions and comparison of decolourization efficiency. *Biochemical Engineering Journal*, **156**, 107501, https://doi.org/10.1016/j.bej.2020.107501

Zhai R., Zhang B., Wan Y., Li C., Wang J. and Liu J. (2013). Chitosan-halloysite hybrid-nanotubes: horseradish peroxidase immobilization and applications in phenol removal. *Chemical Engineering Journal*, **214**, 304–309, https://doi.org/10.1016/j.cej.2012.10.073

doi: 10.2166/9781789063592_099

Chapter 6

Assessing the feasibility of inorganic nanomaterials for nanohybrids formation

Ranjana Pande[1]* *and Anirudh Gururaj Patil*[2]

[1]*School of Basic Sciences and Research, Life Sciences Department, Sharda University, Greater Noida, India*
[2]*Annasaheb Dange College of Engineering & Technology, Department of Food Technology, Ashta, Maharashtra, India*
**Corresponding Author: ranjana.pande@sharda.ac.in*

ABSTRACT

Pollutant removal from industrial effluents is a big challenge for industries and these pollutants can have a great risk to the environment. Traditional methodologies for nanomaterials fabrication have drawbacks such as high energy consumption and use of toxic agents. But there is an option of green nanobiotechnology that synthesizes biogenic metal nanoparticles (Bio-Me NPs) with microbes. Biogenic metal NPs are produced at ambient biosynthesis conditions, and exhibit unique surface properties and crystalline structures, which can eliminate various contaminants from water. Microbes-based nanomaterials exhibit high-quality catalytic performance because of their greater number of active sites and higher surface area as compared to polycrystalline materials. Hence, microorganisms used for nanomaterial production can provide a novel approach for wastewater treatment and decontamination. This chapter focuses on the feasibility of nanomaterials for microbes to fabricate nanohybrids for wastewater treatment and decontamination. Apart from this, the use of microbial fuel cells to not only treat wastewater but also generate electricity in the process is discussed.

Keywords: nanoparticles, biogenic metal, microbial nanohybrid, emerging pollutants, wastewater treatment, green nanotechnology

6.1 INTRODUCTION

6.1.1 Production of nanoparticles

In the last few decades, nanoscience has attracted the attention of the scientific community worldwide for the sustainable production of various nanoparticles (NPs) using innovative techniques, which find applications in the pharmacy, medical diagnostics and disease treatment, energy, electronics, agriculture as well as chemical and space industries (Elfeky *et al.*, 2020; Gahlawat & Choudhury, 2019; Grasso *et al.*, 2020; Khan *et al.*, 2019; Salem & Fouda, 2021). Nanotechnology is an emerging branch of science which works at a size less than 100 nm on the atomic, molecular and supramolecular level for the development of desired properties and functions for diverse applications (Koul *et al.*, 2021). The high surface area-to-volume/mass ratios of nanomaterials greatly improve the adsorption properties of sorbent materials (Thakur & Mukherjee, 2021). It is expected that by 2030, the global

NPs of a desired size, structure, monodispersity and morphology is crucial in terms of their various applications in nanoscience and associated nanobusiness (Dobias *et al.*, 2011). In 2021, Koul *et al.* reviewed recent advancements in nanotechnology are based on the synthesis of NPs/materials using novel techniques. NPs can be fabricated using physical, chemical and biological methods (Figure 6.1). The major physical methods used for the synthesis of NPs include pyrolysis, physical vapor deposition (PVD), lithography, crushing, grinding, milling, and ball processing (Koul *et al.*, 2021) On the other hand, chemosynthesis of NPs consists of sol–gel synthesis, chemical reduction, electrolysis, chemical vapor deposition (CVD), photocatalytic reduction, and microwave-assisted synthesis.

There are some drawbacks of physical and chemical methods for the production of nanomaterials, which are listed below.

(a) Physical methods are not suitable for large-scale production of nanomaterials because of the lower yield of NPs, requirement of high energy and higher input costs.
(b) Chemical methods require less energy consumption and allow production of homogenous NPs with high accuracy (Albanese *et al.*, 2012; Koul *et al.*, 2021).
(c) Both traditional methods are laborious, time consuming and hazardous due to the use of toxic chemicals.
(d) Biomedical applications of NPs produced from chemosynthesis have been limited due to their instability and toxic nature (Koul *et al.*, 2021; You *et al.*, 2013).

Koul *et al.* (2021) described that NP synthesis using biological systems is rapid, feasible and ecofriendly. Moreover, the toxicity and size characteristics of the NPs can be controlled. Microbial cells have the natural ability to grow in diverse habitat, are fast growing and easy to maintain, bacteria and microalgae possess the specific potential to fabricate distinctive nanomaterials, such as exopolysaccharides, nanocellulose, and nanowires (Golmohammadi *et al.*, 2017; He *et al.*, 2008; Li *et al.*, 2017). Figure 6.2 shows a schematic representation of microbe-based biological synthesis of different NPs, their characterization and applications.

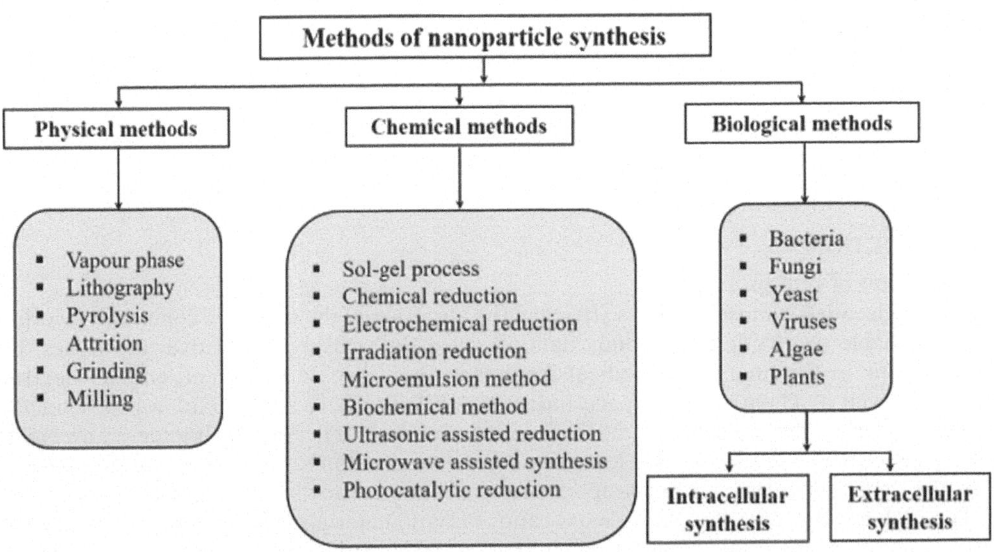

Figure 6.1 Different methods for NPs synthesis. (*Source:* From Koul *et al.* 2021).

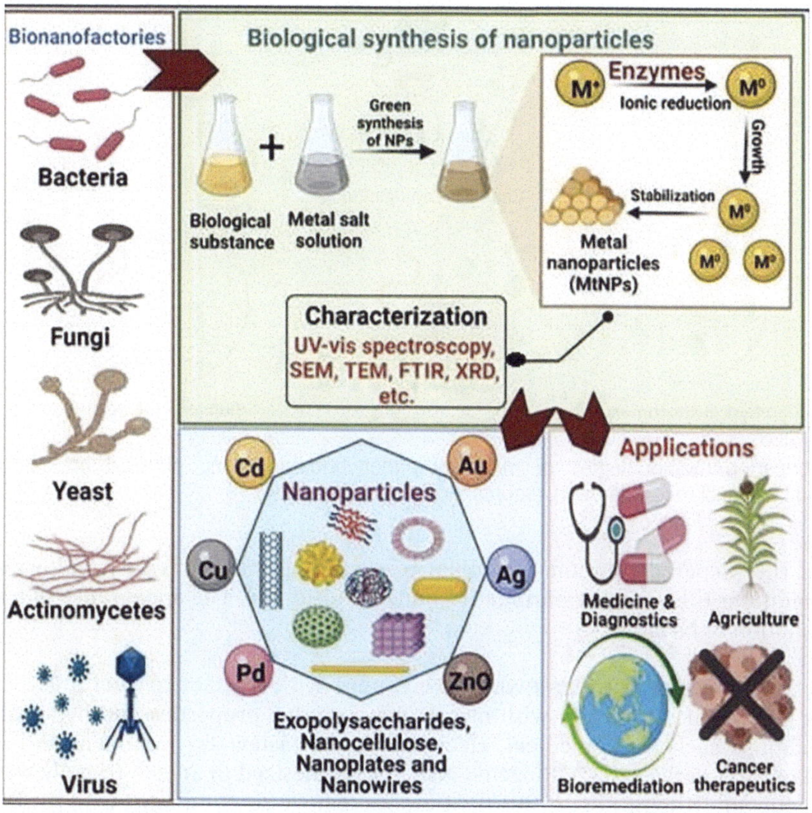

Figure 6.2 Schematic representation of the microbe-based biological synthesis of various NPs, their characterization and applications. (*Source:* From Koul *et al.* 2021).

6.1.2 Microbial nanohybrids

In 2020, Huo *et al.* defined microbial nanohybrids as the conjugates between microorganisms and nanomaterials through specific physiochemical interactions. The fabrication methodologies of microorganism-based nanohybrids can be generally categorized into (1) physical or chemical conjugations of nanomaterials to the extracellular space (ECS), and (2) nanomaterial encapsulations into the intracellular space of the microorganism (Figure 6.3).

Microbes have the potential to reduce inorganic materials into NPs through extracellular or intracellular routes (Fariq *et al.*, 2017; Koul *et al.*, 2021; Salem & Fouda 2021). Microbes absorb metal ions either from the environment or media and convert these metallic ions into elemental form via an enzymatic reduction (Koul *et al.*, 2021; Li *et al.*, 2011). Figure 6.2 explains the extracellular fabrication of NPs. Microbes are grown in suitable media and the broth containing the microbial cells is then centrifuged, and the supernatant containing microbial enzymes is then used for the synthesis of NPs (Al-Dhabi *et al.*, 2018; Das *et al.*, 2014; Koul *et al.*, 2021; Yadav *et al.*, 2015).

In the intracellular fabrication of NPs, the cellular mechanism of microbial cells is used for the synthesis of NPs. The microbial cultures are maintained in appropriate liquid media and the microbial biomass is washed with sterile distilled water followed by centrifugation to obtain the biomass pellet (Castro *et al.*, 2014; Fernández-Llamosas *et al.*, 2017; Koul *et al.*, 2021; Shah *et al.*, 2015). Furthermore, Koul *et al.* (2021) explained the microbial biomass is then allowed to react with an aqueous solution

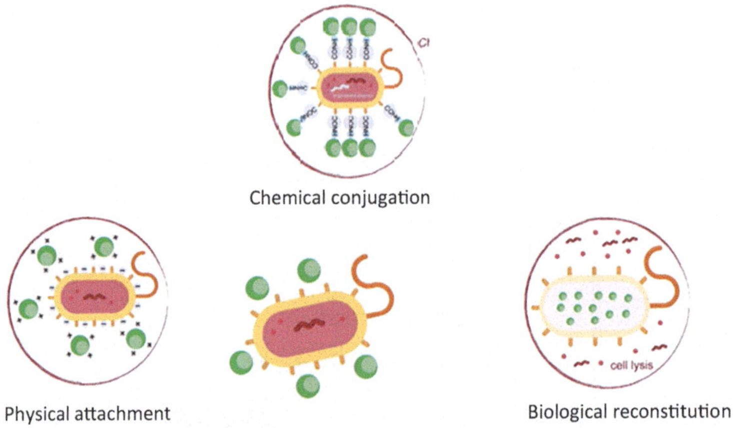

Figure 6.3 Fabrication methodologies of microbial nanohybrids including physical attachments, chemical conjugation, and biological reconstruction. (*Source:* From Huo *et al.* 2020).

of metals and the prepared solution (microbial biomass and metals) is then cultivated at desired incubation conditions till a specific chromatic change is observed. The appearance of a specific color shows the formation of NPs.

6.1.3 Nanohybrid materials for wastewater treatment with respect to microbes

Biogenic NPs (microbial interaction with metals) have unique properties and hence are utilized in various applications such as biomedical, electronic, and wastewater treatment (Ali *et al.*, 2017a, 2017b; Thakur & Mukherjee, 2021). Biogenic NPs are synthesized in an eco-friendly way by utilizing microbes as reductant (Kitching *et al.*, 2015). Microbes remove contaminants with the help of specific functional groups and different reducing substances synthesized by the cells that are charged opposite to the target pollutants via ion-exchange processes and electrostatic interactions (Ahluwalia *et al.*, 2016; Thakur & Mukherjee, 2021). Further for the degradation of toxic and emerging pollutants, biogenic NPs are stimulated by supplying hydrogen gas (De *et al.*, 2012a, 2012b; Suja *et al.*, 2014; Thakur & Mukherjee, 2021). However, biogenic NPs have shown superior performance in the removal of toxic and emerging pollutants, but still certain barriers exist in the marketing of biogenic NPs-based wastewater treatment technologies (Thakur & Mukherjee, 2021).

6.2 BIOSYNTHESIS OF METAL NPS WITH DIFFERENT MICROBES

NPs have gained significant attention due to their unique physicochemical properties, which make them highly valuable for various applications such as catalysis, drug delivery, and electronics. The biosynthesis of NPs using microorganisms offers a sustainable and eco-friendly alternative to traditional chemical methods. In this chapter, the biosynthesis of metal NPs utilizing, different types of microbes, including bacteria, fungi, algae, and yeast, are over reviewed.

6.2.1 Bacteria

Bacteria are commonly employed microorganisms for the biosynthesis of metal NPs due to their high metabolic activity and adaptability to different environmental conditions. Microbes (bacteria) can adapt to higher concentrations of metals and have the capability to reduce inorganic materials into NPs through their extracellular or intracellular routes (Dulinska-Molak *et al.*, 2018; Wells *et al.*, 2017). Microbes (bacteria) absorb metal ions from their surrounding environment/media and convert these metallic ions into elemental form via an enzymatic reduction (Fathima *et al.*, 2017). They can synthesize a wide range of metal NPs and are therefore called as 'the factory of nanomaterials' (Salem & Fouda, 2021). Bacterial species belonging to *Bacillus licheniformis, Klebsiella pneumonia, and Morganella psychrotolerans* were used for Bio-Ag NPs synthesis (John *et al.*, 2020). On the other hand, titanium dioxide NPs were synthesized by *Bacillus subtilis* and *Lactobacillus* sp. (Khan & Fulekar, 2016). Gold NPs were synthesized by *Pseudomonas aeruginosa, Rhodopseudomonas capsulata, Escherichia coli* DH5α, *Bacillus subtilis*, and *Bacillus licheniformis* (Srinath *et al.*, 2018), while *Escherichia coli, Clostridium thermoaceticum*, and *Rhodopseudomonas palustris* were used previously for the synthesis of cadmium NPs (Salem & Fouda, 2021; Sweeney *et al.*, 2004).

6.2.2 Algae

Algae, as photosynthetic microorganisms, have been extensively explored for the biosynthesis of metal NPs owing to their high biomass yield and ability to produce various metabolites. Algae can reduce metal ions to NPs through the action of specific enzymes such as nitrate reductase and hydrogenase. They are capable of synthesizing a wide range of metal NPs including gold, silver, and platinum NPs.

Several studies have reported the biosynthesis of metal NPs using algae. For example, the dried algal cells of *Chlorella vulgaris* were expanded to produce Bio-Au NPs by reduced tetra-chloroaurate ions to form Au NPs (Luangpipat *et al.*, 2011; Salem & Fouda, 2021). *Fucus vesiculosus* is a brown alga capable of bioreduction and biosorption of Au (III) ions (Mata *et al.*, 2009; Salem & Fouda, 2021). *Phaeodactulum tricornatum* possesses phytochelatin-covered CdS nanocrystals fabricated in response to Cd (Scarano & Morelli, 2003). Rapid formation of Au NPs through extracellular biosynthesis has been observed viable in the marine alga *Sargassum wightii* Greville (Singaravelu *et al.*, 2007). Konishi *et al.* (2007) reported that *Shewanella algaehas* has the ability to reduce aqueous $PtCl_6$ to elemental Pt at neutral pH under room temperature. Brayner and coauthors described the synthesis of platinum, gold, palladium, and silver NPs using cyanobacteria (Brayner *et al.*, 2007). Other alga like *Turbinaria conoides* was used for gold NP biosynthesis (Rajeshkumar *et al.*, 2014). On the other hand, four marine macroalgae, viz., *Pterocladia capillacae, Jania rubins, Ulva faciata, and Colpmenia sinusa,* were used for the biosynthesis of Ag NPs (Azizi *et al.*, 2013; El-Rafie *et al.*, 2013; Rajeshkumar *et al.*, 2014; Salem & Fouda, 2021).

6.2.3 Fungi

Fungi have also shown potential for the biosynthesis of metal NPs due to their high efficiencies of fungal metabolites to fabricate different NPs (Dhillon *et al.*, 2012; Fouda *et al.*, 2018; Mohamed *et al.*, 2019). Fungi have been given more attention as they are involved in the study on biological synthesis of metallic nanomaterials due to their tolerance and metal bioaccumulation capability (Sastry *et al.*, 2003) and fungi have the ability to produce various proteins or enzymes (Salem & Fouda, 2021; Spagnoletti *et al.*, 2019). Different species of fungi can be used to produce gold and silver NPs such as *Phanerochaete chrysosporium, Pleurotus sajorcaju, Coriolus versicolor,* and *Schizophyllum commune* (Elamawi *et al.*, 2018; El Domany *et al.*, 2018). Other species including *Aspergillus niger, Aspergillus terreus, Fusarium keratoplasticum, Fusarium oxysporum, and Alternaria alternata* have been reported to biosynthesize zinc oxide and iron oxide NPs (Mohamed *et al.*, 2019; Sarkar *et al.*, 2017). *Fusarium* spp., *Fusarium keratoplasticum, Helminthosporium tetramera,* and *Schizophyllum radiatum* were used for the biosynthesis of Ag NPs (Gaikwad *et al.*, 2013; Mohmed *et al.*, 2017; Shelar & Chavan, 2014). *Penicillium aurantiogriseum, P. waksmanii, P. citrinum, Fusarium oxysporum, and Aspergillus sydowii* were used for Au biosynthesis (Honary *et al.*, 2012; Thakker *et al.*, 2012; Vala, 2015), while *Aspergillus* sp. was used for the biosynthesis of iron NPs (Pavani & Kumar, 2013). *Fusarium oxysporum* can be used to produce zinc sulfide (ZnS), lead sulfide (PbS), cadmium sulfide (CdS), and molybdenum sulfide (MoS) nanomaterials, when the appropriate salt is added to the growth medium (Ahmad *et al.*, 2002; Salem & Fouda, 2021).

These studies exemplify the potential of microorganisms for the biosynthesis of various metal NPs with diverse applications. The biosynthesis of metal NPs using different types of microbes offers several advantages, including their eco-friendly nature, cost-effectiveness, and the ability to produce NPs with unique properties. Moreover, the utilization of microorganisms provides a sustainable approach for NP synthesis, reducing reliance on hazardous chemicals and energy-intensive processes.

In conclusion, the biosynthesis of metal NPs using various microbes represents a promising and exciting field of research. Bacteria, fungi, algae, and yeast offer distinct capabilities for the eco-friendly and sustainable production of metal NPs with a wide range of potential applications. As the field continues to advance, further exploration of different microorganisms and their potential for NP synthesis is expected to contribute to the development of innovative wastewater treatment strategies.

6.3 FEASIBILITY OF MICROBE-BASED BIOGENIC NPS FOR WASTEWATER TREATMENT

6.3.1 Use of biogenic NPs to treat wastewater

The feasibility of microbe-based biogenic NPs for wastewater treatment has attracted significant interest in recent years. These NPs, synthesized by microorganisms, offer a promising approach for efficient and environmentally friendly wastewater remediation. By harnessing the unique properties of biogenic NPs, such as their high surface area, reactivity, and selectivity, novel strategies for the removal of pollutants from wastewater can be developed.

Liu *et al.* (2023) provided a comprehensive overview of the feasibility of microbe-based biogenic NPs in wastewater treatment. The authors discussed various aspects, including the microbial synthesis of NPs, their characterization, and their application in pollutant removal. The review emphasized the need for further research to optimize NP synthesis processes, improve their stability and reactivity, and evaluate their long-term effects on the environment. By addressing these considerations, microbe-based biogenic NPs have the potential to revolutionize wastewater treatment by offering sustainable and efficient solutions.

In conclusion, microbe-based biogenic NPs hold great promise for wastewater treatment due to their unique properties and environmentally friendly synthesis. The studies mentioned above demonstrate the feasibility and considerations associated with the application of microbe-based biogenic NPs for heavy metal, removal of organic pollutant degradation and disinfection in wastewater. These NPs offer advantages such as cost-effectiveness, low energy requirements, and minimal generation of toxic byproducts. However, further research is needed to optimize NP synthesis processes, evaluate their stability and reusability, and assess their long-term impacts on the environment. By addressing these considerations, microbe-based biogenic NPs have the potential to revolutionize wastewater treatment and contribute to sustainable water management practices.

6.3.2 Biogenic inorganic NPs

6.3.2.1 Bio-Fe and Bio-Mn NPs

Biogenic iron (Bio-Fe NPs) and manganese (Bio-Mn NPs) NPs are a type of iron and manganese-based nanomaterials that is synthesized using biological sources such as bacteria, fungi, and plants. The synthesis of Bio-Fe and Bio-Mn NPs using microorganisms has become an attractive alternative to traditional methods due to its eco-friendliness, cost-effectiveness, and potential for large-scale production (Shedbalkar *et al.*, 2014; Yang *et al.*, 2016). The microorganisms used for the synthesis of Bio-Fe NPs are known to produce various enzymes and biomolecules that act as reducing agents, stabilizing agents, and capping agents during the synthesis process. These biomolecules help to regulate the size, shape, and surface properties of the NPs, leading to the formation of highly stable and uniform NPs. Iron (Fe) could be an electron acceptor or donor, depending on the oxidation environment, for example, anoxic or oxic conditions. As a result, Fe can exist with the valency of 0, II, III, and IV in Bio-Fe NPs (Huo *et al.*, 2016; Tuo *et al.*, 2015; Watts *et al.*, 2015). The size of Bio-Fe NPs may range from few nanometers to dozens of nanometers (Du *et al.*, 2022).

The oxidation of Mn in the natural environment is mainly attributed to the microbes since the rate of microbial oxidation of Mn(II) to Mn(III/IV) is several orders of magnitude (up to 105 times) higher than that of abiotic oxidation (Du *et al.*, 2022; Furgal *et al.*, 2015; Tanaka *et al.*, 2010). The synthesis of Bio-Mn NPs using microorganisms has become an attractive alternative to traditional methods due to its eco-friendliness, cost-effectiveness, and potential for large-scale production (Das & Das, 2015).

The unique properties of Bio-Fe and Bio-Mn NPs make them highly desirable for various industrial and environmental applications, including wastewater treatment, bioremediation, and biomedical applications (Yang *et al.*, 2016). The potential applications of Bio-Fe NPs are vast and diverse, ranging from the removal of heavy metals from contaminated water to the treatment of cancer and other diseases. However, further research is needed to fully understand the mechanisms behind the synthesis of Bio-Fe NPs and to optimize their properties for specific applications.

6.3.2.2 Bio-Pd NPs

Biogenic palladium NPs (Bio-Pd NPs) refer to NPs made of palladium (Pd) that have been synthesized through biological methods.

Patel *et al.* (2015) used *Escherichia coli* to synthesize Bio-Pd NPs with a size range of 5–30 nm. The resulting NPs showed excellent catalytic activity in the reduction of nitroarenes to arylamines.

With electron donors supplied, the bioreduction of Pd(II) can proceed through metabolic activity. Pd may precipitate on the cell surface, in the periplasmic or cytoplasm, or independent from biomass (Hennebel *et al.*, 2011). Marine strains with the ability to generate Bio-Pd NPs include both Gram-negative and Gram-positive strains (Du *et al.*, 2022; Hosseinkhani *et al.*, 2014). They also found that Bio-Pd NPs were located in the cytoplasm and periplasmic space of the Gram-negative strains, while in the periplasmic space and on the cell wall of Gram-positive strains. XRD patterns and TEM analysis indicated that Bio-Pd NP are metallic crystalline in nano-sizes. Bio-Pd NP production with an indigenous marine bacterial community was manifested as the conversion from Pd(II) ions to metallic Pd(0) and Pd^0 crystals (Du *et al.*, 2022; Hosseinkhani *et al.*, 2014).

6.3.2.3 Bio-Au and Bio-Ag NPs

Bio-Au and Bio-Ag NPs, also known as bio-functionalized gold and silver NPs, respectively, have garnered significant attention in recent years due to their unique properties and wide-ranging applications. These NPs, which are typically in the range of 1–100 nm, exhibit exceptional stability, high surface-to-volume ratios, and tunable optical properties (Mukherjee *et al.*, 2008). Moreover, their compatibility with biological systems makes them attractive for various biomedical and environmental applications.

Bio-Au and Bio-Ag NPs derived from the Actinomycete *Gordonia amicalis* HS-11 were depicted with a face-centered cubic (fcc) crystalline (Sowani *et al.*, 2016). Bio-Au NPs originated from *Escherichia coli* K12 were strictly controlled in the circular shape of approximately 50 nm with the aid of stabilizing agents (Du *et al.*, 2022; Srivastava *et al.*, 2013).

In conclusion, Bio-Au and Bio-Ag NPs offer a wide range of applications due to their unique properties and compatibility with biological systems. The synthesis and functionalization of these NPs have been extensively investigated, as evidenced by the research papers mentioned above. The bio-functionalization strategies employed in these studies enhance the stability, biocompatibility, and functionality of the NPs, allowing for targeted drug delivery, antimicrobial activity, and environmental remediation. As further research is conducted, the potential of Bio-Au and Bio-Ag NPs is expected to be harnessed in various fields, leading to significant advancements in biomedicine and environmental science.

6.3.2.4 Bio-bimetal NPs

Bio-bimetal NPs refer to NPs made of two different metals that have been synthesized through biological methods. Several studies have reported the successful synthesis of bio-bimetal NPs using various microorganisms, such as bacteria and fungi. For example, Gade *et al.* (2011) used the fungus *Aspergillus fumigatus* to synthesize bio-bimetal NPs consisting of silver (Ag) and gold (Au). The resulting NPs showed excellent antibacterial activity against both Gram-positive and Gram-negative bacteria. De Corte *et al.* also found that Bio-Pd/Au NPs produced by means of decorating Au particles on Bio-Pd NPs can improve the removal of halogenated organics (De Corte *et al.*, 2011, 2012a; Du *et al.*, 2022). Overall, bio-bimetal NPs synthesized through biological methods offer several advantages over traditional chemical methods, including lower toxicity, milder reaction conditions, and better control over particle size and shape. As research in this area continues to grow, it is likely that bio-bimetal NPs will find increasing applications in various fields such as catalysis, biomedicine, and environmental remediation.

6.3.2.5 Composite Bio-Me NPs

Composite Bio-Me NPs refer to NPs made of a combination of a metal and a biopolymer. These NPs have attracted increasing attention in recent years due to their potential applications in various fields such as biomedicine, drug delivery, and biosensing. The use of biopolymers in the synthesis of composite Bio-Me NPs offers several advantages over traditional chemical methods, including lower toxicity, biocompatibility, and the ability to control particle size and shape.

Several biopolymers have been used to synthesize composite Bio-Me NPs, including chitosan, cellulose, and alginate. Chitosan, a naturally occurring polysaccharide, has been widely used in the synthesis of composite Bio-Me NPs due to its biocompatibility and ability to form stable complexes with metal ions. Cellulose, another natural polymer, has also been used to synthesize composite Bio-Me NPs due to its excellent mechanical properties and biodegradability. Alginate, a biopolymer derived from brown seaweed, has been used to synthesize composite Bio-Me NPs due to its ability to form gels and its potential as a drug delivery system.

Several studies have reported the successful synthesis of composite Bio-Me NPs using various biopolymers. For example, Chen *et al.* (2013) used chitosan to synthesize composite Bio-Pd NPs. The resulting NPs showed excellent catalytic activity in the reduction of 4-nitrophenol to 4-aminophenol. Arshad *et al.* (2017) used cellulose to synthesize composite Bio-Au NPs. The resulting NPs showed excellent antibacterial activity against various pathogenic bacteria, including *Staphylococcus aureus* and *Escherichia coli*.

In addition to their potential applications in catalysis and antibacterial activity, composite Bio-Me NPs have also shown promise in drug delivery. For example, Wei *et al.* (2016) used alginate to synthesize composite Bio-Fe NPs as a drug delivery system. The resulting NPs showed excellent biocompatibility and the ability to release drugs in a controlled manner.

6.4 CONCLUSIONS

Millions of people's lives are altered by many sectors and the items they produce. The wastewater quality generated across many sectors is a major source of worry. This chapter has covered multiple kinds of water pollutants, conventional and cutting-edge wastewater treatment techniques, and NPs based on microorganisms. The need for timely water quality characterization is urgent given how poorly untreated water impacts both the environment and human lives. Nanomaterials based on microorganisms are ecologically friendly in contrast to conventional approaches, which are toxic and bad for the environment. Furthermore, microorganism-based nanomaterials improve the removal of certain wastewater contaminants and alter wastewater treatment approaches.

Increasing attention is paid to microbes-based wastewater treatment as it is an inexpensive and ecologically safe method of purifying and treating water. Researchers have become more interested in nanotechnology as a result of its positive impacts, such as the enormous surface area it offers, the possibility for various applications, stability under adverse conditions, simple and effective material manipulation and greater contact. Microorganisms and enzymes are being combined with nanotechnology to provide a greener way for addressing industrial effluents. Microorganisms can be utilized to lessen the harm caused by NPs produced chemically. The remaining residues are either biodegradable or easily separable using straightforward filtration/precipitation procedures.

Microbial cells can be employed for the faster, safer production of NPs having the necessary nature and structure as they can be quick-growing, simple to maintain, and fast-growing. To achieve a consistent shape, symmetry, composition, and size of NPs, which can be affected by environmental factors such as temperature and pH of the medium, is one of the limitations of the biosynthesis of NPs by microbial synthesis. Although recent studies have shown the enormous potential of microbes for the synthesis of novel NPs and their use in biomedicine and cancer treatment, the methods of microorganism-based biosynthesis need to be modified for commercial production, and the synthesis of NPs must be scaled up in comparison with conventional methods.

The commercialization of these nanotechnological features is the more difficult task. Currently, only 1% of these nanotechnological features are being sold commercially. Therefore, the widespread use of these simple and effective microorganism-assisted nanotechnology techniques will be a step ahead for the industries. To foster the potential of nanotechnology for sustainable and affordable manufacturing facilities in industries, this requires for continuous encouragement and verification from scientists and funding from governments. Microorganisms and nanotechnology collectively have developed a green method for bioremediating industrial effluents. It is being suggested that microbes can produce nanomaterials, opening more effective and sustainable solutions for wastewater treatment.

ACKNOWLEDGEMENT

This fruitful effort was possible with the constant support, access to facilities and guidance from resourceful people from the Life Science Department, School of Basic Sciences and Research, Sharda University, Greater Noida (India) and Annasaheb Dange College of Engineering & Technology, Department of Food Technology, Astha (India).

REFERENCES

Ahluwalia S., Prakash N. T., Prakash R. and Pal B. (2016). Improved degradation methyl orange dye using bio-co-catalyst Se nanoparticles impregnated ZnS photocatalyst under UV irradiation. *Chemical Engineering Journal*, **306**, 1041–1048, https://doi.org/10.1016/j.cej.2016.08.028

Ahmad A., Mukherjee P., Mandal D., Senapati S., Khan M. I., Kumar R. and Sastry M. (2002). Enzyme mediated extracellular synthesis of CdS nanoparticles by the fungus, *Fusarium oxysporum*. *Journal of the American Chemical Society*, **124**(41), 12108–12109, https://doi.org/10.1021/ja027296o

Albanese A., Tang P. S. and Chan W. C. (2012). The effect of nanoparticle size, shape, and surface chemistry on biological systems. *Annual Review of Biomedical Engineering*, **14**, 1–16, https://doi.org/10.1146/annurev-bioeng-071811-150124

Al-Dhabi N. A., Mohammed Ghilan A.-K. and Arasu M. V. (2018). Characterization of silver nanomaterials derived from marine *Streptomyces* sp. Al-dhabi-87 and it's in vitro application against multidrug resistant and extended-spectrum beta-lactamase clinical pathogens. *Nanomaterials*, **8**, 279, https://doi.org/10.3390/nano8050279

Ali I., Peng C., Naz I., Khan Z. M., Sultan M. and Islam T. (2017a). Phytogenic magnetic nanoparticles for wastewater treatment: a review. *RSC Advances*, **7**, 40158–40178, https://doi.org/10.1039/C7RA04738J

Ali I., Peng C., Khan Z. M. and Naz I. (2017b). Yield cultivation of magneto tactic bacteria and magnetosomes: a review. *Journal of Basic Microbiology*, **57**, 643–652, https://doi.org/10.1002/jobm.201700052

Arshad T., Iqbal M., Riaz S., Azeem M., Akhtar M. J. and Ali A. (2017). Synthesis and characterization of cellulose supported gold nanoparticles for antibacterial applications. *International Journal of Biological Macromolecules*, **103**, 1146–1151, https://doi.org/10.1016/j.ijbiomac.2017.05.147

Azizi S., Namvar F., Mahdavi M., Ahmad M. and Mohamad R. (2013). Biosynthesis of silver nanoparticles using brown marine macroalga, *Sargassum muticum* aqueous extract. *Materials*, **6**(12), 5942–5950, https://doi.org/10.3390/ma6125942

Brayner R., Barberousse H., Hemadi M., Djedjat C., Yéprémian C., Coradin T., Livage J., Fiévet F. and Couté A. (2007). Cyanobacteria as bioreactors for the synthesis of Au, Ag, Pd, and Pt nanoparticles via an enzyme-mediated route. *Journal of Nanoscience Nanotechnology*, **7**(8), 2696–2708, https://doi.org/10.1166/jnn.2007.600

Castro L., Blázquez M. L., González F. G. and Ballester A. (2014). Mechanism and applications of metal nanoparticles prepared by bio-mediated process. *Reviews in Advanced Sciences and Engineering*, **3**, 199–216, https://doi.org/10.1166/rase.2014.1064

Chen X., Wu S., Cao S. and He Q. (2013). Green synthesis of chitosan-based palladium nanoparticles for catalytic reduction of 4-nitrophenol. *Carbohydrate Polymers*, **93**(1), 41–47, https://doi.org/10.1016/j.carbpol.2012.05.037

Das S. K. and Das A. R. (2015). Mycosynthesis of manganese nanoparticles: a review. *Journal of Applied Microbiology*, **119**(1), 1–14, https://doi.org/10.1111/jam.12808

Das V. L., Thomas R., Varghese R. T., Soniya E., Mathew J. and Radhakrishnan E. (2014). Extracellular synthesis of silver nanoparticles by the *Bacillus* strain cs 11 isolated from industrialized area. *Biotechnology*, **4**, 121–126.

De Corte S., Hennebel T., Verschuere S., Cuvelier C., Verstraete W. and Boon N. (2011). Gold nanoparticle formation using *Shewanella oneidensis*: a fast biosorption and slow reduction process. *Journal of Chemical Technology and Biotechnology*, **86**(4), 547–553, https://doi.org/10.1002/jctb.2549

De Corte S., Hennebel T., De G. B., Verstraete W. and Boon N. (2012a). Bio-palladium: from metal recovery to catalytic applications. *Microbial Biotechnology*, **5**(1), 5–17, https://doi.org/10.1111/j.1751-7915.2011.00265.x

De Corte S., Sabbe T., Hennebel T., Vanhaecke L., De G. B. and Verstraete W. (2012b). Doping of biogenic Pd catalysts with Au enables dechlorination of diclofenac at environmental conditions. *Water Research*, **46**(8), 2718–2726, https://doi.org/10.1016/j.watres.2012.02.036

Dhillon G. S., Brar S. K., Kaur S. and Verma M. (2012). Green approach for nanoparticle biosynthesis by fungi: current trends and applications. *Critical Review of Biotechnology*, **32**(1), 49–73, https://doi.org/10.3109/07388551.2010.550568

Dobias J., Suvorova E. I. and Bernier-Latmani R. (2011). Role of proteins in controlling selenium nanoparticle size. *Nanotechnology*, **22**, 195605, https://doi.org/10.1088/0957-4484/22/19/195605

Du Z., Zhang Y., Xu A., Pan S. and Zhang Y. (2022). Biogenic metal nanoparticles with microbes and their applications in water treatment: a review. *Environmental Science and Pollution Research*, **29**, 3213–3229, https://doi.org/10.1007/s11356-021-17042-z

Dulinska-Molak I., Chlanda A., Li J., Wang X., Bystrzejewski M., Kawazoe N., Chen G. and Swieszkowski W. (2018). The influence of carbon-encapsulated iron nanoparticles on elastic modulus of living human mesenchymal stem cells examined by atomic force microscopy. *Micron*, **108**, 41–48, https://doi.org/10.1016/j.micron.2018.02.006

El Domany E. B., Essam T. M., Ahmed A. E. and Farghali A. A. (2018). Biosynthesis physico-chemical optimization of gold nanoparticles as anti-cancer and synergetic antimicrobial activity using *Pleurotus ostreatus* fungus. *Journal of Applied Pharmaceutical Science*, **2018**, 8.

Elamawi R. M., Al-Harbi R. E. and Hendi A. A. (2018). Biosynthesis and characterization of silver nanoparticles using *Trichoderma longibrachiatum* and their effect on phytopathogenic fungi. *Egyptian Journal of Biological Pest Control*, **28**(1), 28, https://doi.org/10.1186/s41938-018-0028-1

Elfeky A. S., Salem S. S., Elzaref A. S., Owda M. E., Eladawy H. A., Saeed A. M., Awad M. A., Abou-Zeid R. E. and Fouda A. (2020). Multifunctional cellulose nanocrystal/metal oxide hybrid, photo-degradation, antibacterial and larvicidal activities. *Carbohydrate Polymer*, **230**, 115711, https://doi.org/10.1016/j.carbpol.2019.115711

El-Rafie H., El-Rafie M. and Zahran M. (2013). Green synthesis of silver nanoparticles using polysaccharides extracted from marine macro algae. *Carbohydrate Polymer*, **96**(2), 403–410, https://doi.org/10.1016/j.carbpol.2013.03.071

Fariq A., Khan T. and Yasmin A. (2017). Microbial synthesis of nanoparticles and their potential applications in biomedicine. *Journal of Applied Biomedicine*, **15**, 241–248, https://doi.org/10.1016/j.jab.2017.03.004

Fathima J. B., Pugazhendhi A. and Venis R. (2017). Synthesis and characterization of ZrO_2 nanoparticles-antimicrobial activity and their prospective role in dental care. *Microbial Pathogenesis*, **110**, 245–251, https://doi.org/10.1016/j.micpath.2017.06.039

Fernández-Llamosas H., Castro L., Blázquez M. L., Díaz E. and Carmona M. (2017). Speeding up bioproduction of selenium nanoparticles by using *Vibrio natriegens* as microbial factory. *Scientific Reports*, **7**, 16046, https://doi.org/10.1038/s41598-017-16252-1

Fouda A., Saad E., Salem S. S. and Shaheen T. I. (2018). In-vitro cytotoxicity, antibacterial, and UV protection properties of the biosynthesized zinc oxide nanoparticles for medical textile applications. *Microbial Pathogenesis*, **125**, 252–261, https://doi.org/10.1016/j.micpath.2018.09.030

Furgal K. M., Meyer R. L. and Bester K. (2015). Removing selected steroid hormones, biocides and pharmaceuticals from water by means of biogenic manganese oxide nanoparticles in situ at ppb levels. *Chemosphere*, **136**, 321–326, https://doi.org/10.1016/j.chemosphere.2014.11.059

Gade A., Ingle A. P., Whiteley C. G. and Rai M. (2011). Mycogenic metal nanoparticles: progress and applications. *Biotechnology Journal*, **6**(11), 1437–1448.

Gahlawat G. and Choudhury A. R. (2019). A review on the biosynthesis of metal and metal salt nanoparticles by microbes. *RSC Advances*, **9**, 12944–12967, https://doi.org/10.1039/C8RA10483B

Gaikwad S. C., Birla S. S., Ingle A. P., Gade A. K., Marcato P. D., Rai M. and Duran N. (2013). Screening of different *Fusarium* species to select potential species for the synthesis of silver nanoparticles. *Journal of the Brazillian Chemical Society*, **24**(12), 1974–1982.

Golmohammadi H., Morales-Narvaez E., Naghdi T. and Merkoci A. (2017). Nanocellulose in sensing and biosensing. *Chemistry of Materials*, **29**, 5426–5446, https://doi.org/10.1021/acs.chemmater.7b01170

Grasso G., Zane D. and Dragone R. (2020). Microbial nanotechnology: challenges and prospects for green biocatalytic synthesis of nanoscale materials for sensoristic and biomedical applications. *Nanomaterials*, **10**, 11, https://doi.org/10.3390/nano10010011

He S., Zhang Y., Guo Z. and Gu N. (2008). Biological synthesis of gold nanowires using extract of *Rhodopseudomonas capsulata*. *Biotechnology Progress*, **24**, 476–480, https://doi.org/10.1021/bp0703174

Hennebel T., Van Nevel S., Verschuere S., De Corte S., De Gusseme B., Cuvelier C., Fitts J. P., Van Der Lelie D., Boon N. and Verstraete W. (2011). Palladium nanoparticles produced by fermentatively cultivated bacteria as catalyst for diatrizoate removal with biogenic hydrogen. *Applied Microbiology Biotechnology*, **91**(5), 1435–1445, https://doi.org/10.1007/s00253-011-3329-9

Honary S., Barabadi H., Gharaei-Fathabad E. and Naghibi F. (2012). Green synthesis of copper oxide nanoparticles using *Penicillium aurantiogriseum*, *Penicillium citrinum* and *Penicillium waksmanii*. *Digest Journal of Nanomaterials and Biostructures*, **7**(3), 999–1005.

Hosseinkhani B., Hennebel T., Van Nevel S., Verschuere S., Yakimov M. M., Cappello S., Blaghen M. and Boon N. (2014). Biogenic nanopalladium based remediation of chlorinated hydrocarbons in marine environments. *Environment Science Technology*, **48**(1), 550–557, https://doi.org/10.1021/es403047u

Huo Y. C., Li W. W., Chen C. B., Li C. X., Zeng R., Lau T. C. and Huang T. Y. (2016). Biogenic FeS accelerates reductive dechlorination of carbon tetrachloride by *Shewanella putrefaciens* CN32. *Enzyme and Microbial Technology*, **95**, 236–241, https://doi.org/10.1016/j.enzmictec.2016.09.013

Huo H., Wang L., Chen Y. and Shi S. (2020). Nanomaterials/microorganism-integrated microbiotic nanomedicine. *Nano Today*, **32**, 100854, https://doi.org/10.1016/j.nantod.2020.100854

John M. S., Nagoth J. A., Ramasamy K. P., Mancini A., Giuli G., Natalello A., Ballarini P., Miceli C. and Pucciarelli S. (2020). Synthesis of bioactive silver nanoparticles by a *Pseudomonas* strain associated with the Antarctic psychrophilic protozoon *Euplotes focardii*. *Marine Drugs*, **18**(1), 38, https://doi.org/10.3390/md18010038

Khan R. and Fulekar M. (2016). Biosynthesis of titanium dioxide nanoparticles using *Bacillus amyloliquefaciens* culture and enhancement of its photocatalytic activity for the degradation of a sulfonated textile dye Reactive Red 31. *Journal of Colloid Interface Science*, **475**, 184–191, https://doi.org/10.1016/j.jcis.2016.05.001

Khan I., Saeed K. and Khan I. (2019). Nanoparticles: properties, applications and toxicities. *Arabian Journal of Chemistry*, **12**, 908–931, https://doi.org/10.1016/j.arabjc.2017.05.011

Kitching M., Ramani M. and Marsili E. (2015). Fungal biosynthesis of gold nanoparticles: mechanism and scaleup. *Microbial Biotechnology*, **8**(6), 904–917, https://doi.org/10.1111/1751-7915.12151

Konishi Y., Ohno K., Saitoh N., Nomura T., Nagamine S., Hishida H., Takahashi Y. and Uruga T. (2007). Bioreductive deposition of platinum nanoparticles on the bacterium *Shewanella* algae. *Journal of Biotechnology*, **128**(3), 648–653, https://doi.org/10.1016/j.jbiotec.2006.11.014

Koul B., Poonia A. K., Yadav D. and Jin J. O. (2021). Microbe-Mediated Biosynthesis of Nanoparticles: Applications and Future Prospects. *Biomolecules*, (11), **886**, 1–33.

Li C., Zhou L., Yang H., Lv R., Tian P., Li X., Zhang Y., Chen Z. and Lin F. (2017). Self-assembled exopolysaccharide nanoparticles for bioremediation and green synthesis of noble metal nanoparticles. *ACS Applied Materials and Interfaces*, **9**, 22808–22818, https://doi.org/10.1021/acsami.7b02908

Li X., Xu H., Chen Z.-S. and Chen G. (2011). Biosynthesis of Nanoparticles by Microorganisms and Their Applications. *Journal of Nanomaterials*, 1–16.

Luangpipat T., Beattie I. R., Chisti Y. and Haverkamp R. G. (2011). Gold nanoparticles produced in a microalga. *Journal of Nanoparticle Research*, **13**(12), 6439–6445, https://doi.org/10.1007/s11051-011-0397-9

Mata Y., Torres E., Blazquez M., Ballester A., González F. and Munoz J. (2009). Gold (III) biosorption and bioreduction with the brown alga *Fucus vesiculosus*. *Journal of Hazard Material*, **166**(2–3), 612–618, https://doi.org/10.1016/j.jhazmat.2008.11.064

Mohamed A. A., Fouda A., Abdel-Rahman M. A., Hassan SE.-D., ElGamal M. S., Salem S. S. and Shaheen T. I. (2019). Fungal strain impacts the shape, bioactivity and multifunctional properties of green synthesized zinc oxide nanoparticles. *Biocatalysis and Agricultural Biotechnology*, **19**, 101103, https://doi.org/10.1016/j.bcab.2019.101103

Mohmed A. A., Fouda A., Elgamal M. S., Hassan S., Shaheen T. I. and Salem S. S. (2017). Enhancing of cotton fabric antibacterial properties by silver nanoparticles synthesized by new Egyptian strain *Fusarium keratoplasticum* A1-3. *Egyptian Journal of Chemistry*, **60**, 63–71 (conference issue (the 8th international conference of the Textile Research Division (ICTRD 2017), National Research Centre, Cairo 12622, Egypt.).

Mukherjee P., Ahmad A. and Mandal D. (2008). Bioreduction of $AuCl_4^-$ ions by the fungus, *Verticillium* sp. and surface trapping of the gold nanoparticles formed. *Angewandte Chemie International Edition*, **47**(31), 5836–5839, https://doi.org/10.1002/anie.200801435

Patel K., Kumar V. and Krishnan V. (2015). *Escherichia coli* mediated synthesis of palladium nanoparticles and their catalytic activity towards nitroarenes reduction. *Journal of Nanoscience and Nanotechnology*, **15**(7), 4999–5006, https://doi.org/10.1166/jnn.2015.10423

Pavani K. and Kumar N. S. (2013). Adsorption of iron and synthesis of iron nanoparticles by *Aspergillus species kvp* 12. *American Journal of Nanomaterials*, **1**(2), 24–26.

Rajeshkumar S., Malarkodi C., Paulkumar K., Vanaja M., Gnanajobitha G. and Annadurai G. (2014). Algae mediated green fabrication of silver nanoparticles and examination of its antifungal activity against clinical pathogens. *International Journal of Metals*, **2014**, 8.

Salem S. S. and Fouda A. (2021). Green synthesis of metallic nanoparticles and their prospective biotechnological applications: an overview. *Biological Trace Element Research*, **199**, 344–370, https://doi.org/10.1007/s12011-020-02138-3

Sarkar J., Mollick M. M. R., Chattopadhyay D. and Acharya K. (2017). An eco-friendly route of γ-Fe_2O_3 nanoparticles formation and investigation of the mechanical properties of the HPMC-γ-Fe_2O_3 nanocomposites. *Bioprocess Biosystem Engineering*, **40**(3), 351–353, https://doi.org/10.1007/s00449-016-1702-x

Sastry M., Ahmad A., Khan M. I. and Kumar R. (2003). Biosynthesis of metal nanoparticles using fungi and actinomycete. *Current Science*, **85**(2), 162–170.

Scarano G. and Morelli E. (2003). Properties of phytochelatin-coated CdS nanocrystallites formed in a marine phytoplanktonic alga (*Phaeodactylum tricornutum*, Bohlin) in response to Cd. *Plant Science*, **165**(4), 803–810, https://doi.org/10.1016/S0168-9452(03)00274-7

Shedbalkar S. U. U., Jadhav J. P. and Govindwar S. P. (2014). Biogenic synthesis of manganese oxide nanoparticles by fungus *Aspergillus* sp. *Journal of Nanoscience and Nanotechnology*, **14**(5), 3640–3648.

Shelar G. B. and Chavan A. M. (2014). Fungus-mediated biosynthesis of silver nanoparticles and its antibacterial activity. *Archives of Applied Science Research*, **6**, 111–114.

Singaravelu G., Arockiamary J., Kumar V. G. and Govindaraju K. (2007). A novel extracellular synthesis of monodisperse gold nanoparticles using marine alga, *Sargassum wightii Greville*. *Colloids Surface B: Biointerfaces*, **57**(1), 97–101, https://doi.org/10.1016/j.colsurfb.2007.01.010

Sowani H., Mohite P., Damale S., Kulkarni M. and Zinjarde S. (2016). Carotenoid stabilized gold and silver nanoparticles derived from the Actinomycete *Gordonia amicalis* HS-11 as effective free radical scavengers. *Enzyme and Microbial Technology*, **95**, 164–173, https://doi.org/10.1016/j.enzmictec.2016.09.016

Spagnoletti F. N., Spedalieri C., Kronberg F. and Giacometti R. (2019). Extracellular biosynthesis of bactericidal Ag/AgCl nanoparticles for crop protection using the fungus *Macrophomina phaseolina*. *Journal of Environment and Management*, **231**, 457–466, https://doi.org/10.1016/j.jenvman.2018.10.081

Srinath B., Namratha K. and Byrappa K. (2018). Eco-friendly synthesis of gold nanoparticles by *Bacillus subtilis* and their environmental applications. *Advance Science Letters*, **24**(8), 5942–5946, https://doi.org/10.1166/asl.2018.12224

Srivastava S. K., Yamada R., Ogino C. and Kondo A. (2013). Biogenic synthesis and characterization of gold nanoparticles by *Escherichia coli* K12 and its heterogeneous catalysis in degradation of 4-nitrophenol. *Nanoscale Research Letters*, **8**(70), 1–9.

Suja E., Nancharaiah Y. V. and Venugopalan V. P. (2014). Biogenic nanopalladium production by self-immobilized granular biomass: application for contaminant remediation. *Water Research*, **65**, 395–401, https://doi.org/10.1016/j.watres.2014.08.005

Sweeney R. Y., Mao C., Gao X., Burt J. L, Belcher A. M., Georgiou G. and Iverson B. L. (2004). Bacterial biosynthesis of cadmium sulfide nanocrystals. *Chemical Biology*, **11**(11), 1553–1559, https://doi.org/10.1016/j.chembiol.2004.08.022

Tanaka K., Tani Y., Takahashi Y., Tanimizu M., Suzuki Y., Kozai N. and Ohnuki T. (2010). A specific Ce oxidation process during sorption of rare earth elements on biogenic Mn oxide produced by *Acremonium* sp. Strain KR21-2. *Geochimica et Cosmochimica Acta*, **74**(19), 5463–5477, https://doi.org/10.1016/j.gca.2010.07.010

Thakker J. N., Dalwadi P. and Dhandhukia P. C. (2012). Biosynthesis of gold nanoparticles using *Fusarium oxysporum f.* sp. *cubense* JT1, a plant pathogenic fungus. *Hindawi Publishing Corporation ISRN Biotechnology*, **2013**, 5.

Thakur P. and Mukherjee G. (2021). Chapter title: microbes-based nanomaterials for the wastewater treatment and decontamination of water. In: Wastewater Treatment Cutting Edge Molecular Tools, Techniques and Applied Aspects, M. P. Shah, A. Sarkar and S. Mandal (eds.), Elsevier, pp. 311–325. ISBN 978-0-12-821881-5.

Tuo Y., Liu G., Dong B., Zhou J., Wang A., Wang J., Jin R., Lv H., Dou Z. and Huang W. (2015). Microbial synthesis of Pd/Fe_3O_4, Au/Fe_3O_4 and PdAu/Fe_3O_4 nanocomposites for catalytic reduction of nitroaromatic compounds. *Scientific Reports*, **5**, 13515, https://doi.org/10.1038/srep13515

Vala A. K. (2015). Exploration on green synthesis of gold nanoparticles by a marine-derived fungus *Aspergillus sydowii*. *Environmental Progress and Sustainable Energy*, **34**(1), 194–197, https://doi.org/10.1002/ep.11949

Watts M. P., Coker V. S., Parry S. A., Pattrick R. A. D., Thomas R. A. P., Kalin R. and Lloyd J. R. (2015). Biogenic nano-magnetite and nano-zero valent iron treatment of alkaline Cr(VI) leachate and chromite ore processing residue. *Applied Geochemistry*, **54**, 27–42, https://doi.org/10.1016/j.apgeochem.2014.12.001

Wei X., Luo M., Li W., Yang J. and Zhou C. (2016). Synthesis of alginate-stabilized Fe nanoparticles and their potential application as a drug delivery system. *Journal of Nanoparticle Research*, **18**(5), 142.

Wells J., Kazakova O., Posth O., Steinhoff U., Petronis S., Bogart L. K., Southern P., Pankhurst Q. and Johansson C. (2017). Standardisation of magnetic nanoparticles in liquid suspension. *Journal of Physics D Applied Physics*, **50**(38), 383003, https://doi.org/10.1088/1361-6463/aa7fa5

Yadav A., Kon K., Kratosova G., Duran N., Ingle A. P. and Rai M. (2015). Fungi as an efficient mycosystem for the synthesis of metal nanoparticles: progress and key aspects of research. *Biotechnology Letter*, **37**, 2099–2120, https://doi.org/10.1007/s10529-015-1901-6

Yang X., Li C. and Wan L. (2016). Biosynthesis of iron nanoparticles by *T. reesei*: synthesis mechanism and optimization of the synthesis process. *RSC Advances*, **6**(54), 48887–48894, https://doi.org/10.1039/C6RA05176F

You H., Yang S., Ding B. and Yang H. (2013). Synthesis of colloidal metal and metal alloy nanoparticles for electrochemical energy applications. *Chemical Society Reviews*, **42**, 2880–2904, https://doi.org/10.1039/C2CS35319A

doi: 10.2166/9781789063592_0113

Chapter 7

Sustainable wastewater treatment using magnetic nanohybrids

Anirudh Gururaj Patil[1], Vivekanandan Alangadu Kothandan[2], A. B. Hemavathi[3], Shweta Suri[4], Ranjana Pande[5], Sneha Kagale[6], Snehal Yadav[1] and Gouri Patil[3]

[1]Annasaheb Dange College of Engineering & Technology, Department of Food Technology, Ashta, Maharashtra, India
[2]Annasaheb Dange College of Engineering & Technology, Department of Aeronautical Engineering, Ashta, Maharashtra, India
[3]Sri Jayachamarajendra College of Engineering (SJCE), Jagadguru Sri Shivarathreeshwara Science and Technology University, Department of Polymer Science and Technology, Mysore, India
[4]Amity University, Amity Institute of Food Technology, Noida, India
[5]School of Basic Sciences and Research, Life Sciences Department, Sharda University, Noida, India
[6]Kolhapur Institute of Technology's College of Engineering (Autonomous), Department of Biotechnology Engineering, Kolhapur, India

ABSTRACT

Water pollution by various toxic contaminants has become one of the most serious problems across the globe. Various technologies have been used since ages to treat water and waste water that include filtration (membrane and gravel) and chemical (precipitation, coagulation, flocculation, adsorption, ion exchange, and electrochemical) methods. Although it requires an expensive material, commercial adsorbents are widely used in the industries for waste water treatment. Adsorption is generally used because of its adaptability in design and operation, reversibility, and feasibility. The choice of right and environmentally friendly adsorbent material for especially heavy metals has become the need of the hour. The nanomaterial/composites have demonstrated remarkable capacity for elimination of heavy metals from water. Different magnetic nanohybrid adsorbents have different adsorbing efficiencies for heavy metals due to their intrinsic properties such as surface area, functional group, zeta potential, and pore structure. Magnetic nanohybrid materials not only possess the properties of nanomaterials but also have magnetic properties that enable them as excellent adsorbents or catalysts in water pollution remediation. Also, extrinsic factors such as pH, temperature, and contact time influence the adsorption. In the last few decades, nanotechnology has received special attention and different nanomaterials or composites have been developed for water purification. Among which nanohybrid materials, combined with two or more compositions together possess a great advantage with new properties, diverse functionalities, and unlimited possibilities. In this chapter, sustainable waste water treatment with focus on nanohybrids is discussed that include sources of pollutants, methods of magnetic nanohybrid preparation, and factors influencing adsorption followed by applications and challenges.

Keywords: adsorption, carbon-based nanoparticle, dyes and radionuclides, graphene, nanoparticle.

7.1 INTRODUCTION

Of the total water present on the earth only 1% of it is potable for human consumption (Adeleye *et al.*, 2016; Grey *et al.*, 2013). It was reported by WHO in 2015 that about 1.1 billion people lack supply of sufficient drinking water due to the rise in the cost of potable water, uncontrolled human growth, climatic changes, and environmental issues (Adeleye *et al.*, 2016). One of the major challenges in the supply of fresh water is the contamination of water resources by a variety of pollutants from mining, pharmaceutical wastes, dyes, and pigment wastes (Schwarzenbach *et al.*, 2006). The pollution of water resources is one of the most alarming problems among the many global issues, which is causing great amounts of destruction to both terrestrial and aquatic life. The primary cause of water resource contamination is the human activities that release untreated wastewater into the resources and cause water pollution. Wastewater is the used water that has been affected by domestic, industrial, or commercial use. It contains loads of oxygen demanding wastes, pathogens, inorganic and organic materials, minerals, and biomedical wastes that release tons of toxic compounds in the environment (Tchobanoglus *et al.*, 2003). Wastewater treatment carried out over the decades focused primarily on the reduction of suspended solids, inorganic waste, pathogens and oxygen-demanding materials. However, only recently have we realized that the issue is much bigger than assumed and the means of disposal must be addressed too.

The basic method of wastewater treatment occurs in three stages: a primary stage which involves grit removal, screening, grinding, and sedimentation, in the secondary stage the oxidation of dissolved organic matter is carried out using a biologically activated sludge which is then filtered off and the tertiary treatment involves elimination of nitrogen through biological means or physical–chemical methods such as granular filtration and activated carbon absorption (Pratap *et al.*, 2014). These physical processes remove pollutants by transforming one phase to the other by producing a highly concentrated sludge which is toxic and very difficult to dispose. However, these traditional methods of wastewater treatment do not efficiently remove the pollutants and even the modern technologies have drawbacks such as huge energy consumption, partial pollutant removal and release of toxic sludge (Ferroudj *et al.*, 2013). Some of the other used technologies include ozone treatment, irradiation and chlorination, activated carbon and plasma technology, which efficiently remove bacterial contaminations but the removal of chemical contamination is tough.

Looking at the efficiencies of the conventional methods, we can clearly say that there is a need for a more efficient and powerful technology to treat the wastewaters (Ferroudj *et al.*, 2013; Singh *et al.*, 2023). This is exactly when nanomaterials come into the picture with great advantages over the conventional methods. Nanomaterials are promising materials for wastewater treatment because of their potential characteristics such as large surface area yet small size, high porosity, hydrophobicity, ease of separation, and regeneration with tunable molecular size and catalytic nature, making them one of the most suitable pollutant absorbent materials. These nanomaterials are made into several kinds of nanocomposites using engineering nanotechnology to achieve great features.

The heavy metal contamination of soil and water resources remains a growing global issue due to rapid industrialization and careless modernization (Jia *et al.*, 2017). Metals, specifically lead and mercury, are carcinogenic and bio-accumulating with no signs of degradability (Yang *et al.*, 2018). Hence, it becomes necessary to carry out a systematic removal of wastewater contaminated with heavy metals. A plethora of methodologies have already been employed for wastewater purification such as ion-exchange, solvent extraction, chemical precipitation, adsorption, and membrane separation (Wu *et al.*, 2015). Amongst the mentioned methods, the adsorption technology is easy for operation with good recyclability and does not generate toxic sludge (Li *et al.*, 2018).

This chapter gives a brief introduction about the water pollutants such as anthropogenic and pharmaceutical wastes, dyes, and radionuclides. It discusses the sustainable or green approach for the remediation of water pollutants using magnetic-based nanohybrid materials with a thorough explanation of their preparation, the principle involved for their application in the water treatment arena, types of magnetic nanohybrids, different factors affecting their efficacy with a brief discussion of the prospects and the challenges that may arise.

7.2 SOURCE OF POLLUTANTS

7.2.1 Ore extraction

Ore extraction or mining is the process of obtaining minerals and metals from the earth (Gosselin *et al.*, 2013). These metals and minerals are used in agriculture, telecommunication, space studies, construction, and housing facilities (Sumi & Thomsen, 2001). Mining industry provides employment opportunities to the people, contributing to the nation's economy. Ore extraction is turbulent to the environment as it produces enormous quantities of waste that can impact the globe for decades mainly as a result of inappropriate working and waste disposal practices (Kitula, 2006; Mondal *et al.*, 2014; Sumi & Thomsen, 2001). Some of the environmental issues caused by mining are soil degradation, noise pollution, major loss of biodiversity, and water pollution (Dasagupta *et al.*, 2012; Sumi & Thomsen, 2001). Surface and groundwater contamination is one of the most alarming impacts (Hudson, 2016).

The mined ores require further processing to concentrate the desired minerals. In case of metal ores, these concentrates must be separated from other elements present in the ore for purification through a process called smelting. Therefore, smelting is a process of separating metal from its impurities by heating the ore with a reducing agent such as charcoal, coke, or other purifying agents. Smelting the concentrate produces a metallic mixture alongside a solid waste called slag which can impact the environment if left untreated. Slag can be differentiated into two main types: one from steel manufacturing and the other from smelting. Steel manufacturing slag has the lowest toxin levels, whereas the slag produced from smelters contains high levels of hazardous elements. This resembles a volcanic rock with a significantly high ratio of toxic metals such as copper, lead, arsenic, zinc, cadmium, and barium; which when released into the environment cause soil pollution as well as surface and groundwater contamination (Keller & Gunn, 1995). Smelting of sulfide ores cause emission of sulfur dioxide gas, which reacts in the atmosphere to form a sulfuric acid mist that falls on earth as acid rain, which increases the acidity of water reservoirs and soils affecting the aquatic and terrestrial life (Gregory *et al.*, 2014).

Other than slag and acid rain, acid drainage is another environmental concern caused by the mining industry. Acid drainage or acid mine drainage (AMD) is the outflow of acidic water from mines. It usually occurs naturally due to rock weathering but is exaggerated by mining activities. When the earth is disturbed during construction and mining activities, a liquid drains out of the mine stocks which is highly toxic due to the presence of heavy metals. This liquid when combined with reduced pH causes detrimental impact on water reservoirs and aquatic life (Ferguson & Morin, 1991). As discussed so far, these are a few impacts of mining on the water resources. Although it is difficult to overcome these problems within a short period of time, implementing modern technologies and consistent efforts will save our blue planet from the negative consequences of mining.

7.2.2 Electroplating

Electroplating is the process of application of a metal layer or coating on another metal/conducting material with the help of electrochemical processes. The material to be electroplated is treated as the cathode of an electrolysis cell and the electric current is passed through the anode. It is performed to modify the surface features of an object such as corrosion resistance, lubrication, and abrasion properties or to elevate the aesthetics. After plating, the electroplated item is washed with water and dried. This process involves use of different metal salts and chemicals that generate environmental pollution. The most hazardous heavy metal ions such as copper, nickel, chromium, and lead are discarded during electroplating in the wastewater.

To prevent environmental pollution from these substances, many conventional methods such as chemical precipitation, membrane filtration, and ion-exchange are being employed, but due to high capital investment and disposal problems, these methods are ineffective in use (Ye *et al.*, 2019).

7.2.3 Water pollution

7.2.3.1 Pharmaceutical waste

Pharmaceuticals are substances that are used to detect, treat, reverse, and/or prevent illnesses. The term is broadened to include veterinary substances, and it also applies to illicit substances, drugs in specific (Blair, 2016). Thousands of tonnes of human pharmaceuticals, some of which are used every year, are created and consumed, including statins, hormones, anti-inflammatories and cytotoxins (Metcalfe *et al.*, 2003). Medications have raised significant concerns, after ingestion, residues or metabolites are eliminated or excreted, either directly or indirectly after ineffective treatment, which enters the water supplies (Kümmerer, 2001). Although pharmaceutical residue concentrations in surface waters are very modest, their existence and persistence pose a hazard to both aquatic and terrestrial species, and their impacts should not be disregarded. Estimating long-term impacts is still quite challenging (Asghar *et al.*, 2018). The sources of pharmaceutical waste includes landfills which contaminate the ground water, freshwater aquaculture waste, animal waste, hospital waste, domestic waste, and industrial waste (Li, 2014). There are various studies on the occurrence of pharmaceuticals in water, one of which is trimethoprim found in the United States, and Hebei (China) in the range of 0–0.64 and 0.64–2.15 μg/kg, respectively (Chen *et al.*, 2011; Kinney *et al.*, 2008).

7.2.3.2 Dyes

Dye effluent from dye industries are serious water pollutants as they may lead to mutagenicity, cancer, dysfunction of reproductive system, kidney, and liver (Kaykhaii *et al.*, 2018). The source of the dyes in the aquatic environment are the tannery, paper, and textile industries (Malik *et al.*, 2012). To date, dyes are present in river water, ponds, drinking water, and waste water along with it in the soil and wild fish. The dyes like red 60 and disperse blue 26 were identified in Canada in river water (Tkaczyk *et al.*, 2020). In an earlier study, azo dye was found in the effluents from a processing plant of the dye, as well as the dye disperse blue 373 in the range 57.9–316 μg/L (Oliveira *et al.*, 2007).

7.2.4 Radionuclides

While the pollution from radionuclides is not well targeted in wastewater treatment, these facilities must eliminate the pollutants for improving the water quality and also protect the health of the public. The source of radionuclides in waste water and water treatment plants include nuclear power plants, nuclear medicines from hospitals, mining uranium, leaching from building materials, oil and natural gas production, industrial activities, ocean disposal, research and diagnostic laboratory (Hossain, 2020). The occurrences are evident from various studies, one of which include the Chernobyl disaster, the groundwater was contaminated by ^{137}Cs with a maximum concentration of 50 MBq/L, ^{134}Cs, ^{131}I, ^{90}Sr, ^{239}Pu, ^{240}Pu, ^{106}Ru and also Americium-241 (^{241}Am). Till 2001, the highest groundwater concentration of ^{137}Cs, ^{239}Pu and ^{240}Pu was 200 Bq/L, 7 and 7 Bq/L, respectively. The source of taint of radioisotopes was because of the fallout isotopes and contaminated waste water discharge (Bugai, 2014).

7.3 SUSTAINABLE WASTEWATER TREATMENT WITH NANOHYBRIDS

Due to the catastrophic impact on both human and environmental health, the removal of hazardous components from wastewater discharge has received a lot of attention. Rapid urbanization and industrial development in underdeveloped nations have accelerated environmental devastation in recent years. Due to its negative effects on the environment and sustainability, the majority of wastewater effluents are released either untreated or insufficiently treated, this has become a serious problem. Innovative and creative wastewater treatment techniques with modest investment requirements must be put into practice. By using unusual water sources, nanohybrid adsorbents offer the potential to enhance wastewater treatment and boost water supply. Examples of nanoadsorbents that are utilized to remove pollutants include carbon nanotubes, activated carbon (AC), titanium oxide, magnesium oxide, graphene, ferric oxides, manganese oxide, and zinc oxide. These showed the efficient removal of pollutants, such as chemicals (see Section 7.2.3.1), coloring agents (see Section 7.2.3.2) and metals (see Sections 7.2.1 and 7.2.2) present in wastewater. The potential usefulness of nanohybrids for specialized applications has garnered enormous interest for sustainable treatment of wastewater.

7.4 MAGNETIC NANOHYBRIDS MATERIALS FOR WATER CONTAMINANT REMOVAL

7.4.1 Preparation of magnetic nanohybrid materials

The most often utilized materials for the source of magnetic property are nano-Fe_3O_4 particles or the Fe–M (M = Co, Mn, Ni, Zn) complex (Li *et al.*, 2021). The easiest way to create spherical nano-Fe_3O_4 with somewhat uniform size is through chemical coprecipitation. Additionally, the reaction duration, temperature, pH of the solution, and mechanical rotational speed may all be changed to regulate the particle size (Thanh *et al.*, 2018). As an illustration, we created magnetic nano-Fe_3O_4 with somewhat equal particle size using this technique. In addition, using the alkanolamines isopropanolamine and disopropanolamine as an alkaline agent, Pereira *et al.* (2012) synthesized superparamagnetic ferrite nanoparticles (MFe_2O_4, M = Fe, Co, Mn) by a unique one-step chemical coprecipitation process (Schwaminger *et al.*, 2019). Since the relative high temperature (100–200°C) and pressure (>10 MPa) conditions are suitable for the improvement of purity and magnetic properties, the hydrothermal approach offers certain benefits for producing magnetic nanoparticles. $Co_xNi_{1-x}Fe_2O_4$ ($x = 0$, 0.3, 0.7, 0.9, 1.0) magnetic nanoparticles are with good magnetic characteristics and a high capacity for adsorbing Congo Red. They were prepared quickly using a hydrothermal method, as described by Chen *et al.* (2014). Additionally, the pyrolysis approach (Li, 2014) and the sol–gel method (Li, 2014; Zheng *et al.*, 2014) are frequently used to create magnetic nanoparticles. To create magnetic nanohybrid materials, inorganic or organic functional substances, microorganisms, or both, are hybridized with the magnetic nanosubstance by chemical or physical alterations, considerably expanding the variety and applications of magnetic nanomaterials.

Numerous magnetic nanohybrid materials have been created and used. Three groups may be made based on their composition and structure. One method involves employing magnetic nanoparticles as hosts that have been functionally changed using materials like magnetic nanohybrids modified with polyethylenimine (PEI) or polyacrylic acid (PAA). Another option is to add magnetic nanoparticles as a dopant or guest component to the primary composition. For example, by dispersing them on nanotubes or microorganisms. The third method involves equally distributing magnetic nanoparticles and hybrid materials, such as magnetic biochar and polymetallic nanohybrid materials. Additionally, the one-pot approach may be used to create magnetic nanohybrid materials (Mahmoud *et al.*, 2022; Nizamuddin *et al.* 2019). For instance, $CuSO_4{\cdot}5H_2O$ and $FeSO_4{\cdot}7H_2O$ may be combined easily to create CuO-Fe_3O_4 nanohybrid material while maintaining an air purging flow rate of 53 L/min at 80°C and a pH of 8. To make graphene oxide (GO), reduced graphene oxide (rGO), and Fe^{2+} and Fe^{3+}, the solution's pH was adjusted from 9 to 11, and stirring continuously for 30 min at 80°C. In a similar manner, this technique may also be used to create Fe_3O_4/CNTs or Fe_3O_4/biochar (Li, 2014; Shen *et al.*, 2012).

However, due to some compositions being unable to form under the same reaction circumstances, its usefulness is restricted. Another common approach for making magnetic nanohybrid materials is the multistep process (Khan *et al.*, 2022; Pereira *et al.*, 2012). For instance, $Fe(NO_3)_3{\cdot}9H_2O$ and melamine were used as precursors, followed by ultrasonication, drying, and carbonization to produce nitrogen-doped magnetic carbon materials at 800°C in a N_2 environment (Chen *et al.*, 2010).

Additionally, our team used the chemical coprecipitation approach to create nano-Fe_3O_4 particles, and then polyethylenimine (PEI) was added to their surface using 2% glutaric dialdehyde as a cross-linking agent (Xu *et al.*, 2007). When selecting an appropriate technique for a certain application, the desired production, the precursor, the cost, and the operating circumstances should be considered.

The shape and physical–chemical characteristics of the magnetic nanohybrid materials must be explored after preparation. The most popular instruments are scanning electron microscopy (SEM), transmission electron microscopy (TEM), x-ray powder diffraction (XRD), Fourier transform infrared spectroscopy (FTIR), and x-ray photoelectron spectroscopy (XPS). To study the morphology and physical structure, SEM and TEM are employed. Crystal structure may be described using XRD. The potential functional groups of materials are found using FTIR. Additionally, XPS is a helpful technique for examining the surface chemical state of materials at the atomic level, which might yield important quantitative data of the substance under study (Asuha *et al.*, 2012).

7.4.2 Magnetic nanohybrid development

Owing to their minimal cost, non-toxic nature, and easy modification/separation, the magnetic nanoparticles for example Fe^0, Fe_3O_4, γ-Fe_2O_3, and $CuFe_2O_4$ have been extensively researched and utilized in the form of catalysts, adsorbents as well as in the form of a carrier (Feng *et al.*, 2010; Liu *et al.*, 2018; Xu *et al.*, 2014). However, these magnetic nanoparticles have some disadvantages in wastewater treatment when used alone, thus they are often used together with another material for nano-hybridization. The hybrid magnetic nanomaterials not only possess the properties of nanomaterials but also contain magnetic properties, making them excellent adsorbents in the remediation of water contamination. The hybridization process can establish novel compositions and modify the chemical constitution of the initial material, giving the material improved properties. Modification of natural nanomaterials with various functional groups, for instance amino, hydroxyl, carboxyl and thiol groups, to enhance the efficient adsorption sites is an approach for developing nanohybrid magnetic adsorbents. The use of magnetic biosorbents offers distinct advantages, as nanomaterials with magnetic potent are swiftly and effortlessly removed from the treated water by employing an external magnetic field. Furthermore, iron oxide-based nanostructures are widely used for wastewater treatment as it has mild ecological harm (Soares *et al.*, 2017; Sousa *et al.*, 2015).

The chemical co-precipitation technique is favorable for the formulation of Fe_3O_4 magnetic nanohybrids of relatively uniform size. Furthermore, the particle size of these nanohybrids can be managed by modifying the reaction time, pH of the solution, temperature conditions, as well as mechanical rotational speed. Fe_3O_4-based magnetic nanohybrids with uniform particle sizes were synthesized (Feng *et al.*, 2010; Xu *et al.* 2014). Besides, the chemical co-precipitation technique using alkanolamine isopropanolamine and disopropanolamine as alkaline agents were also employed for the preparation of nanohybrids (Sousa *et al.*, 2015). Usually, the chemical-based functionalization of nanoparticles is performed to improve the selection and absorption capacity of the magnetic nanoparticles concerning target contaminants. A group of scientists described the elimination of methylene blue from wastewater by employing κ-carrageenan or organic/inorganic hybrid silica shell-coated with magnetic ferrous oxide-based nanoparticles as an adsorbent. Using a new surface alteration method, the researchers confirmed the massive grafting of κ-carrageenan onto the siliceous fields of the hybrid material and demonstrated a high adsorption capacity (530 mg/g). After cleansing with an aqueous KCl solution, the adsorbent exhibited recyclability with eminent colloidal stability. The robustness and regenerative capability of the developed hybrid sorbents were demonstrated by the methylene blue removal efficiency (>97%) in six successively repeated trials. The efficient removal of organic dyes from wastewater was demonstrated using magnetic separation technology together with this newly developed sorbent (Liu *et al.*, 2018). Other researchers proposed the development of a silica-carrageenan bioadsorbent to eliminate toxic heavy metal ions (Cd^{2+} and Pb^{2+}). The findings of the study revealed that the bioadsorbent material consisted of ash (81.70%), moisture (2.10%), density

($1.10 g/cm^3$), and 21.80% adsorbent iodine. FTIR spectra revealed the presence of the hydroxyl, silanol as well as siloxane functional groups of carrageenan. SEM micrographs showed uniformly spread spherical nanoparticles (<100 nm). Heavy metal ions cannot be eluted on a thin layer chromatography (TLC) plate by all organic eluents (Pérez-Page *et al.*, 2016).

Likewise, hydrothermal treatment has its benefits for the preparation of magnetic nanoparticles as the relatively high temperature (100–200°C), as well as pressure (0.10 MPa) settings improve the purity as well as magnetic properties. The hydrothermal method was used for the prompt formulation of $Co_x Ni_{1-x} Fe_2O_4$ magnetic nanoparticles ($x = 0, 0.3, 0.7, 0.9, 1.0$) with exceptional magnetic attributes as well as superior adsorption capacity for Congo Red (Soares *et al.*, 2017). Besides, the researchers recognized a sequence of polyethylenimine (PEI)-modified nanohybrid magnetic adsorbents for the removal of heavy metals. PEI-grafted magnetic porous adsorbents were produced for taking up heavy metals like copper, zinc, and cadmium from water with the highest adsorption capacity of 157.8, 138.8, and 105.2 mg/g, correspondingly, as per the Langmuir adsorption model. The adsorption choice between the heavy metals is ascribed to the heavy metal radius as well as the stability constant among amino groups and heavy metals, indicating that the intrinsic attributes of heavy metals determine the adsorption behavior.

The biomass-based magnetic nanohybrid adsorbent is an additional favorable approach for cost-effective adsorbent development (Soares *et al.*, 2017; Sousa *et al.*, 2015). This type of adsorbent is largely developed by doping or else adding magnetic nanoparticles on biomass. Biomass feedstocks such as bacteria, fungi, algae as well as agricultural wastes have multiple functional groups that are appropriate for the development of inexpensive, along with effective magnetic nanoadsorbents. For example, chitosan-coated magnetic nanoparticles are economical and effective nanoadsorbents to remove mercury from industrial wastewater (Soares *et al.*, 2017). Magnetic carbon nanoadsorbents obtained from coconut shells help remove toxic dyes (Xu *et al.*, 2014). All through the operation, the adsorbents could be easily separated by operating an external magnetic field and the liquid became clear within minutes. Lastly, various nanohybrid magnetic adsorbents differ in their heavy metal adsorption performance depending on internal properties for example surface area, functional groups, zeta potential, assembly of the pore, and adsorption parameters like solution pH, temperature, and contact time.

7.4.3 Mechanism of adsorptive removal of pollutants using magnetic nanohybrid materials

The potential of magnetic nanohybrid materials in water contamination removal has gained a lot of interest in recent years. The adsorption and catalytic capabilities of the nanohybrids are enhanced by the synergic effect of magnetic nanoparticles and other components such as carbon, silica, or polymers. Commonly, magnetic nanohybrids are considered as an effective adsorbent for removing a diverse array of pollutants from aqueous solutions, including heavy metal ions, organic pollutants, and bacterial contaminants. The large surface area and chemical reactivity of nanohybrids allow for a larger number of adsorption sites and improved affinity for diverse contaminants. Magnetic responsiveness enables separation from water using a magnetic field, simplifying the water treatment process (Bulin *et al.*, 2023). The reactivity can be tailored for specific contaminants such as pharmaceutical waste, dyes, and radionuclides, enabling the effective absorption of a diverse spectrum of contaminants from water.

The magnetic nanohybrid, iron oxide nanoparticles functionalized with graphene oxide ($GO\text{-}CuFe_2O_4$) exhibits effective removal of the CR dye from aqueous solution. The absorption of the CR dye depends on the initial concentration and rises with treatment time. The adsorption process followed pseudo-second-order kinetics and the Freundlich isotherm model, as shown in the following equation (Zourou *et al.*, 2022):

$$\frac{1}{q_t} = \frac{1}{K_2 q_{2e}^2} + \frac{1}{q_{2e}} t \tag{7.1}$$

where q_t is the absorbed amount of CR (mg/g), t is the time of absorption (min), K_2 is the pseudo-second-order constant (g/mg/min), and q_{2e} is the amount of adsorbed CR at equilibrium for a pseudo-second-order kinetic, K_2 and q_{2e} can be determined using a linear plot of t/q_t versus t. The GO-$CuFe_2O_4$ absorption of CR dye is related to several fundamental factors, such as basal plane and edges of GO sheets, which are responsible for hydrogen bonding among -OH and -NH_2 groups of CR molecules, and the pi–pi stacking of the GO plane and the aromatic ring of CR. GO enhances the surface area of the iron oxide nanoparticles, increasing their adsorption capacity and providing a catalytic surface for the degradation of organic pollutants. GO-$CuFe_2O_4$ nanohybrid materials have strong adsorption and catalytic capabilities, facile separability, and adaptability in terms of composition and functionalization, which make them promising for applications in water pollutant removal.

7.5 FACTORS INFLUENCING ADSORPTION BY MAGNETIC NANOHYBRID ADSORBENT

Magnetic nanohybrids have emerged as promising materials for removing heavy metals from water due to their high surface area, high reactivity, and magnetic properties. These materials are composed of a magnetic core (such as iron oxide) and a functional shell (such as a polymer or carbon) that can selectively bind to heavy metal ions (Sharma *et al.*, 2022). The magnetic core allows for easy separation of the nanohybrids from water using a magnet, making the process more efficient and cost-effective. Despite their promising properties, there are key challenges associated with the use of magnetic nanohybrid materials for heavy metal removal from water. It is crucial to optimize the performance of magnetic nanohybrid materials for specific contaminants, and the effectiveness of these materials can be impacted by the following variables:

(1) pH of the water: The adsorption capacity of magnetic nanohybrid adsorbents for heavy metal removal is significantly influenced by the pH of the solution, owing to its potential to alter the surface charge of the adsorbent and the speciation of metal ions in solution, which subsequently affects the electrostatic interactions and affinity between the adsorbent and heavy metal ions (Lan Huong *et al.*, 2016).
(2) Treatment time: The contact time of the magnetic nanohybrid adsorbent with the aqueous solution has a significant effect on its adsorption capacity. As the contact time increases, the adsorption capacity rises until an equilibrium is attained (Huong *et al.*, 2016).
(3) Adsorbent concentration: A linear relationship exists between the quantity of adsorbent and the adsorption capacity until saturation is reached (Kumar *et al.*, 2014).
(4) Heavy metal concentration: The adsorption capacity of the nanohybrid adsorbent can be influenced by the initial concentration of heavy metal ions present in the solution. Higher initial concentrations may lead to a limited adsorption capacity, as the adsorption sites may become saturated. Therefore, the heavy metal concentration is an important parameter to consider when designing an efficient adsorption system using nanohybrid adsorbents (Kumar *et al.*, 2014).
(5) Temperature: The temperature of the solution has a crucial impact on the rate and extent of adsorption of metal ions onto the adsorbent. An increase in temperature usually leads to an accelerated adsorption process, as the energy of the adsorbate and the adsorbent surface is increased, leading to higher collision frequencies and more effective mass transfer. However, it should be noted that excessively high temperatures can lead to thermal degradation of the adsorbent and destabilization of the metal ions, leading to a decrease in the adsorption capacity. Hence, optimizing the temperature is crucial for obtaining efficient adsorption of heavy metal ions in a nanohybrid adsorption system (Lan Huong *et al.*, 2016).
(6) Presence of interfering ions: The adsorption capacity of a nanohybrid adsorbent can be affected by the presence of interfering ions in the solution. Interfering ions have the potential to reduce the overall adsorption capacity by competing for adsorption sites. Therefore, it is essential to consider the effect of interfering ions while designing an adsorption process using nanohybrid adsorbents (Mohan *et al.*, 2017).

Understanding these factors and their interactions can assist in optimizing the use of magnetic nanohybrid adsorbents for the removal of heavy metals from water treatment.

7.6 REMOVAL OF WATER POLLUTANTS BASED ON MAGNETIC NANOHYBRID CATALYST

7.6.1 Carbon-based magnetic nanohybrid adsorbents

Carbon-based materials are powdered or bulk non-metallic solid materials that contain carbon as the main ingredient, such as activated carbon/biochar (AC/BC), carbon nanotubes (CNTs), and graphene (GN) analogues (Duan *et al.*, 2020). Carbon-based materials have been investigated as better adsorbents for the removal of inorganic and organic contaminants among the many nanomaterials-based adsorbents (Santhosh *et al.*, 2016).

7.6.1.1 Activated charcoal/biochar-based materials

Rice husk carbon is created by combining dried rice husk and sulfuric acid and heating the mixture for a length of time. It was discovered that Cr(VI) is entirely adsorbed after 120 min of adsorption with rice husk carbon. Adsorption tests with Cd (II) and Ni (II) with rice husk carbon were similarly successful, with ion exchange adsorption and precipitation being the primary adsorption mechanisms.

After three adsorption–desorption cycles, maize straw porous carbon demonstrated the highest adsorption capacity for Cr(VI), with a removal efficiency of 70.65% (Duan *et al.*, 2020). The adsorption process includes several processes, including ion exchange. The removal of Pb(II), Cu(II), Cd(II), and Ni(II) from discarded mushroom-stick biochar was explored (Duan *et al.*, 2020). At higher temperatures, more aromatic structures are present and a larger number of initial mineral ions (e.g., Na^{+}, Ca^{2+}, and Mg^{2+}) to exchange with heavy metal ions were present (Duan *et al.*, 2020).

7.6.1.2 Carbon nanotubes

Carbon nanotubes (CNTs) absorption ability is effective across a wide pH range. The pH range of 7–10 was shown to give optimal performance (Santhosh *et al.*, 2016). CNTs are classified into two types: single-walled carbon nanotubes (SWCNT) and multiwalled carbon nanotubes (MWCNT). CNT have hydrophobicity, an abundant pore structure and functional groups, and performs a broad-spectrum adsorption of heavy metals, such as Pb, Cu, Cd, and Zn in water (Duan *et al.*, 2020). CNT surface modification boosts its total adsorption activity. Various surface modification procedures, such as acid treatment, metal impregnation and functional molecules or group grafting, have been described by various researchers. Modified carbon nanotubes grafted with different functional groups enable removal of heavy metal from wastewater. Furthermore, CNTs treated with metal/metal oxides such as MnO_2 and iron oxide show promising results in the removal of heavy metals from wastewater (Anjum *et al.*, 2019).

7.6.1.3 Graphene-based nanoadsorbents

GO is a two-dimensional carbon nanomaterial formed by the chemical oxidation of a graphite sheet. The presence of these hydroxyl and carboxyl groups as functional groups in GO increases heavy metal adsorption. When compared to other nanomaterials such as CNTs, GO offers two distinct advantages. First, a single layer GO features two-dimensional basal planes for maximum heavy metal adsorption. Second, it has a simple synthesis procedure that can be accomplished by chemical exfoliation of graphite without the use of a metallic catalyst or sophisticated equipment. Furthermore, because GO has already hydrophilic functional groups, it does not require any further acid treatment to improve its adsorption ability (Anjum *et al.*, 2019). Graphene is a CNT alternative, and rGO may be rapidly synthesized by chemical exfoliation of graphite without the use of sophisticated equipment or metallic catalysts, making it appropriate for water treatment (Santhosh *et al.*, 2016). rGO (Duan *et al.*, 2020) and GO have been demonstrated to have higher Pb(II) adsorption than virgin graphene sheets. A composite containing GO and metal oxide typically has distinct properties and has been employed as efficient adsorbents for the elimination of various contaminants. The removal of Zn^{2+}, Cd^{2+} and Pb^{2+} ions from water has been accomplished using flower-like TiO_2 on GO hybrid (GO-TiO_2) (Santhosh *et al.*, 2016).

7.6.1.4 Chitosan-based magnetic nanohybrid catalyst

Chitosan and iron oxide nanoparticle-based magnetic nanocomposites remove heavy metals from water. Chitosan is a naturally occurring biopolymer derived from chitin that has a high surface area and can absorb heavy metals for use in water treatment. The incorporation of iron oxide nanoparticles into chitosan results in the formation of a nanohybrid with enhanced magnetic responsiveness, thereby allowing the heavy metal removal from water. Additionally, the iron oxide nanoparticles provide a catalytic surface for the oxidation of the heavy metals. The chitosan-iron oxide nanohybrids have been widely used in the literature for the effective removal of various heavy metals such as cobalt, mercury, copper, zinc, chromium, cadmium, lead, nickel, and arsenic from water bodies (Vakili *et al.*, 2019). The performance of the nanohybrids can be influenced by the pH of the water, initial concentration of the metal ions, and predominately the treatment time. Moreover, the chitosan-iron oxide hybrids show potential as a low-cost and environmentally friendly metal removal method from water.

7.6.2 Metal-based magnetic nanohybrid catalyst

7.6.2.1 Zeolites

Among the materials used in the adsorption of heavy metals from water, zeolite magnetic hybrids are indispensable due to their porosity, ion exchange, and high surface area properties. When zeolites are incorporated with iron oxides, they display magnetic characteristics that aid in the separation and recovery of the adsorbent after use. These materials are effective for the removal of heavy metals such as arsenic, cadmium, and lead from water (Loiola *et al.*, 2022). Using magnetic separation techniques, the materials can be readily separated from the water, making them suitable for large-scale applications. In addition to heavy metal removal, zeolite magnetic hybrid materials are good at absorbing phosphorous from wastewater. Phosphorus is a nutrient that can cause eutrophication in water bodies, which can have negative effects on aquatic ecosystems. Zeolite magnetic nanohybrid materials effectively remove phosphorus from wastewater, which can help mitigate the effects of eutrophication.

7.6.2.2 Multi-metals-based magnetic nanohybrid catalyst

Magnetic nanohybrid catalysts based on multi-metals, as well as magnetic CuO/Fe_3O_4 nanocatalysts for phenol degradation, can be used to activate persulfate, in which persulfate (PS) acts as catalyst for phenol degradation. The removal of phenol was revealed to be mostly attributable to the radicals formed by the interaction of PS and Cu(II). The binary metal catalyst was stable and reusable, and the presence of bicarbonate ions assists in the removal of phenol, making this magnetic nanohybrid catalyst beneficial for practical applications. A magnetic nano-$CoFe_2O_4$/titanate nanotubes ($CoFe_2O_4$/TNTs) catalyst was created using an impregnation calcination method (Pang *et al.*, 2019). Furthermore, the cobalt (Co) leakage was substantially lower than in regular $CoFe_2O_4$ or $CoFe_2O_4$/SBA-15, which might be attributed to Co's unique ion-exchange properties (Pang *et al.*, 2019).

7.7 FUTURE PROSPECTIVES WITH CHALLENGES

Greener nanotechnology has the potential to make industrial water treatment more efficient where the contaminants can be removed but this field has some pros and cons. Carbon and graphene quantum dots and graphene-based nanomaterials may emerge as effective, cost-effective, and environmentally friendly substitutes for the prevailing treatment supplies. There is a need to evaluate toxicity in brief for the modification of CNTs. Applying electrochemical assistance can meaningfully improve their adsorption rate and capacities for the removal of contaminants, and may classify as an innovative strategy in environmental remediation (Nasrollahzadeh *et al.*, 2021).

To fabricate advanced and suitable catalytic powered nanomaterials such as carbon nanotubes, carbon and graphene quantum dots and graphene-based nanomaterials, more attention should be given on the following issues in future investigations.

Novel and innovative synthetic methods are required to reduce the cost of production of nanomaterials and improve their catalytic power, and reusability. Utilization of biowaste-derived nanomaterials and use of mild, water soluble, and low-cost reductants would reduce the exploitation of natural materials. There is a chance of manipulation of animal residues such as bones, eggshell, and bristles for the synthesis of nanomaterials and application in water treatment. There is a need to do toxicity evaluations and comprehensive assessments of the risks and the potential impact of nanomaterials to human health and ecosystems.

Photocatalysis of organic contaminants in water by noble metal-free-doped graphitic carbon nitride-based nanohybrids has several challenges such as doping mechanism, site of attachment of dopant and assessment of visible light absorption (Hasija *et al.*, 2019). In accordance to the future aspects there is a need to set a novel model high quantum degradation efficiency. It is also important towards the oxidative and reductive nature of g-C_3N_4 which is responsible for photodegradation of pollutants and evolution of H_2 and O_2, respectively. Hence, for designing and fabrication of robust and efficient noble metal-free-doped g-C_3N_4-based photocatalytic systems, following key directions must be taken into consideration suggested by Hasija *et al.* (2019):

(1) Researchers should focus on extending the light-harvesting range from UV to NIR spectrum to avail extensive renewable solar energy to deteriorate environmental implications via biofriendly photocatalytic systems. For this, noble metal-free g-C_3N_4 should be combined with plasmonic photocatalysts or metal/carbon quantum dots.

(2) The role of various morphologies such as nanofibers, nanotubes, nanosheets, nanorods, and mesoporous structures of g-C_3N_4 must be explored for achieving a higher surface area. 2D photocatalysts have more catalytic reaction sites for anchoring of pollutants, thereby enhancing their photocatalytic ability and are preferred for designing noble metal-free doped g-C_3N_4-based photocatalysts.

(3) The recovery of noble metal-free-doped g-C_3N_4 photocatalysts is highly needed for higher recyclability of photocatalysts. For this, researchers should dope g-C_3N_4 with magnetic dopants like iron oxide and ferrites exhibiting paramagnetic/superparamagnetic nature.

(4) The development of innovative, feasible, facile, and cost-effective fabrication methods should be adopted for their synthesis. Non-metal doping of g-C_3N_4 is preferred due to less thermal variation, high cost and scarcity of metallic dopants. Moreover, researchers should try to find a more innovative way to improve the overall efficacy of synthesis methods using microwave radiation-assisted polycondensation methods.

ACKNOWLEDGEMENTS

This endeavor was possible with the constant support, access to facilities and guidance from resourceful people from the Annasaheb Dange College of Engineering & Technology (Astha, India), the Department of Polymer Science and Technology, Sri Jayachamarajendra College of Engineering (SJCE), JSSSTU (Mysore, India), Amity Institute of Food Technology, Amity University, Noida, India, School of Basic Sciences and Research, Sharda University (Noida, India) and Department of Biotechnology Engineering, Kolhapur Institute of Technology (Kolhapur, India).

REFERENCES

Adeleye A. S., Conway J. R., Garner K., Huang Y., Su Y. and Keller A. A. (2016). Engineered nanomaterials for water treatment and remediation: costs, benefits, and applicability. *Chemical Engineering Journal*, **286**, 640–662, https://doi.org/10.1016/j.cej.2015.10.105

Anjum M., Miandad R., Waqas M., Gehany F. and Barakat M. A. (2019). Remediation of wastewater using various nanomaterials. *Arabian Journal of Chemistry*, **12**(8), 4897–4919, https://doi.org/10.1016/j.arabjc.2016.10.004

Asghar M. A., Zhu Q., Sun S. and Shuai Q. (2018). Suspect screening and target quantification of human pharmaceutical residues in the surface water of Wuhan, China, using UHPLC-Q-Orbitrap HRMS. *Science of the Total Environment*, **635**, 828–837, https://doi.org/10.1016/j.scitotenv.2018.04.179

Asuha S., Wan H. L., Zhao S., Deligeer W., Wu H. Y., Song L. and Tegus O. (2012). Water-soluble, mesoporous Fe_3O_4: synthesis, characterization, and properties. *Ceramics International*, **38**(8), 6579–6584, https://doi.org/10.1016/j.ceramint.2012.05.042

Blair B. D. (2016). Potential upstream strategies for the mitigation of pharmaceuticals in the aquatic environment: a brief review. *Current Environmental Health Reports*, **3**, 153–160, https://doi.org/10.1007/s40572-016-0088-x

Bugai D. (2014). Groundwater contamination following the Chernobyl accident: overview of monitoring data, assessment of radiological risks and analysis of remedial measures. IAEA TM Conference on Groundwater Contamination following Fukushima Accident, Vienna, Austria.

Bulin C., Zheng R., Song J., Bao J., Xin G. and Zhang B. (2023). Magnetic graphene oxide-chitosan nanohybrid for efficient removal of aqueous Hg (*II*) and the interaction mechanism. *Journal of Molecular Liquids*, **370**, 121050, https://doi.org/10.1016/j.molliq.2022.121050

Chen M. A., Bin L. U. O., Song H. H. and Zhi L. J. (2010). Preparation of carbon-encapsulated metal magnetic nanoparticles by an instant pyrolysis method. *New Carbon Materials*, **25**(3), 199–204, https://doi.org/10.1016/S1872-5805(09)60028-7

Chen F., Ying G. G., Kong L. X., Wang L., Zhao J. L., Zhou L. J. and Zhang L. J. (2011). Distribution and accumulation of endocrine-disrupting chemicals and pharmaceuticals in wastewater irrigated soils in Hebei, China. *Environmental Pollution*, **159**(6), 1490–1498, https://doi.org/10.1016/j.envpol.2011.03.016

Chen R., Wang W., Zhao X., Zhang Y., Wu S. and Li F. (2014). Rapid hydrothermal synthesis of magnetic $CoxNi_{1-x}Fe_2O_4$ nanoparticles and their application on removal of Congo red. *Chemical Engineering Journal*, **242**, 226–233, https://doi.org/10.1016/j.cej.2013.12.016

Dasgupta A., Sastry K. L. N. and Dhinwa P. S. (2012). Impact of mining on rural environment and economy – a case study, Kota district, Rajasthan. *International Journal of Remote Sensing and Geoscience*, **2**(3), 21–26.

Duan C., Ma T., Wang J. and Zhou Y. (2020). Removal of heavy metals from aqueous solution using carbon-based adsorbents: a review. *Journal of Water Process Engineering*, **37**, 101339, https://doi.org/10.1016/j.jwpe.2020.101339

Feng Y., Gong J. L., Zeng G. M., Niu Q. Y., Zhang H. Y., Niu C. G. and Yan M. (2010). Adsorption of Cd(II) and Zn(II) from aqueous solutions using magnetic hydroxyapatite nanoparticles as adsorbents. *Chemical Engineering Journal*, **162**(2), 487–494, https://doi.org/10.1016/j.cej.2010.05.049

Ferguson K. D. and Morin K. A. (1991). The prediction of acid rock drainage—lessons from the database. Second International Conference on the Abatement of Acidic Drainage. Vol. **3**, pp. 83–106.

Ferroudj N., Nzimoto J., Davidson A., Talbot D., Briot E., Dupuis V. and Abramson S. (2013). Maghemite nanoparticles and maghemite/silica nanocomposite microspheres as magnetic Fenton catalysts for the removal of water pollutants. *Applied Catalysis B: Environmental*, **136**, 9–18, https://doi.org/10.1016/j.apcatb.2013.01.046

Gosselin D. C., Manduca C., Bralower T. and Mogk D. (2013). Transforming the teaching of geoscience and sustainability. *Eos, Transactions American Geophysical Union*, **94**(25), 221–222, https://doi.org/10.1002/2013EO250002

Gregory S. J., Anderson C. W. N., Arbestain M. C. and McManus M. T. (2014). Response of plant and soil microbes to biochar amendment of an arsenic-contaminated soil. *Agriculture, Ecosystems & Environment*, **191**, 133–141, https://doi.org/10.1016/j.agee.2014.03.035

Grey D., Garrick D., Blackmore D., Kelman J., Muller M. and Sadoff C. (2013). Water security in one blue planet: twenty-first century policy challenges for science. *Philosophical Transactions of the Royal Society A: Mathematical, Physical and Engineering Sciences*, **371**(2002), 20120406, https://doi.org/10.1098/rsta.2012.0406

Hasija V., Sudhaik A., Raizada P., Hosseini-Bandegharaei A. and Singh P. (2019). Carbon quantum dots supported AgI/ZnO/phosphorus doped graphitic carbon nitride as Z-scheme photocatalyst for efficient photodegradation of 2, 4-dinitrophenol. *Journal of Environmental Chemical Engineering*, **7**(4), 103272, https://doi.org/10.1016/j.jece.2019.103272

Hossain F. (2020). Natural and anthropogenic radionuclides in water and wastewater: sources, treatments and recoveries. *Journal of Environmental Radioactivity*, **225**, 106423, https://doi.org/10.1016/j.jenvrad.2020.106423

Hudson T. (2016). *Living with Earth: An introduction to environmental geology*, Routledge.

Huong P. T. L., Huy L. T., Phan V. N., Huy T. Q., Nam M. H., Lam V. D. and Le A. T. (2016). Application of graphene oxide-$MnFe_2O_4$ magnetic nanohybrids as magnetically separable adsorbent for highly efficient removal of arsenic from water. *Journal of Electronic Materials*, **45**, 2372–2380, https://doi.org/10.1007/s11664-015-4314-3

Jia F., Wang Q., Wu J., Li Y. and Song S. (2017). Two-dimensional molybdenum disulfide as a superb adsorbent for removing Hg^{2+} from water. *ACS Sustainable Chemistry & Engineering*, **5**(8), 7410–7419, https://doi.org/10.1021/acssuschemeng.7b01880

Kaykhaii M., Sasani M. and Marghzari S. (2018). Removal of dyes from the environment by adsorption process. *Chemical and Materials Engineering*, **6**(2), 31–35, https://doi.org/10.13189/cme.2018.060201

Keller W. and Gunn J. M. (1995). Lake water quality improvements and recovering aquatic communities. In: Restoration and Recovery of an Industrial Region: Progress in Restoring the Smelter-Damaged Landscape Near Sudbury, J. M. Gunn (ed.), Canada, Springer New York, New York, NY, pp. 67–80.

Khan B., Nawaz M., Price G. J., Hussain R., Waseem M., Haq S. and Rehman W. (2022). Effect of pH on the morphology of magnetite nanoparticles for adsorption of Cr (VI) ions from aqueous medium. *Journal of Dispersion Science and Technology*, **44**, 1–8.

Kinney C. A., Furlong E. T., Kolpin D. W., Burkhardt M. R., Zaugg S. D., Werner S. L. and Benotti M. J. (2008). Bioaccumulation of pharmaceuticals and other anthropogenic waste indicators in earthworms from agricultural soil amended with biosolid or swine manure. *Environmental Science & Technology*, **42**(6), 1863–1870, https://doi.org/10.1021/es702304c

Kitula A. G. N. (2006). The environmental and socio-economic impacts of mining on local livelihoods in Tanzania: a case study of Geita District. *Journal of Cleaner Production*, **14**(3–4), 405–414, https://doi.org/10.1016/j.jclepro.2004.01.012

Kumar S., Nair R. R., Pillai P. B., Gupta S. N., Iyengar M. A. R. and Sood A. K. (2014). Graphene oxide–$MnFe_2O_4$ magnetic nanohybrids for efficient removal of lead and arsenic from water. *ACS Applied Materials & Interfaces*, **6**(20), 17426–17436, https://doi.org/10.1021/am504826q

Kümmerer K. (2001). Drugs in the environment: emission of drugs, diagnostic aids and disinfectants into wastewater by hospitals in relation to other sources – a review. *Chemosphere*, **45**(6–7), 957–969, https://doi.org/10.1016/S0045-6535(01)00144-8

Lan Huong P. T., Huyen N. T., Giang C. D., Tu N., Phan V. N., Quy N. V. and Le A. T. (2016). Facile synthesis and excellent adsorption property of GO-Fe_3O_4 magnetic nanohybrids for removal of organic dyes. *Journal of Nanoscience and Nanotechnology*, **16**(9), 9544–9556, https://doi.org/10.1166/jnn.2016.12161

Li W. C. (2014). Occurrence, sources, and fate of pharmaceuticals in aquatic environment and soil. *Environmental Pollution*, **187**, 193–201, https://doi.org/10.1016/j.envpol.2014.01.015

Li Z., Wang L., Meng J., Liu X., Xu J., Wang F. and Brookes P. (2018). Zeolite-supported nanoscale zero-valent iron: new findings on simultaneous adsorption of Cd (II), Pb (II), and As (III) in aqueous solution and soil. *Journal of Hazardous Materials*, **344**, 1–11, https://doi.org/10.1016/j.jhazmat.2017.09.036

Li H., Tang C., Wang M., Mei C. and Liu N. (2021). Decolorization of azo dyes in a heterogeneous persulfate system using FeS as the activator. *Water Science and Technology*, **83**(7), 1703–1713, https://doi.org/10.2166/wst.2021.085

Liu N., Ding F., Weng C. H., Hwang C. C. and Lin Y. T. (2018). Effective degradation of primary color direct azo dyes using FeO aggregates-activated persulfate process. *Journal of Environmental Management*, **206**, 565–576, https://doi.org/10.1016/j.jenvman.2017.11.006

Loiola A. R., Bessa R. A., Oliveira C. P., Freitas A. D., Soares S. A., Bohn F. and Pergher S. B. (2022). Magnetic zeolite composites: classification, synthesis routes, and technological applications. *Journal of Magnetism and Magnetic Materials*, **560**, 169651, https://doi.org/10.1016/j.jmmm.2022.169651

Mahmoud M. H., Hassan A. M., Said A. E. A. A. and Taha T. A. (2022). Structural, magnetic, and catalytic studies of microwave-combustion/ball-mill synthesized zinc ferrite nanoparticles. *Inorganic Chemistry Communications*, **144**, 109932, https://doi.org/10.1016/j.inoche.2022.109932

Malik A., Rahman M., Ansari M. I., Masood F. and Grohmann E. (2012). Environmental protection strategies: an overview. *Environmental Protection Strategies for Sustainable Development*, 1–34.

Metcalfe C. D., Miao X. S., Koenig B. G. and Struger J. (2003). Distribution of acidic and neutral drugs in surface waters near sewage treatment plants in the lower Great Lakes, Canada. *Environmental Toxicology and Chemistry: an International Journal*, **22**(12), 2881–2889.

Mohan S., Kumar V., Singh D. K. and Hasan S. H. (2017). Effective removal of lead ions using graphene oxide-MgO nanohybrid from aqueous solution: isotherm, kinetic and thermodynamic modeling of adsorption. *Journal of Environmental Chemical Engineering*, **5**(3), 2259–2273, https://doi.org/10.1016/j.jece.2017.03.031

Mondal S., Bandayopadhyay J. and Chakravatry D. (2014). Scientific investigation of the environmental impact of mines using geospatial techniques over a small part of Keonjhar district of Orissa. *International Journal of Scientific and Research Publications*, **4**(1), 33.

Nasrollahzadeh M., Sajjadi M., Iravani S. and Varma R. S. (2021). Carbon-based sustainable nanomaterials for water treatment: state-of-art and future perspectives. *Chemosphere*, **263**, 128005, https://doi.org/10.1016/j.chemosphere.2020.128005

Nizamuddin S., Siddiqui M. T. H., Mubarak N. M., Baloch H. A., Abdullah E. C., Mazari S. A. and Tanksale A. (2019). Iron oxide nanomaterials for the removal of heavy metals and dyes from wastewater. *Nanoscale Materials in Water Purification*, 447–472, https://doi.org/10.1016/B978-0-12-813926-4.00023-9

Oliveira D. P., Carneiro P. A., Sakagami M. K., Zanoni M. V. B. and Umbuzeiro G. A. (2007). Chemical characterization of a dye processing plant effluent – identification of the mutagenic components. *Mutation Research/Genetic Toxicology and Environmental Mutagenesis*, **626**(1–2), 135–142, https://doi.org/10.1016/j.mrgentox.2006.09.008

Pang Y., Yu J., Tang L., Zeng G., Zhu C. and Wei X. (2019). Magnetic nanohybrid materials for water-pollutant removal. In: Nanohybrid and Nanoporous Materials for Aquatic Pollution Control, L. Tang, Y. Deng, J. Wang, J. Wang and G. Zeng (eds.), Elsevier, Changsha, P. R. China, pp. 1–30.

Pereira C., Pereira A. M., Fernandes C., Rocha M., Mendes R., Fernández-García M. P., Guedes A., Tavares P. B., Grenèche J. M., Araújo J. P. and Freire C. (2012). Superparamagnetic MFe_2O_4 (M = Fe, Co, Mn) nanoparticles: tuning the particle size and magnetic properties through a novel one-step coprecipitation route. *Chemistry of Materials*, **24**(8), 1496–1504, https://doi.org/10.1021/cm300301c

Pérez-Page M., Yu E., Li J., Rahman M., Dryden D. M., Vidu R. and Stroeve P. (2016). Template-based syntheses for shape controlled nanostructures. *Advances in Colloid and Interface Science*, **234**, 51–79, https://doi.org/10.1016/j.cis.2016.04.001

Pratap N., Sharma S., Rani R. and Singh P. (2014). The negative and dangerous effects of water pollution and storm water runoff on the environment and the surrounding community. *International Journal of Engineering Technologies and Management Research*, **1**, 185–196.

Santhosh C., Velmurugan V., Jacob G., Jeong S. K., Grace A. N. and Bhatnagar A. (2016). Role of nanomaterials in water treatment applications: a review. *Chemical Engineering Journal*, **306**, 1116–1137, https://doi.org/10.1016/j.cej.2016.08.053

Schwaminger S. P., Blank-Shim S. A., Scheifele I., Pipich V., Fraga-García P. and Berensmeier S. (2019). Design of interactions between nanomaterials and proteins: a highly affine peptide tag to bare iron oxide nanoparticles for magnetic protein separation. *Biotechnology Journal*, **14**(3), 1800055, https://doi.org/10.1002/biot.201800055

Schwarzenbach R. P., Escher B. I., Fenner K., Hofstetter T. B., Johnson C. A., Von Gunten U. and Wehrli B. (2006). The challenge of micropollutants in aquatic systems. *Science*, **313**(5790), 1072–1077, https://doi.org/10.1126/science.1127291

Sharma P., Nanda K., Yadav M., Shukla A., Srivastava S. K., Kumar S. and Singh S. P. (2022). Remediation of noxious wastewater using nanohybrid adsorbent for preventing water pollution. *Chemosphere*, **292**, 133380, https://doi.org/10.1016/j.chemosphere.2021.133380

Shen H., Pan S., Zhang Y., Huang X. and Gong H. (2012). A new insight on the adsorption mechanism of amino-functionalized nano-Fe_3O_4 magnetic polymers in Cu (II), Cr (VI) co-existing water system. *Chemical Engineering Journal*, **183**, 180–191, https://doi.org/10.1016/j.cej.2011.12.055

Singh A., Garg P., Srivastava P. and Sharma V. P. (2023). Technological interventions for wastewater treatment: monitoring and management. In: Advanced Application of Nanotechnology to Industrial Wastewater, M. P. Shah (ed.), Ankleshwar, India, Springer Nature Singapore, Singapore, pp. 353–363.

Soares S. F., Simoes T. R., Trindade T. and Daniel-da-Silva A. L. (2017). Highly efficient removal of dye from water using magnetic carrageenan/silica hybrid nano-adsorbents. *Water, Air, & Soil Pollution*, **228**, 1–11, https://doi.org/10.1007/s11270-017-3281-0

Sousa F. L., Daniel-da-Silva A. L., Silva N. J. O. and Trindade T. (2015). Bionanocomposites for magnetic removal of water pollutants. *Eco-Friendly Polymer Nanocomposites: Chemistry and Applications*, **74**, 279–310, https://doi.org/10.1007/978-81-322-2473-0_9

Sumi L. and Thomsen S. (2001). Mining in remote areas, issues and impacts. Minind Watch Canada/Mines Alerte.

Tchobanoglus G., Burton F. and Stensel H. D. (2003). Wastewater engineering: treatment and reuse. *American Water Works Association Journal*, **95**(5), 201.

Thanh D. N., Novák P., Vejpravova J., Vu H. N., Lederer J. and Munshi T. (2018). Removal of copper and nickel from water using nanocomposite of magnetic hydroxyapatite nanorods. *Journal of Magnetism and Magnetic Materials*, **456**, 451–460, https://doi.org/10.1016/j.jmmm.2017.11.064

Tkaczyk A., Mitrowska K. and Posyniak A. (2020). Synthetic organic dyes as contaminants of the aquatic environment and their implications for ecosystems: a review. *Science of the Total Environment*, **717**, 137222, https://doi.org/10.1016/j.scitotenv.2020.137222

Vakili M., Deng S., Cagnetta G., Wang W., Meng P., Liu D. and Yu G. (2019). Regeneration of chitosan-based adsorbents used in heavy metal adsorption: a review. *Separation and Purification Technology*, **224**, 373–387, https://doi.org/10.1016/j.seppur.2019.05.040

Wu Q., Cui Y., Li Q. and Sun J. (2015). Effective removal of heavy metals from industrial sludge with the aid of a biodegradable chelating ligand GLDA. *Journal of Hazardous Materials*, **283**, 748–754, https://doi.org/10.1016/j.jhazmat.2014.10.027

Xu J., Yang H., Fu W., Du K., Sui Y., Chen J. and Zou G. (2007). Preparation and magnetic properties of magnetite nanoparticles by sol–gel method. *Journal of Magnetism and Magnetic Materials*, **309**(2), 307–311, https://doi.org/10.1016/j.jmmm.2006.07.037

Xu J., Sun J., Wang Y., Sheng J., Wang F. and Sun M. (2014). Application of iron magnetic nanoparticles in protein immobilization. *Molecules*, **19**(8), 11465–11486, https://doi.org/10.3390/molecules190811465

Yang Q., Li Z., Lu X., Duan Q., Huang L. and Bi J. (2018). A review of soil heavy metal pollution from industrial and agricultural regions in China: pollution and risk assessment. *Science of the Total Environment*, **642**, 690–700, https://doi.org/10.1016/j.scitotenv.2018.06.068

Ye Z., Yin X., Chen L., He X., Lin Z., Liu C. and Wei Y. (2019). An integrated process for removal and recovery of Cr(VI) from electroplating wastewater by ion exchange and reduction–precipitation based on a silica-supported pyridine resin. *Journal of Cleaner Production*, **236**, 117631, https://doi.org/10.1016/j.jclepro.2019.117631

Zheng X., Wu D., Su T., Bao S., Liao C. and Wang Q. (2014). Magnetic nanocomposite hydrogel prepared by ZnO-initiated photopolymerization for La (III) adsorption. *ACS Applied Materials & Interfaces*, **6**(22), 19840–19849, https://doi.org/10.1021/am505177c

Zourou A., Ntziouni A., Adamopoulos N., Roman T., Zhang F., Terrones M. and Kordatos K. (2022). Graphene oxide-$CuFe_2O_4$ nanohybrid material as an adsorbent of Congo red dye. *Carbon Trends*, **7**, 100147, https://doi.org/10.1016/j.cartre.2022.100147

doi: 10.2166/9781789063592_0129

Chapter 8

Feasibility of nanomaterials to support electroactive microbes in nanobiohybrids

Zheng Chen[1,2], Yanqiong Zeng[1], Feng Wang[1], Peng Huang[1], Jian Li[1] and Yibin Chen[2]*

[1]*Wenzhou Medical University, School of Public Health & Management, Wenzhou, People's Republic of China*
[2]*Fujian Provincial Key Lab of Coastal Basin Environment, Fujian Polytechnic Normal University, Fuqing, People's Republic of China*
**Corresponding author: chenzheng_new@163.com*

ABSTRACT

Electrochemically active bacteria (EAB) harboring extracellular electron transfer capabilities display great potential for applications in wastewater treatment and energy recovery. However, limited biocatalytic efficiency can curtail their practical engineering applications. Building on the merits of nanomaterials (NMs) that are biocompatible with EAB, engineered nanobiohybrids that are synergistically integrated with specific bacterial components and NMs provide a novel design for engineering an advanced biocatalyst complex. In this chapter, we summarize the efficacy of NMs for their suitability in constructing nanobiohybrids. Favorable engineered nanobiohybrids attributes include high electrical conductivity, large specific surface area, desirable photocatalytic capability, stimulated production of cellular components related to electron transfer, and cytoprotection capabilities. Considering the mechanisms of interfacial electron transport and electron flux generation, the configuration of nanobiohybrids with NMs can be implemented through both endogenous and exogenous bioaugmentation, wherein nanobiohybrids are constructed at a single cell-based or biofilm-based scale with different interfacial connections. Finally, we summarize the current challenges for nanobiohybrid construction and envision the future prospects for nanobiohybrid development and applications.

Keywords: electrochemically active bacteria, nanobiohybrids, nanomaterials, bioaugmentation, wastewater treatment, energy recovery.

8.1 INTRODUCTION

Electrochemically active bacteria (EAB) harboring extracellular electron transfer (EET) capabilities display great potential for environmental applications, including bioremediation of pollutants and renewable/carbon-neutral energy production (Chen *et al.*, 2022; Zhao *et al.*, 2022). Several recent advances for wastewater treatment and energy recovery via biocatalytic approaches have been achieved; however, these approaches face numerous challenges. For example, a primary challenge results from inefficient EET capabilities of natural EAB and high energy barriers for reactions under real-world environmental conditions (Zhao *et al.*, 2022). These limitations become more prominent when applied to practical industrial applications. To address this challenge, it is imperative to develop effective approaches to engineer advanced biocatalysts that possess finely tuned activation energies to promote the forward reaction and accommodate energetically intensive processes.

In the past decade, the rapid development of nanotechnology has allowed the advancement of novel biomaterials. Nanomaterials (NMs) featuring (semi)conductance, photosensitivity and redox properties are able to form favorable associations with EAB (Chen *et al.*, 2022; Dong *et al.*, 2020a). These advances make possible the synthesis of reliable and efficient biocatalyst complexes incorporating novel nanobiohybrids that synergistically integrate specific bacterial components (e.g., individual cell or biofilm) and one or more types of NMs (Guo *et al.*, 2020). These coupled processes provide the potential to expand extrinsic functional properties that surpass their natural capabilities, thereby contributing to enhanced reaction efficacy to overcome several of the inefficiencies associated with sequentially independent processes (Dong *et al.*, 2020a, 2020b). Correspondingly, their intimate cooperation might broaden the substrate ranges for microbial assimilation and decreased reaction times, as well as eliminate the harsh conditions required for several conventional chemical or biological catalysis. Thus, specific NMs could be applied as effective green synthetic practice capable of functioning under milder reaction conditions with EAB cells.

In this chapter, recent advances in engineered nanobiohybrids to enhance or enable new biological functions are systematically summarized, including the types of NMs and the related methodologies employed in designing and optimizing NM-biointerfaces. Further, the different scales (e.g., single cell-based versus biofilm-based) of engineered nanobiohybrids are illustrated. Therein, the choice of both biological and synthetic components has a crucial impact on the biofunctionality of engineered nanobiohybrids. Overall, nanobiohybrids could greatly improve the efficacy of NMs used for environmental applications and correspondingly enhance nanocatalytic reactions to achieve efficient wastewater treatment and energy recovery technologies.

8.2 INHERENT BOTTLENECKS FOR ELECTRON TRANSFER IN NATURAL EAB CELLS

In terms of EET processes occurring in natural EAB, there are two main pathways consisting of direct and indirect electron transport processes (Logan *et al.*, 2019). Electron transport occurring in the direct pathway is launched through conduction with the conductive and nano electrical pili (e-pili) and c-type cytochromes (c-Cyts) (Logan *et al.*, 2019). In contrast, electron transport occurring in the indirect pathway is manipulated through foreign mediators (e.g., special electrically conductive media or electron shuttles) (Logan *et al.*, 2019). Notably, the path of transmembrane electron transfer in natural EAB cells across the cell envelope follows the inner membrane → periplasm → outer membrane pathway (Logan *et al.*, 2019). The complex redox environment in the periplasmic space, persistent electron escape and sluggish electron-export induce an inefficient export of intracellular electrons, thereby limiting the electron flux to extracellular acceptors. Thus, appropriate optimizations aimed at intracellular and/or extracellular interfacial modifications are essential to minimize the energy loss at the junction sites between intracellular proteins or the outer-membrane with extracellular electron acceptors.

Rapid development of nanotechnology in the past few decades has provided alternative modification strategies for enhancing or exploiting completely new functions within biological systems. Especially, the size of artificial NMs can be effectively matched with that of subcellular and molecular components of EAB (Chen *et al.*, 2020b). Considering the various mechanisms associated with EET processes, the merits of NMs are expected to diminish the inherent bottleneck of electron transfer in natural EAB cells. For example, incorporation of specific functional NMs could enable a closer electrical contact and ensure a low-energy barrier charge injection. The concept of integrating bacteria with NMs offers an alternative pathway for engineering nanobiohybrids for wastewater treatment and energy recovery. Therefore, integration of functional NMs with living EAB for constructing nanobiohybrid systems is expected to promote a decoupling of cellular electron-transport machineries and achieve a synergistic bioelectrocatalytic effect.

8.3 NANOMATERIAL SELECTION FOR CONSTRUCTING EFFICIENT NANOBIOHYBRIDS

8.3.1 Favorable electrical conductivity of NMs

8.3.1.1 Metal/metal oxide-based NPs and conductive carbon-based NMs

Due to redox potential discrepancies among substances, electron flow between different substances is thermodynamically permitted (Logan *et al.*, 2019). Thus, electron transfer is theoretically permitted across the microbial cyt-protein complex and the conduction band of (semi)conductive NMs. This is due to the wide range of redox potentials between the cyt-protein complex in the cytoplasmic membrane of microbial cells and (semi)conductive NMs. Considering the mechanism associated with the electrical conductivity-derived EET process, NMs with a higher electrical conductivity are commonly able to more easily transmit the electrons delivered from microbes to terminal electron acceptors (profiled in Figure 8.1a). Thus far, a variety of electrically conductive metal/metal oxides-based NPs (e.g., Au, Pd, and Fe_3O_4) and conductive carbon-based NMs (e.g., carbon dots [CDs], graphene and carbon nano tubes [CNTs]) have been extensively used for the configuration of nanobiohybrids (Chen *et al.*, 2022).

Previously, Malvankar *et al.* (2012) confirmed that the e-pili of *Geobacter sulfurreducens* featured a prominent metal-like electrical conductivity (up to 188 Ms/cm) due to the dense configuration of the OmcS protein in e-pili. This implies that certain metal-based conductive NMs might display an equivalent electrically conductive effect with multi-heme *c*-Cyt OmcS protein. Furthermore, Liu *et al.* (2015) demonstrated that a syntrophism was originally unable to participate between OmcS-defected *G. sulfurreducens* and *G. metallireducens*, but was restarted after supplementing with conductive

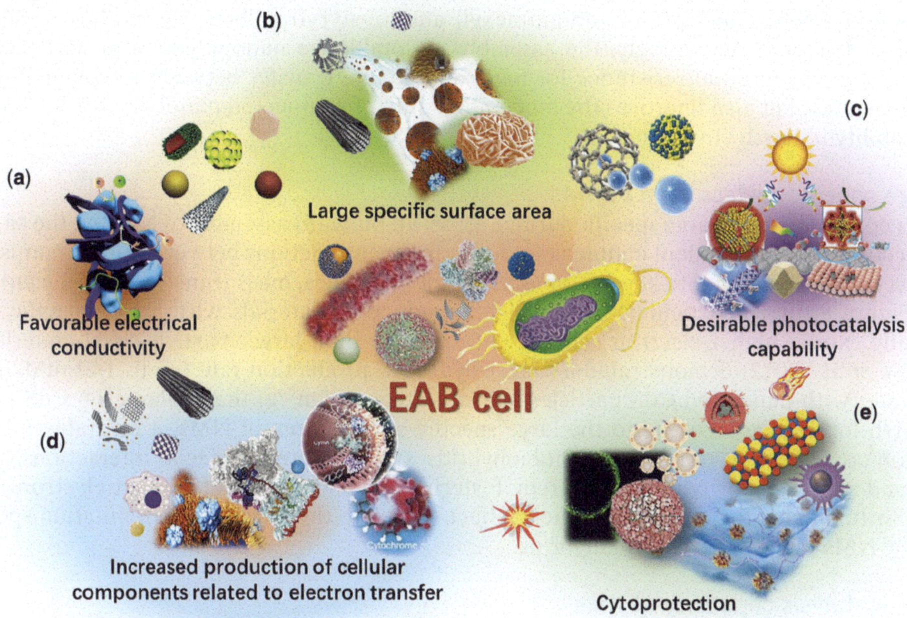

Figure 8.1 Important properties of NMs for optimizing engineered nanobiohybrids, including favorable electrical conductivity (a), large specific surface area (b), desirable photocatalysis capability (c), increased production of cellular components related to electron transfer (d), and special cytoprotective functions used for protecting engineered nanobiohybrids (e).

magnetite NPs. This demonstrates that Fe_3O_4 NPs could serve as a substitute for the OmcS protein to participate in EET processes. Analogous to conductive metal-based NMs, several conductive NMs such as CDs and CNTs can be strongly bound to the periplasm and/or outer membrane. For example, Yang *et al.* (2020) found more than a 6.5-fold increase in the maximum power density of bioelectricity generation by *S. oneidensis* MR-1 cells that were densely aligned with CDs compared to that of the wild strain. Hence, effective interfacial intercalation with conductive metal/metal oxides or carbon-based NMs are effective avenue for improving the performance of electron export across intracellular interfaces.

8.3.1.2 Conductive organic nanopolymers

Soluble organic nanopolymers are an appealing alternative to improve the orbital mixing for enhancing interfacial conductivity of EAB. One approach to execute this strategy is through doping with macromolecular nanopolymers (e.g., poly(3, 4-ethylenedioxythiophene) [PEDOT], polypyrrole [PPy] and polydopamine [PDA]) on the surface of EAB cells to form a large heterojunction complex to connect the interface between the outer membrane and extracellular electrodes (Chen *et al.*, 2022). The tailored structures greatly broaden the channels for charge transfer. Another approach is intercalating micromolecular conductive oligomers into the lipid bilayer of the cell envelope that is inherently electrically nonconductive (Apetrei *et al.*, 2019). This modulation effectively activates the conductive orbits for electron transport and improves the charge transfer orientation, resulting in a considerable boost in transmembrane electron transfer. For example, Sivakumar *et al.* (2014) reported a 1.2-fold higher enhancement, compared to the wild strain, for extracellular reduction of ferrihydrite by a biosystem with *S. oneidensis* MR-1 cells that were inward assembled with the oligoelectrolyte 4,4′-bis (4′-N,N-bis (6″-(N,N,N-trimethyl ammonium)hexyl) amino)-styryl) stilbene tetraiodide (DSSN+, an analogue for oligomer). Accordingly, the assembly of conductive nanopolymers in nanobiohybrids not only reduces the internal resistance for mass and charge transfer between nanobiohybrids and electrode receptors, but also improves the biocompatibility of engineered nanobiohybrids to sustain a longer durability in harsh environments.

8.3.2 Large specific surface area of NMs

Since NMs commonly have large specific surface areas, an outward assembly of NMs on the surface of microbial cells provides abundant contact sites for intimate interactions between microorganisms and electron acceptors that facilitate adsorption of reactant molecules, mass transfer, and efficient charge transfer (Figure 8.1b). In particular, smaller-sized conductive materials with larger specific surface areas usually exhibit higher electron transfer performance than larger particle-sized materials. For instance, Yu *et al.* (2012) demonstrated a higher hydrogen production rate (up to 1.21 mM/h) when splitting water with nanosized CdS particles (surface area = 46 m^2/g) than submicron CdS particles (<0.35 mM/h for 17 m^2/g). Owing to the large specific surface area of NMs, a favorable interfacial construction configuration of NMs in nanobiohybrids will promote interfacial interactions between microbes and substrates or reactants, thereby fostering an efficient direct-contact electron transfer process. This highlights the notable interfacial effect related to the affinity and orientation specificity mediated by NMs in designed nanobiohybrids.

8.3.3 Photocatalysis capability of NMs

8.3.3.1 Metal-based semiconductor NPs

The electron flux of most conventional bioelectrochemical systems is extremely limited because the supply of electrons mainly relies on microbial degradation of low-molecular-weight organic substrates. To increase the electron flux and ensure more efficient extracellular reduction or energy production processes, living photosensitive nanobiohybrids are engineered through coating specific metal oxide or metal sulfide semiconductor NPs on the surface of a given EAB to harvest photoelectrons under illuminated conditions. Therein, photoelectrons are generated from photoelectron–hole pairs in the illuminated semiconductor NPs (Figure 8.1c), accompanied by the capture of oxidizing holes scavenged by reductive sacrificial compounds, such as cysteine (Cestellos-Blanco *et al.*, 2020). Since photoelectrons are highly reductive and their potentials are more negative than those of most biological compounds (Dong *et al.*, 2020b), the photoelectrons excited from illuminated semiconductor NPs are readily accepted by EAB to drive specific reduction processes. Currently, a variety of photosensitive nanobiohybrids have been engineered through pairing various host EAB with specific metal-based semiconductor NPs (e.g., TiO_2, Cu_2O, CdS, and ZnS) that contribute an additional electron flux to desired reduction processes under illuminated conditions (Chen *et al.*, 2022; Zhao *et al.*, 2022). Among them, CdS NPs are a highly effective engineered photosensitive nanobiohybrid (Table 8.1). This is mainly attributed to CdS NPs having more favorable biocompatibility with a variety of cellular units,

Table 8.1 Summary of representative photosensitive nanobiohybrids used for wastewater treatment and energy recovery.

NM	Strain/Enzyme	Substrates/ Products	Remarkable Results	References
CdS NPs	*G. sulfurreducens*	Methyl orange (MO)/ $CO_2 + H_2O$	Nanobiohybrid achieved a maximum kinetic rate constant of 1.441 h^{-1} for methyl orange biodecolorization	Huang *et al.* (2019)
CdS NPs	*Moorella thermoacetica*	CO_2/acetate	Key enzymes involved in WLP and ATP production were upregulated in illuminated nanobiohybrid cells	Zhang *et al.* (2020)
CdS NPs	*Clostridium autoethanogenum*	CO_2/acetate	Expression of flavin-binding proteins increased with the activation of genes associated with the WL pathway and energy conservation system	Jin *et al.* (2021)
CdS NPs	*Methanosarcina barkeri*	CO_2/CH_4	CH_4 generation (0.19 μmol/h) occurred at a high quantum efficiency (0.34%), which is comparable to the quantum efficiency of plants	Ye *et al.* (2019)
g-C_3N_4	*R. eutropha*	CO_2/PHB	A 1.2-fold increase in PHB production achieved by a C_3N_4–*R. eutropha* nanobiohybrid in the absence of an external supply of electron donor	Xu *et al.* (2019)
NH_2-doped CDs	*S. oneidensis* MR–1	Cr(VI)/Cr(III)	Catalytic reduction capacity for Cr(VI) from nanobiohybrid was 40-fold that of wild strain	Li *et al.* (2022)
Red-emission CDs (RCDs)	*S. xiamenensis* (*S.x.*)	*S.x.*-based electrode/ RCDs-*S.x.* nanobiohybrid-based electrode	Illumination of nanobiohybrid electrode lead to a 4.8-fold increase in current density compared to *S.x*-based electrode	Liu *et al.* (2020)

WLP: Wood–Ljungdahl pathway; ATP: adenosine triphosphate; C_3N_4: carbon nitride; PHB: polyhydroxybutyrate.

as well as their desirable photocatalysis effect under visible-light illumination, which could circumvent the adverse effects resulting from ultraviolet light that are prominent among other competing metal-based semiconductor NPs (Dong *et al.*, 2020b).

8.3.3.2 Carbon-based semiconductor NPs

The leaching of toxic metals resulting from photocorrosion of metal-based semiconductor NP-assembled nanobiohybrids creates a potential toxicity for EAB cells. In addition, cellular damages might result from high-intensity illumination (e.g., UV-irradiation) required for effective excitation of some photosensitive biohybrids (Svitkova *et al.*, 2017). In contrast, several carbon-based semiconductor NMs (e.g., C_3N_4 and CDs, shown in Table 8.1) represent a more bio-friendly photosensitizer for photocatalysis applications that might be more suitable for engineered photosensitive nanobiohybrids. Favorable attributes of carbon-based semiconductors NMs include a higher water solubility, photostability, and tunability compared to that of metal-based semiconductors (Chen *et al.*, 2022). CDs-coated *S. xiamenensis* cells possessed a higher regulation of intracellular flavins responsible for boosting EET, which generated a higher interfacial conductivity between EAB and extracellular electrodes in a cyclic voltammetry test (Liu *et al.*, 2020). This corroborates that CDs not only sustain the common photocatalytic function found for conventional metal-based semiconductors, but also offer favorable biological and physical associations between EAB cells and extracellular acceptors. Thus, the merits of carbon-based semiconductors are worthy of greater exploitation in engineering more advanced nanobiohybrids in the future.

8.3.4 NMs stimulate production of cellular components related to electron transfer

8.3.4.1 Increased production of c-Cyts in the presence of NMs

Electron-transporting c-Cyts serve as important transmembrane electron-export carriers (Heidary *et al.*, 2020). Evidence suggests that the responses of e-pili and c-Cyts from EAB might be altered by the addition of (semi)conductive NMs to establish an effective communication between EAB and extracellular electron acceptors (Figure 8.1d). For instance, a previous investigation demonstrated that up-regulation of c-Cyts (e.g., OmcJ, PgcA, and OmcK) and pili-relating PilA in *G. sulfurreducens* was stimulated by direct contact with (semi)conductive NMs (Fe_3O_4, Fe_2O_3 and TiO_2 NPs) (Dong *et al.*, 2020a). Similarly, recent studies examining engineered nanobiohybrids confirmed a more than 3-fold up-regulation of c-Cyts in a system of *S. oneidensis* MR-1 cells that were configured with NH_2-doped CDs (Li *et al.*, 2022; Yang *et al.*, 2020). The increased production of c-Cyts with the addition of (semi)conductive NMs manifested a positive response for electron-transporting cellular components to the foreign stimuli of (semi)conductive NMs.

8.3.4.2 Increased EPS production in the presence of NMs

Several conductive NMs (e.g., Fe_3O_4 NPs, and CDs) appear to stimulate extracellular polymeric substance (EPS) production (Yang *et al.*, 2020; Ye *et al.*, 2018). EPS is often comprised of a large amount of extracellular redox enzymes and electron-shuttling compounds (Xiao *et al.*, 2017). This highlights a facilitated EET process administrated by secreted EPS components (Figure 8.1d). Recent studies demonstrated that direct contact with Fe_3O_4 NPs led to the formation of augmented the EPS in biofilms, thereby propagating long-distance electron transfer across larger scales (Wang *et al.*, 2018). Likewise, our group previously reported a higher production (increased by 0.7 times versus wild strain) of flavins in nanobiohybrids comprised of *S. oneidensis* MR-1, NH_2-doped CDs carbon dots, and polydopamine (Li *et al.*, 2022). The increased flavin production enhanced shuttling of inward-to-outward delivery of electrons in the nanobiohybrid. These observations offer new insights into the enhanced electron transfer performance resulting from beneficial molecular and physiological responses to coexisting (semi)conductive NMs.

8.3.5 Special functionalized NMs used for cytoprotection in engineered nanobiohybrids

Exploitation of living biomaterials in the past decade has witnessed great advancements in the use of NMs for cytoprotection in engineered nanobiohybrids. According to the inherent properties of specific materials, functionalized NMs used for cytoprotection are divided into four characteristic types: biomimetic inorganic NPs, nano-hydrogels, hybrid coordination NMs, and nanoenzymes (illustrated in Figure 8.1e). These NMs exert their protective effects by first constructing a unique shell as a shield for EAB cells, and then establishing a collaboration with various cellular functions to block the external stressor, such as reactive oxygen species (ROS), heat, and pH conditions.

8.3.5.1 Biomimetic inorganic NPs

To sustain the reliable functioning of a given nanobiohybrid system, biomimetic inorganic NMs (e.g., $CaCO_3$, SiO_2, and iron oxide) are often exploited in bioaugmentation as a protective coating for biological systems (Chen *et al.*, 2022; Wei *et al.*, 2018). These inorganic biomineral NPs are abundant in nature making them popular as an artificial cell-coating material. Because of their biocompatibility properties and well-ordered architectures, these inorganic NMs are ideal for coating nanobiohybrids. Wei *et al.* (2018) engineered nanobiohybrids of CdS NPs-*E. coli* that were firstly immersed with polyelectrolyte sodium polystyrene sulfonate and poly (diallyldimethylammonium chloride) solutions, and thereafter coated with silicic acid. The resulting SiO_2 NP-coated nanobiohybrid resisted ROS damage and produced highly efficient H_2 generation for 96 h under natural aerobic conditions. Thus, in-situ surface coating with biomimetic inorganic NMs has excellent potential for constructing a defensive system as a protective function for nanobiohybrids.

8.3.5.2 Nano-hydrogels

As the outer-membranes of EAB are negatively charged, in-situ encapsulation with positively charged nano-hydrogels via electrostatic attraction provides an antioxidant protective shell for nanobiohybrids (Cestellos-Blanco *et al.*, 2019). Multiple examples demonstrate that nanobiohybrids with ROS-defensive abilities can be fabricated through LBL-coating technology, in which a compact polymeric shell consisting of nano-hydrogels (such as hyaluronic acid, poly-l-lysine (PLL) and PVA) encapsulates an EAB cell (Liu *et al.*, 2019). Creation of such a defensive protection system allows nanobiohybrid functionalization to withstand harsh environmental conditions.

8.3.5.3 Hybrid coordination NMs

Porous coordination network materials [e.g., metal-organic frameworks (MOFs) and metal-phenolic networks (MPNs)] with complex structural topologies, multiple adjustable pores and tunability provide superior attributes when incorporated into living nanobiohybrids, such as resistance to heat, UV radiation, osmotic pressure, and mechanical pressure (Guo *et al.*, 2020). In this approach, MOFs and MPNs are formed around an EAB cell as a cell wall-like shell. EAB are fully retained inside the hybrid coordination materials by generating macromolecular defects during crystallization, hence eliminating the pore size limitation of the pre-synthesized MOF/MPN materials. The highly efficient heterogeneous nucleation sites on the cell surface lead to the rapid buildup of MOF coatings in which the living cells remain viable. For instance, Li *et al.* (2015) created MPN-wrapped yeast cells to protect against UV irradiation and ROS damage. Likewise, Ji *et al.* (2018) developed a special nanobiohybrid consisting of a *M. thermoacetica* cell coated with a MOF monolayer (a composite of $Zr_6O_4(OH)_4(BTB)_2(OH)_6(H_2O)_6$) that effectively resisted ROS damage and enabled acetate production by CO_2 fixation.

8.3.5.4 Artificial nanoenzymes

Nanozymes are specific NMs with enzyme-like characteristics. Among various NMs, artificial nanoenzymes display highly effective defensive functions comparable to natural enzymes, such as superoxide dismutase (SOD), peroxidase (POD), and catalase (CAT), that can resist ROS damage

(Ighodaro & Akinloye, 2018). Nanoenzymes involve relatively simple synthesis procedures, stable morphology and tunable specific catalytic activities, which render them desirable compared to hybrid coordination materials (Section 8.3.5.3) that are complicated to synthesize and incur high manufacturing costs (Wang *et al.*, 2020). Hence, coating nanobiohybrids with nanoenzymes is appealing for creation of high-performance nanobiohybrids. Nanoenzyme-coated nanobiohybrids comprised of EAB and semiconductor NPs have ROS radicals separated from photoelectron–hole pairs that are rapidly blocked and caged by nanoenzymes upon illumination, along with an intermediate release of photoelectrons. Therein, artificial nanosized minerals (e.g., Fe_3O_4, Mn_3O_4, Co_3O_4, and V_2O_5 NPs) could serve as enzyme mimics and are capable of displaying POD/SOD/CAT-mimicking activities to resist ROS attack (Jo *et al.*, 2020). For example, the successful encapsulation of CdS NPs-*Thiobacillus denitrificans* nanobiohybrids with Mn_3O_4 artificial nanosized minerals not only ensured highly efficient (>90%) ROS decomposition, but also generated an appreciable denitrification effect (>28% via photoelectrotrophic denitrification coupled with a 78% decrease in N_2O emission) (Chen *et al.*, 2020a).

8.4 ASSEMBLY PROTOCOLS AND SYNTHETIC STRATEGIES EMPLOYED FOR DIFFERENT FUNCTIONAL NANOBIOHYBRID SYSTEMS

Bioaugmentation at the molecular and cellular level is an emerging concept, largely owing to the rapid progress in materials research and nanotechnology in the past decade. Hitherto, advanced functional materials have been applied at the molecular, single cellular, and simple multicellular levels for enhancing biofunctions, such as biostability and biocatalytic efficiency, and for introducing extrinsic functionalities to living systems, such as new metabolic pathways, electrical conductivity, and optical properties. Different hybrid patterns (internal or external) are created by introducing extrinsic NMs to construct specific nanobiohybrids for enhancing distinct biofunctions.

Considering the functionality, stability, and durability of nanobiohybrids, construction of nanobiohybrids is carried out through endogenous (internal) and exogenous (external) bioaugmentation approaches at different interfacial connections (Chen *et al.*, 2022; Guo *et al.*, 2020). The selected NMs are targeted to site-specific cellular locations on engineered nanobiohybrids using specialized modifications as described in Table 8.2. According to the desired type of configured living biointerface, the assembly objects are focused on internal bioaugmentation on individual cells, external bioaugmentation on individual cells and/or external bioaugmentation on biofilms (Chen *et al.*, 2022). Therein, the technology of acquiring precise biointerfaces through NM modification comprises the approaches of layer-by-layer (LBL) deposition, surface coating, immersion, and direct microbial uptake (shown in Figure 8.2).

8.4.1 Internal bioaugmentation on an individual cell scale

The very limited electron flux delivered from a single EAB cell to extracellular acceptors is an inherent bottleneck incurred by natural EAB systems. The lack of imperative conductive c-Cyts in the periplasm (Dong *et al.*, 2020b; Zhao *et al.*, 2022) causes a disruption of the intracellular electron flow and impedes the efficient electron transmission across the cell envelope at the single EAB cell level. Correspondingly, engineered conductive NMs can be intercalated as an interfacial modification across the periplasm and cell envelopes of a natural EAB cell, aiming to enhance the cytomembrane's permeability and stimulate microbial metabolism at the single cell level. This is accomplished by introducing engineered conductive NMs (such as Pd, Au, and CNTs, as well as conductive oligomers such as DSSN+) into a wild EAB cell through conductive NMs-intercalation (Zhang *et al.*, 2018). Owing to these conductive NM-anchors, an efficient pathway for electron transport is thereafter established across the periplasmic space that was originally nonconductive due to the lack of essential conductive c-Cyts (Apetrei *et al.*, 2019; Li *et al.*, 2022; Sun *et al.*, 2010; Wu *et al.*, 2011). The intercalation techniques consist of microbial mineralization and intracellular internalization for

Table 8.2 Summary of different types of nanobiohybrids engineering from different interfacial connections at different scales.

Interface	NM	Living Component	Remarkable Results	References
Internal bioaugmentation on individual cells	Pd NPs	*Desulfovibrio desulfuricans*	A current increase of more than 232 ± 65 nA was achieved in a high Pd^0 loading bio-complex	Wu *et al.* (2011)
	Au NPs	*G. sulfurreducens*	Current density and substrate removal increased by 44% and by 2.2 times, respectively	Chen *et al.* (2018)
	DSSN+	*S. oneidensis* MR-1	A greater than 14% increase in extracellular ferrihydrite reduction occurred with cell-DSSN+ interaction	Sivakumar *et al.* (2014)
	CAT and SOD-coated MOF	HeLa cells	HeLa cells resist toxic ROS for up to a week and retain cellular activity at low pH condition	Lian *et al.* (2017)
External bioaugmentation on individual cells	PPy	*S. oneidensis* MR-1	Power density produced by PPy-coated cell was 14.1 time higher than that of wild strain	Song *et al.* (2017)
	CNTs	*S. oneidensis* MR-1	Nitrobenzene reduction efficiency increased by 74% compared to wild strain	Yan *et al.* (2014)
	CDs	*S. oneidensis* MR-1	CDs-fed cells increased maximum current value by more than 7-fold	Yang *et al.* (2020)
	MOF	*M. thermoacetica*	MOF-wrap reduced strain survival rate by 5-fold and enables CO_2-to-acetate conversion under oxidative stress	Ji *et al.* (2018)
External bioaugmentation on biofilms	Graphene–CNT	*S. oneidensis* MR-1	A greater than 15-fold increase of maximum power density (97.9 μW/cm²) was obtained with hybrid biofilms	Zhao *et al.* (2015)
	GO–PVA	*S. xiamenensis*	Reaction time for Cr(VI) bioreduction decreased by 39 h when amended with composited biofilms	Luo *et al.* (2019)

Note: DSSN+ [oligoelectrolyte (4,4′-bis (4′-N,N-bis (6″-(N,N,N-trimethyl ammonium)hexyl) amino)-styryl) stilbene tetraiodide] is an analogue for the specific oligomer.

in-situ assembly of conductive NMs at the intracellular periplasmic outer-space membrane junction sites (Chen *et al.*, 2022).

Internal interfacial optimization not only enables an EAB cell to generate greater intracellular electron fluxes, but also induces higher ATP levels, higher organic substrate utilization and more efficient transmembrane transfer of intracellular electrons (Yang *et al.*, 2020). Moreover, the innovative work of Lian *et al.* (2017) revealed that specific enzyme-loaded MOF NMs incorporating

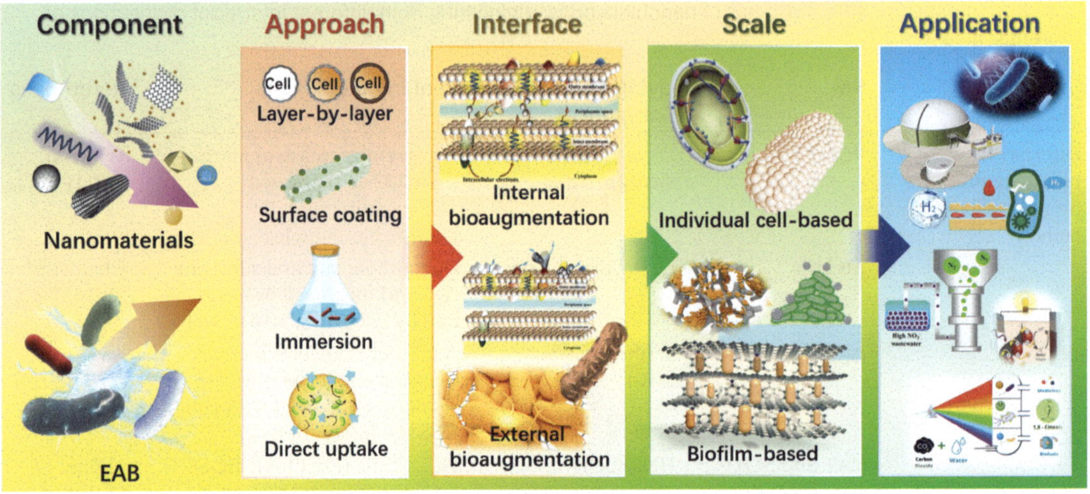

Figure 8.2 Functional nanobiohybrids integrating NMs and EAB cells are synthesized by different approaches using internal/external bioaugmentation at different scales (individual cell versus biofilm).

CAT and SOD could be taken up by living cells to develop advanced nanobiohybrid systems with high efficiency and long-term intracellular activity. These engineered nanobiohybrids were also capable of sustaining their protection capacity by mediating ROS detoxification for up to one week. Internal bioaugmentation methods that implant conductive NMs and functional MOFs into EAB cells can effectively eliminate the intrinsic limitation of inefficient intracellular electron export of natural EAB, while also initiating an efficient electron-exporting capacity for extracellular reactions.

8.4.2 External bioaugmentation on an individual cell scale

External bioaugmentation has been employed to (1) increase membrane–electrode interfaces to increase electrical conductivity, (2) harvest more photoelectrons, and (3) strengthen cytoprotection during fabrication of specific functional nanobiohybrids (Chen *et al.*, 2022). A key feature of nanobiohybrids is the ability to transmit electrical signals over long distances, which is often influenced by a wide range of directional and dynamic interactions. To achieve the above-stated objectives, a series of strategies have been demonstrated as effective controllable assembly mechanisms, such as hydrogen bonding, electrostatic interactions through π–π stacking or ionic interactions and micelle formation (Tibbitt & Langer, 2017). Herein, the methodology of LBL deposition is notable as it is able to precisely control shell thickness and build site-specific and core–shell platforms (Chen *et al.*, 2022; Zhao *et al.*, 2022).

Because of the specificity and functionality of NMs in supramolecular chemistry and colloid science, NMs could provide an operable approach and flexible framework for precisely engineering intermolecular interactions and multiscale structures in nanobiohybrids. To increase the electrical conductivity at the membrane–electrode interface, conductive NMs are selected initially based on their mechanical properties that could restore function while minimizing any adverse effects. For instance, an outward coating with positive charge and conductive NMs (like CNTs and CDs) on the surface could be attached to the negatively charged EAB cell, thereby providing the nanobiohybrids with a location for storing charge while also improving the electrical conductivity and mechanical strength of the nanobiohybrid (Zhao *et al.*, 2022). Thus, external bioaugmentation with conductive NMs confers several desirable attributes to nanobiohybrids, such as developing a reliable architecture and multifaceted functionality.

Considering the limited electron flux supplied from microbial metabolism, incorporating semiconductor NMs (as described in Section 8.3.3) on the surface of an EAB cell to create a photosensitive nanobiohybrid is advantageous. This imparts EAB with emergent properties outside their biotic scope and harnesses solar–chemical pathways to enhance functions (Dong *et al.*, 2020a). In particular, illuminated nanobiohybrids coated with semiconductor NPs give a strong boost to non-photosynthetic bacteria and enable non-photosynthetic bacteria to make full use of light energy to supplement their metabolism (Chen *et al.*, 2019; Dong *et al.*, 2021). Currently, engineered photosensitive living nanobiohybrids show a promising potential for use in the synthesis of acetic acid, bioplastics and biofuels (Fang *et al.*, 2020b). Moreover, to advance the stability of photosensitive nanobiohybrids, LBL cell encapsulation may be used to eliminate ROS-induced cellular damage (as illustrated in Section 8.3.5) in nanobiohybrids. This approach mainly consists of configurations employing coatings of charged polymers, inorganic materials, hydrogel, MOF, and nanoenzyme (Zhao *et al.*, 2022). Thus, suitable external bioaugmentation methods can be used to impart enhanced survivability and new functionalities to EAB beyond what natural EAB evolution offers.

8.4.3 External bioaugmentation on the biofilm scale

As EET performance relies on dispersed single cell-based nanobiohybrids, it is usually less efficient and more time-consuming than an integrated cell network (Fang *et al.*, 2020a). Hence, a successful tactic is fabrication of bulk biofilm-based nanobiohybrids incorporating the synergistic efforts of individual cells to maximize EET efficiency. Biofilms developed under natural conditions are commonly thin and poorly aggregated resulting in an overall poor conductivity (Zhao *et al.*, 2022). Thus, improving the formation/stability of biofilms to enhance electron transfer efficiency is an increasing focus of nanobiohybrid research.

Engineering biofilm-based nanobiohybrids by grafting biofilms with biocompatible and electrically conductive nano-polymers (e.g., PVA and PDA) and NM-based scaffolds (e.g., GO, rGO, and special nanofibers) enables in-situ formation of highly conductive biocomposites at a larger scale (Chen *et al.*, 2022; Guo *et al.*, 2020). A series of investigations on abiotic/biotic coupling heterojunctions (e.g., nanostructured polymers and/or graphene–biofilms) revealed that overlapping heterojunction structures favored microbial substrate utilization and charge transfer, thereby boosting bio-electrical generation (Chen *et al.*, 2022; Zhao *et al.*, 2022). The orbits constructed in these overlapping heterojunction architectures are mostly layer-ordered and multidirectional, favoring the transport of microbial substrates and boosting surface/interface dynamic reactions (Wang *et al.*, 2016). The unique multi-dimensional architecture of these biofilm-based nanobiohybrids prominently facilitates electron transfer across larger spatial and temporal ranges, consequently decreasing energy losses associated with EET processes.

Luo *et al.* (2019) fabricated a unique biofilm-based nanobiohybrid-derived biocomposite consisting of *S. xiamenensis* biofilms, GO and PVA assembled in a commingled pattern for application in Cr(VI) reduction from wastewaters. The beneficial bioaugmentation effect resulted from the enlarged specific surface area of GO and the favorable biocompatibility of PVA to facilitate formation of a large biofilm mass/volume. The biocomposite was highly effective for ten successive reaction cycles and achieved a strong Cr(VI) reduction performance. Further, Song *et al.* (2017) reported that PPy-coated carbon cloth promoted formation of electroactive biofilm-derived nanobiohybrids with a high bacterial cell density that greatly enhanced the EET efficiency. Hence, these findings provide compelling evidence that engineered biofilm-based nanobiohybrids can enhance the functional utility of biocomposites for use in wastewater treatment and energy recovery.

8.5 FUTURE DIRECTIONS

8.5.1 Present challenges for nanobiohybrid development

To alleviate our dependence on fossil fuels and decrease the cost associated with pollution remediation, there is excellent potential to synthesize an increasing amount of bulk chemicals and remediate several pollutants through incorporating microbial technologies. For example, nanobiohybrids can harvest energy more efficiently from biochemical sources, which provides a great advantage for the utilization of living biomaterials. In particular, engineered nanobiohybrids exhibit great potential for revolutionizing the field of environmental pollution remediation and energy recovery. However, to ensure the practical, economic and sustainable use of nanobiohybrids, several challenges must be overcome in the design and optimization of nanobiohybrids to fulfill commercial needs. For instance, EAB cell-laden hydrogels could respond to changing oxygen concentrations, water tensions, and even chemical gradients by secreting growth cytokines to fulfill the metabolic demand of cells. With the integration of synthetic biology technologies, EAB could be genetically engineered to perform more complex functions with greater efficiencies. Correspondingly, manufacturing technologies need to be improved to allow the intimate integration of EAB and NMs at a low cost. It should be highlighted that current research based on nanobiohybrids for use in wastewater treatment and energy recovery is still primarily a laboratory-based study occurring on a temporal scale of days, rather than the long-term (years). Accordingly, field-scale studies under real-world environmental conditions are essential for advancing the emerging technologies to commercial operations.

8.5.2 Outlook for nanobiohybrid development

Future key directions targeting the design of advanced nanobiohybrids will benefit from a focus on five principle aspects as shown in Figure 8.3: (1) broadening the spectrum of organic substrates across microbial metabolic pathways and decoupling the cellular electron-transport machineries will ensure that EAB are able to liberate intracellular electrons through assimilating more diverse substrates. (2) inactivation/inhibition of EAB needs to be prevented to ensure the stability of living nanobiohybrids. Development of a self-healing system (i.e., self-perpetuating/sustainable) will provide cytoprotection and sustain the working lifespan of the nanobiohybrids even under inhospitable conditions, thereby leading

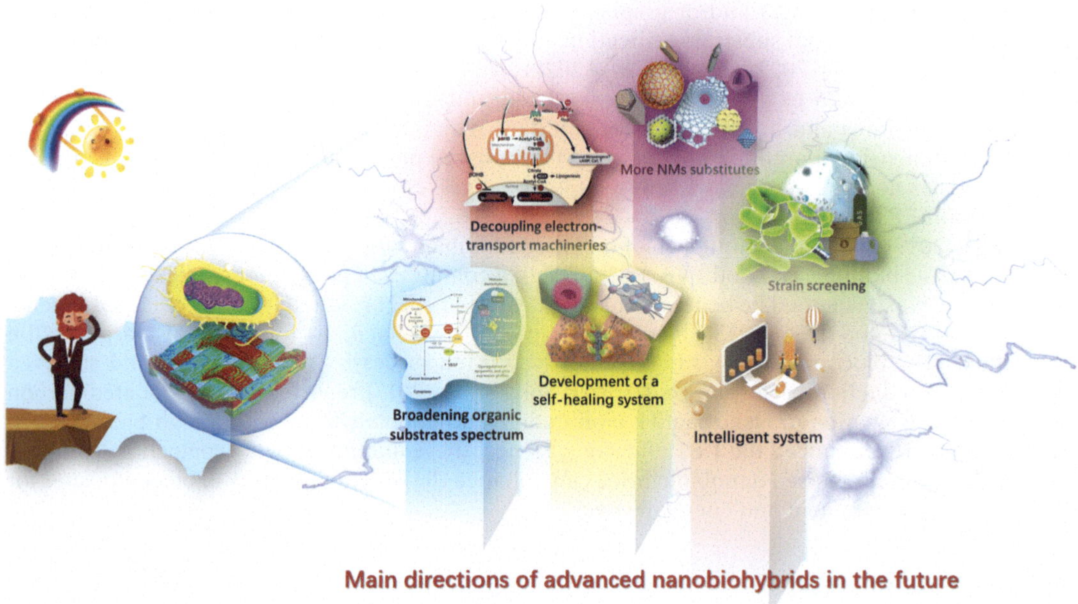

Figure 8.3 Overview of future design directions for advanced living nanobiohybrids.

to wider commercialization potential. (3) To achieve a high value/high efficiency nanobiohybrid, strain screening for more broad and diverse EAB is strongly warranted to create nanobiohybrids with novel functional attributes. (4) Moreover, the selection of NMs at present generally involves a single species, which limits the application scope. Adding substitutes might allow delicate optimization and designed features (e.g., oxygen permeability, porosity, and stability) that provide a favorable environment for EAB to perform their desired functions. (5) Finally, the future development of advanced nanobiohybrids could be directed to smart, controllable, and intelligent living biomaterials that contain an externally controllable switch-off function to meet remediation or manufacturing requirements. We envisage that these synthetic recombination methods will pave the way for highly efficient, resilient nanobiohybrids that can supersede conventional materials with more innovative and effective applications.

ACKNOWLEDGMENTS

The authors sincerely acknowledge the financial support received from Zhejiang Province Public Welfare Technology Application Research Project (LGF22E080002) and the National Natural Science Foundation of China (41807035). We thank Professor Randy A. Dahlgren (University of California - Davis, USA) for his careful review of the manuscript.

REFERENCES

Apetrei R.-M., Carac G., Ramanaviciene A., Bahrim G., Tanase C. and Ramanavicius A. (2019). Cell-assisted synthesis of conducting polymer - polypyrrole - for the improvement of electric charge transfer through fungal cell wall. *Colloids and Surfaces B: Biointerfaces*, **175**, 671–679, https://doi.org/10.1016/j.colsurfb.2018.12.024

Cestellos-Blanco S., Zhang H. and Yang P. D. (2019). Solar-driven carbon dioxide fixation using photosynthetic semiconductor bio-hybrids. *Faraday Discussions*, **215**, 54–65, https://doi.org/10.1039/C8FD00187A

Cestellos-Blanco S., Zhang H., Kim J. M., Shen Y. X. and Yang P. D. (2020). Photosynthetic semiconductor biohybrids for solar-driven biocatalysis. *Nature Catalysis*, **3**(3), 245–255, https://doi.org/10.1038/s41929-020-0428-y

Chen M., Zhou X., Liu X., Zeng R. J., Zhang F., Ye J. and Zhou S. (2018). Facilitated extracellular electron transfer of *Geobacter sulfurreducens* biofilm with in situ formed gold nanoparticles. *Biosensors and Bioelectronics*, **108**, 20–26, https://doi.org/10.1016/j.bios.2018.02.030

Chen Z., Liu Y., Zhang C., Pan Y., Han R., Chen Y., Shang X., Dong G. and Zhang J. (2019). Titanium dioxide nanoparticles induced an enhanced and intimately coupled photoelectrochemical-microbial reductive dissolution of As(V) and Fe(III) in flooded arsenic-enriched soils. *ACS Sustainable Chemistry & Engineering*, **7**(15), 13236–13246, https://doi.org/10.1021/acssuschemeng.9b02494

Chen X., Feng Q., Cai Q., Huang S., Yu Y., Zeng R. J., Chen M. and Zhou S. (2020a). Mn_3O_4 nanozyme coating accelerates nitrate reduction and decreases N_2O emission during photoelectrotrophic denitrification by *Thiobacillus denitrificans*-CdS. *Environmental Science & Technology*, **54**(17), 10820–10830, https://doi.org/10.1021/acs.est.0c02709

Chen Z., Zhou J., Wang Y. and Wang Y. (2020b). Nano on micro: tuning microbial metabolisms by nano-based artificial mediators to enhance and expand production of biochemicals. *Current Opinion in Biotechnology*, **64**, 161–168, https://doi.org/10.1016/j.copbio.2020.03.006

Chen Z., Zhang J., Lyu Q., Wang H., Ji X., Yan Z., Chen F., Dahlgren R. A. and Zhang M. (2022). Modular configurations of living biomaterials incorporating nano-based artificial mediators and synthetic biology to improve bioelectrocatalytic performance: a review. *Science of the Total Environment*, **824**, 153857, https://doi.org/10.1016/j.scitotenv.2022.153857

Dong G., Chen Y., Yan Z., Zhang J., Ji X., Wang H., Dahlgren R. A., Chen F., Shang X. and Chen Z. (2020a). Recent advances in the roles of minerals for enhanced microbial extracellular electron transfer. *Renewable and Sustainable Energy Reviews*, **134**, 110404, https://doi.org/10.1016/j.rser.2020.110404

Dong G., Wang H., Yan Z., Zhang J., Ji X., Lin M., Dahlgren R. A., Shang X., Zhang M. and Chen Z. (2020b). Cadmium sulfide nanoparticles-assisted intimate coupling of microbial and photoelectrochemical processes: mechanisms and environmental applications. *Science of the Total Environment*, **740**, 140080, https://doi.org/10.1016/j.scitotenv.2020.140080

Dong G., Han R., Pan Y., Zhang C., Liu Y., Wang H., Ji X., Dahlgren R. A., Shang X., Chen Z. and Zhang M. (2021). Role of MnO_2 in controlling iron and arsenic mobilization from illuminated flooded arsenic-enriched soils. *Journal of Hazardous Materials*, **401**, 123362, https://doi.org/10.1016/j.jhazmat.2020.123362

Fang K., Park O.-J. and Hong S. H. (2020a). Controlling biofilms using synthetic biology approaches. *Biotechnology Advances*, **40**, 107518, https://doi.org/10.1016/j.biotechadv.2020.107518

Fang X., Kalathil S. and Reisner E. (2020b). Semi-biological approaches to solar-to-chemical conversion. *Chemical Society Reviews*, **49**(14), 4926–4952, https://doi.org/10.1039/C9CS00496C

Guo Z., Richardson J. J., Kong B. and Liang K. (2020). Nanobiohybrids: materials approaches for bioaugmentation. *Science Advances*, **6**(12), 330–345.

Heidary N., Kornienko N., Kalathil S., Fang X., Ly K. H., Greer H. F. and Reisner E. (2020). Disparity of cytochrome utilization in anodic and cathodic extracellular electron transfer pathways of *Geobacter sulfurreducens* biofilms. *Journal of the American Chemical Society*, **142**(11), 5194–5203, https://doi.org/10.1021/jacs.9b13077

Huang S., Tang J., Liu X., Dong G. and Zhou S. (2019). Fast light-driven biodecolorization by a *Geobacter sulfurreducens*–CdS biohybrid. *ACS Sustainable Chemistry & Engineering*, **7**(18), 15427–15433, https://doi.org/10.1021/acssuschemeng.9b02870

Ighodaro O. and Akinloye O. (2018). First line defence antioxidants-superoxide dismutase (SOD), catalase (CAT) and glutathione peroxidase (GPX): their fundamental role in the entire antioxidant defence grid. *Alexandria Journal of Medicine*, **54**(4), 287–293, https://doi.org/10.1016/j.ajme.2017.09.001

Ji Z., Zhang H., Liu H., Yaghi O. M. and Yang P. (2018). Cytoprotective metal-organic frameworks for anaerobic bacteria. *Proceedings of the National Academy of Sciences*, **115**(42), 10582–10587, https://doi.org/10.1073/pnas.1808829115

Jin S., Jeon Y., Jeon M. S., Shin J., Song Y., Kang S., Bae J., Cho S., Lee J.-K., Kim D. R. and Cho B.-K. (2021). Acetogenic bacteria utilize light-driven electrons as an energy source for autotrophic growth. *Proceedings of the National Academy of Sciences*, **118**(9), e2020552118, https://doi.org/10.1073/pnas.2020552118

Jo S.-M., Zhang K. A. I., Wurm F. R. and Landfester K. (2020). Mimic of the cellular antioxidant defense system for a sustainable regeneration of nicotinamide adenine dinucleotide (NAD). *ACS Applied Materials & Interfaces*, **12**(23), 25625–25632, https://doi.org/10.1021/acsami.0c05588

Li W., Bing W., Huang S., Ren J. and Qu X. (2015). Mussel byssus-like reversible metal-chelated supramolecular complex used for dynamic cellular surface engineering and imaging. *Advanced Functional Materials*, **25**(24), 3775–3784, https://doi.org/10.1002/adfm.201500039

Li J., Wang F., Zhang J., Wang H., Zhao C., Shu L., Huang P., Xu Y., Yan Z., Dahlgren R. A. and Chen Z. (2022). Inward-to-outward assembly of amine-functionalized carbon dots and polydopamine to *Shewanella oneidensis* MR-1 for high-efficiency, microbial-photoreduction of Cr(VI). *Chemosphere*, **307**, 135980, https://doi.org/10.1016/j.chemosphere.2022.135980

Lian X., Erazo-Oliveras A., Pellois J.-P. and Zhou H.-C. (2017). High efficiency and long-term intracellular activity of an enzymatic nanofactory based on metal-organic frameworks. *Nature Communications*, **8**(1), 2075, https://doi.org/10.1038/s41467-017-02103-0

Liu F., Rotaru A.-E., Shrestha P. M., Malvankar N. S., Nevin K. P. and Lovley D. R. (2015). Magnetite compensates for the lack of a pilin-associated c-type cytochrome in extracellular electron exchange. *Environmental Microbiology*, **17**(3), 648–655, https://doi.org/10.1111/1462-2920.12485

Liu T., Wang Y., Zhong W., Li B., Mequanint K., Luo G. and Xing M. (2019). Biomedical applications of layer-by-layer self-assembly for cell encapsulation: current status and future perspectives. *Advanced Healthcare Materials*, **8**(1), 1800939, https://doi.org/10.1002/adhm.201800939

Liu S., Yi X., Wu X., Li Q. and Wang Y. (2020). Internalized carbon dots for enhanced extracellular electron transfer in the dark and light. *Small*, **16**(44), 2004194, https://doi.org/10.1002/smll.202004194

Logan B. E., Rossi R., Ragab A. and Saikaly P. E. (2019). Electroactive microorganisms in bioelectrochemical systems. *Nature Reviews Microbiology*, **17**(5), 307–319, https://doi.org/10.1038/s41579-019-0173-x

Luo Q., Chen Z., Li Y., Wang Y., Cai L., Wang L., Liu S., Wang Z., Peng Y. and Wang Y. (2019). Highly efficient and recyclable *Shewanella xiamenensis*-grafted graphene oxide/poly(vinyl alcohol) biofilm catalysts for increased Cr(VI) reduction. *ACS Sustainable Chemistry & Engineering*, **7**(14), 12611–12620.

Malvankar N. S., Tuominen M. T. and Lovley D. R. (2012). Biofilm conductivity is a decisive variable for high-current-density *Geobacter sulfurreducens* microbial fuel cells. *Energy & Environmental Science*, **5**(2), 5790–5797, https://doi.org/10.1039/c2ee03388g

Sivakumar K., Wang V. B., Chen X., Bazan G. C., Kjelleberg S., Loo S. C. J. and Cao B. (2014). Membrane permeabilization underlies the enhancement of extracellular bioactivity in *Shewanella oneidensis* by a membrane-spanning conjugated oligoelectrolyte. *Applied Microbiology and Biotechnology,* **98**(21), 9021–9031, https://doi.org/10.1007/s00253-014-5973-3

Song R.-B., Wu Y., Lin Z.-Q., Xie J., Tan C. H., Loo J. S. C., Cao B., Zhang J.-R., Zhu J.-J. and Zhang Q. (2017). Living and conducting: coating individual bacterial cells with in situ formed polypyrrole. *Angewandte Chemie International Edition,* **56**(35), 10516–10520, https://doi.org/10.1002/anie.201704729

Sun M., Zhang F., Tong Z.-H., Sheng G.-P., Chen Y.-Z., Zhao Y., Chen Y.-P., Zhou S.-Y., Liu G., Tian Y.-C. and Yu H.-Q. (2010). A gold-sputtered carbon paper as an anode for improved electricity generation from a microbial fuel cell inoculated with *Shewanella oneidensis* MR-1. *Biosensors and Bioelectronics,* **26**(2), 338–343, https://doi.org/10.1016/j.bios.2010.08.010

Svitkova V., Blaskovicova J., Tekelova M., Kallai B. M., Ignat T., Horackova V., Skladal P., Kopel P., Adam V., Farkasova D. and Labuda J. (2017). Assessment of CdS quantum dots effect on UV damage to DNA using a DNA/quantum dots structured electrochemical biosensor and DNA biosensing in solution. *Sensors and Actuators B: Chemical,* **243**, 435–444, https://doi.org/10.1016/j.snb.2016.11.160

Tibbitt M. W. and Langer R. (2017). Living biomaterials. *Accounts of Chemical Research,* **50**(3), 508–513, https://doi.org/10.1021/acs.accounts.6b00499

Wang W., You S., Gong X., Qi D., Chandran B. K., Bi L., Cui F. and Chen X. (2016). Bioinspired nanosucker array for enhancing bioelectricity generation in microbial fuel cells. *Advanced Materials,* **28**(2), 270–275, https://doi.org/10.1002/adma.201503609

Wang T., Zhang D., Dai L. L., Dong B. and Dai X. H. (2018). A scavenger for the blockage of electron transfer in anaerobic digestion of high-solids sewage sludge. *Environmental Science & Technology,* **52**(12), 7160–7169, https://doi.org/10.1021/acs.est.8b00891

Wang X., Li S., Wang C., Mujuni C. J., Yue T. and Huang F. (2020). Supramolecular construction of biohybrid nanozymes based on the molecular chaperone GroEL as a promiscuous scaffold. *ACS Biomaterials Science & Engineering,* **6**(2), 833–841, https://doi.org/10.1021/acsbiomaterials.9b00997

Wei W., Sun P., Li Z., Song K., Su W., Wang B., Liu Y. and Zhao J. (2018). A surface-display biohybrid approach to light-driven hydrogen production in air. *Science Advances,* **4**(2), eaap9253, https://doi.org/10.1126/sciadv.aap9253

Wu X., Zhao F., Rahunen N., Varcoe J. R., Avignone-Rossa C., Thumser A. E. and Slade R. C. T. (2011). A role for microbial palladium nanoparticles in extracellular electron transfer. *Angewandte Chemie International Edition,* **50**(2), 427–430, https://doi.org/10.1002/anie.201002951

Xiao Y., Zhang E. H., Zhang J. D., Dai Y. F., Yang Z. H., Christensen H. E. M., Ulstrup J. and Zhao F. (2017). Extracellular polymeric substances are transient media for microbial extracellular electron transfer. *Science Advances,* **3**(7), 1–9, https://doi.org/10.1126/sciadv.1700623

Xu M. Y., Tremblay P. L., Jiang L. L. and Zhang T. (2019). Stimulating bioplastic production with light energy by coupling *Ralstonia eutropha* with the photocatalyst graphitic carbon nitride. *Green Chemistry,* **21**(9), 2392–2400, https://doi.org/10.1039/C8GC03695K

Yan F.-F., He Y.-R., Wu C., Cheng Y.-Y., Li W.-W. and Yu H.-Q. (2014). Carbon nanotubes alter the electron flow route and enhance nitrobenzene reduction by *Shewanella oneidensis* MR-1. *Environmental Science & Technology Letters,* **1**(1), 128–132, https://doi.org/10.1021/ez4000093

Yang C., Aslan H., Zhang P., Zhu S., Xiao Y., Chen L., Khan N., Boesen T., Wang Y., Liu Y., Wang L., Sun Y., Feng Y., Besenbacher F., Zhao F. and Yu M. (2020). Carbon dots-fed *Shewanella oneidensis* MR-1 for bioelectricity enhancement. *Nature Communications,* **11**(1), 1379, https://doi.org/10.1038/s41467-020-14866-0

Ye J., Hu A. D., Ren G. P., Chen M., Tang J. H., Zhang P. Y., Zhou S. G. and He Z. (2018). Enhancing sludge methanogenesis with improved redox activity of extracellular polymeric substances by hematite in red mud. *Water Research,* **134**, 54–62, https://doi.org/10.1016/j.watres.2018.01.062

Ye J., Yu J., Zhang Y., Chen M., Liu X., Zhou S. and He Z. (2019). Light-driven carbon dioxide reduction to methane by *Methanosarcina barkeri*-CdS biohybrid. *Applied Catalysis B: Environmental,* **257**, 117916, https://doi.org/10.1016/j.apcatb.2019.117916

Yu J. G., Yu Y. F. and Cheng B. (2012). Enhanced visible-light photocatalytic H_2-production performance of multi-armed CdS nanorods. *RSC Advances,* **2**(31), 11829–11835, https://doi.org/10.1039/c2ra22019a

Zhang P., Liu J., Qu Y. P., Li D., He W. H. and Feng Y. J. (2018). Nanomaterials for facilitating microbial extracellular electron transfer: recent progress and challenges. *Bioelectrochemistry,* **123**, 190–200, https://doi.org/10.1016/j.bioelechem.2018.05.005

Zhang R. T., He Y., Yi J., Zhang L. J., Shen C. P., Liu S. J., Liu L. F., Liu B. H. and Qiao L. (2020). Proteomic and metabolic elucidation of solar-powered biomanufacturing by bio-abiotic hybrid system. *Chem,* **6**(1), 234–249, https://doi.org/10.1016/j.chempr.2019.11.002

Zhao C.-E., Wu J., Ding Y., Wang V. B., Zhang Y., Kjelleberg S., Loo J. S. C., Cao B. and Zhang Q. (2015). Hybrid conducting biofilm with built-in bacteria for high-performance microbial fuel cells. *Chemelectrochem*, **2**(5), 654–658, https://doi.org/10.1002/celc.201402458

Zhao C., Zhang J., Chen Z., Li J., Shu L. and Ji X. (2022). Effective constructions of electro-active bacteria-derived bioelectrocatalysis systems and their applications in promoting extracellular electron transfer process. *Progress in Chemistry*, **34**(2), 397–410.

Part 3

Environmental Remediation Using NBs

doi: 10.2166/9781789063592_0147

Chapter 9

Nanobiohybrids: a promising approach for sensing diverse environmental water pollutants

Angelene Hannah Jebarani D.[1*] *and Sruthi Ann Alex*[2**]

[1*]*National Institute of Technology (NIT) Andhra Pradesh, Department of Biotechnology, Tadepallidueden, Andhra Pradesh, India*
[2**]*Centre for Nano Science and Technology, Anna University, Chennai, Tamil Nadu, India*

ABSTRACT

Water resource management has become the need of the hour with continuous depletion of good quality water. Sensors are important tools for monitoring the environmental water quality and quantifying the presence of various harmful pollutants (pesticides, heavy metals, and micro-organisms). Biosensors designed with functional biological molecules as recognition elements use a transducer, which converts a biological response into electrochemical, optical, gravimetric, magnetic, or thermal signals that can be easily quantified. The advent of nanotechnology has revolutionized material design and properties of sensing platforms. Materials in the nanoscale dimension possess interesting characteristic features such as high surface-area-to-volume ratio, high conductivity, functionalizability, shock-bearing ability, color tunability, photo-reactivity, and magnetization. Nanomaterial-integrated biomolecules with enhanced functional properties are termed as nanobiohybrids. The amalgamation of the tuneable characteristics of nanomaterials with functional biological molecules by means of functionalization and/or immobilization creates sophisticated transducers for enhancing the analytical performance of sensors. The resultant nanobiohybrid sensor systems are robust, exhibit relatively short response time with improved specificity and have higher sensitivity enabling detection of even trace volumes of analyte molecules. Recent advancements in research and engineering also promote the possibilities of device miniaturization, portability, and incessant monitoring of water quality. This chapter intends to discuss the impact of various toxic pollutants in water, the importance of nanomaterials in design of novel nanobiohybrids, the different nanomaterials relevant for use in design of sensors for environmental analysis, and finally the different transduction mechanisms that enable the detection and quantification of toxins in water.

Keywords: biological recognition, biomolecules, biosensors, environmental monitoring, nanobiohybrids, nanomaterials, sensors, water pollution.

9.1 INTRODUCTION

The ever-growing demand from manufacturing industries, agricultural and irrigational activities, overpopulation together with improper treatment of contaminated effluents and consistent changes in temperature and weather patterns has resulted in the deterioration of quality of the natural water sources. Some of the potential toxicants and environmentally relevant materials include a diverse range of heavy metals, industrially significant chemicals with their derivatives, and pathogenic disease-causing microorganisms. Heavy metals such as lead, mercury, arsenic, cadmium, and copper, which are major components in industrial wastewater streams, are non-biodegradable and dangerous even at low concentrations. Pathogenic organisms like *Salmonella* sp., *Escherichia coli*, *Streptococcus faecalis*, and enteroviruses in polluted waters can cause life-threatening diseases such as cholera, salmonellosis, hepatitis A, and typhoid polio in humans. Nitrites used in fertilizers can impair the functioning of human hemoglobin irreversibly. Chemicals in pesticides including parathion, paraoxon, carbaril, diuron, paraquat, aldrin, chlordane, DDT (dichlorodiphenyltrichloroethane), dieldrin, endrin, and hexachlorobenzene have variable toxicity depending on the levels of exposure and may result in impairment of functional body systems (nervous system and immune system), compromise on the growth and development or cause mutations making it potentially teratogenic and carcinogenic. Phenolic compounds can be released naturally into water due to the decomposition of dead plants/animals, but the quantity has increased exponentially due to the activities of a lot of industries. Their long-term toxicity effects include shortened lifespan, fertility issues, liver/kidney damage, skin burns, tremor, and body weakness. Due to the lethality of many of these pollutants at higher doses, they have been used as chemical and biological warfare agents that can kill, harm or incapacitate humans and other living beings. Hence, monitoring the presence and concentration of these pollutants needs to be performed meticulously to ensure the maintenance of optimum water quality standards and to evaluate the effectiveness of water treatment methods. The continuous monitoring of water quality using conventional analytical instruments becomes difficult due to expensive instrumentation, tedious sample preparation, and requirement for skilled technical operators.

With the advent of the sensor technology, the monitoring of environmentally significant toxins in-situ and ex-situ has become relatively easier and hence, has continuously been preferred over conventional analytical methods. The consistent research in the development of novel materials for sensing eventually led to the emergence of biological material-derived sensing devices. Biosensors can be used to detect specific analytes through some physical, chemical, biological, or electrical signals that arise from the analyte interaction with a biological recognition element; the signal gets transduced into an electrical signal, which can be processed and quantified. Over the years, these biosensors have found extensive use in healthcare, therapeutics, environmental monitoring, and food safety applications. However, the direct loading of these bioelements onto bare transducer surfaces tend to destabilize/denature the bioreceptor molecule and have inherently reduced their ability to functionalize and their availability on the transducer surface, causing lower sensitivity, limited shelf-life and activity, and thereby affecting their reusability. Nanomaterials offer several distinctive features that make their integration into biosensors highly beneficial. Thus, the integration of biomolecules with nanomaterials to provide improved functional properties gives rise to the class of materials termed as 'Nanobiohybrids'.

9.2 IMPORTANCE OF NANOMATERIALS IN THE NANOBIOHYBRIDS

Nanobiohybrids are by definition a class of materials composed of one or more materials in the nanoscale dimension combined with biological components (ranging from simple biological molecules to unicellular microorganisms to complex multicellular organisms). The importance of using nanomaterials in the development of nanobiohybrids relates to the capability of the nanomaterial to impart exceptional properties to biosystems making them functional components in many practical applications, including biomedical, catalysis, sensing, membrane technologies, energy conversion and storage, and coatings. Biosystems have the innate ability to adapt to changes that help augment their performance, a skill of survival from an evolutionary point of view. This ability of biosystems has been exploited to achieve modification by incorporation of suitable nanomaterials creating new sophisticated bio-functional materials for various applications. The current advances in materials engineering, design and processing have promoted the progress of nanobiohybrid generation by tailoring the physicochemical properties of the nanomaterials enabling them to be successfully coupled with the desired biosystem. Nanobiohybrids have been shown to demonstrate exceptional bio-functional properties surpassing years of evolutionary transitions or even genetically mutated or engineered biosystems (Guo *et al.*, 2020).

Nanomaterials offer a large fraction of surface atoms, higher surface energy, and ease of surface modification with multiple functional groups. Thereby, they permit efficient biomolecule immobilization on the transducer. Their reduced imperfections, spatial confinement, and precise crystallographic direction can facilitate uniform bioreceptor attachment on the transducer. These biocompatible nanomaterials can also stabilize the biomolecules through electrostatic and covalent interactions and improve the bioactivity of the sensing element. Moreover, due to their enhanced catalytic and conductive properties and the directed arrangement of the biomolecule on the nanomaterial, they can also behave like a mediator for the bio-recognition events by improving the electron transfer rate to the sensor surface and provide improved integration and amplification of the signals. The usage of nanomaterials in combination with the biological element has improved sensor efficiency in terms of sensitivity, response time, selectivity, and detection range and provides other benefits such as cost-effectiveness, miniaturization, requirement of low sample volumes, and environmental friendliness (Kuswandi, 2019). By modifying the synthesis conditions, different features of the nanomaterial such as size, shape, grain structure, porosity, pore size, and number of layers [for graphene sheets, carbon nanotubes (CNTs)] can be tuned to bring changes in their functional properties. Some of the special tunable properties of nanomaterials that are important for different types of functions are listed below:

(1) Higher catalytic efficacy as a result of the higher surface-area-to-volume ratio.
(2) Improved electrical conductivity in ceramics and magnetic nanocomposites and increased electrical resistance to metals.
(3) Increased magnetic coercivity up to a critical grain size and superparamagnetic behavior.
(4) Improved mechanical properties like higher hardness and toughness for metals and alloys and higher ductility and superplasticity for ceramics.
(5) Optical properties like localized surface plasmon resonance (LSPR) for metallic nanoparticles, spectral shift in optical absorption and fluorescence based on nanoparticle size, increased quantum efficiency for semiconductor nanomaterials.
(6) Sterically significant properties like improved selectivity based on pore size for mesoporous structures.
(7) Better biological applicability due to their improved biocompatibility, and improved ability for nanodrug carriers to penetrate cell membranes and the blood–brain barrier.

A wide spectrum of materials in the nanoscale has been successfully used in the design of biosensors (Kuswandi, 2019). These nanomaterials include, but are not limited to, metal/metal oxide, carbon-based, semiconductor, and composite nanoparticles. By modifying the nanoparticle morphology and composition through controlled synthesis and by employing the appropriate sensing strategy with standard optimization of process parameters, specific nanobiohybrids can be designed to sense varied organic or inorganic environmental pollutants of interest (Ramezani *et al.*, 2020). Miniaturization of existing sensors using nanotechnology has paved the way for developing highly sensitive and selective systems for facile detection and continuous monitoring of environmental pollutants on-site.

9.3 CHOICE OF NANOMATERIAL

Over the years, considerable effort has been made in the design and development of nanobiohybrids containing nanostructures (synthetic component) combined with structures of biological origin (biological component). The careful selection of both biological and synthetic components is of high priority as the properties of the two ultimately influence the biofunctionality of the nanobiohybrid

Table 9.1 Classification of materials in the nanoscale based on the restriction of electron movement along the axes (X, Y, Z) with examples.

Dimensionality of the Material in the Nanoscale	Directions of Electron Confinement	Examples
Zero dimensional (0D)	X-, Y- and Z-axis	Quantum dots, fullerenes
One dimensional (1D)	X- and Y-axis	Nanowires, nanotubes, nanorods
Two dimensional (2D)	X-axis	Films, plates, sheets
Three dimensional (3D)	No confinement	Nanocomposites

system, which may either result in the enhancement of selective existing traits of the biosystem or bestow unique new properties to the biosystem. The synthetic component of a nanobiohybrid assembly implicating the nanomaterial can be chosen from a wide spectrum of structurally versatile materials, including zero-dimensional quantum nanostructures, one-dimensional wires or rods, two-dimensional sheets, and three-dimensional networks (Table 9.1). The choice of specific-type of nanomaterials to design a nanobiohybrid differs primarily with the purpose of application, example nanomaterials with photosensitization properties can be used to induce photocatalytic properties, materials with magnetic properties are chosen for biomedical imaging, while polymeric materials used in constructing the hybrids can manipulate the properties of organisms (Guo *et al.*, 2020; Huang *et al.*, 2014).

For the development of bioelectronic sensors, nanomaterials have been used as elements of transduction in combination with biological recognition molecules either to achieve improved device performance or to aid in design of novel advanced detection methodologies. Plasmonic nanostructures such as gold or silver coupled with biological recognition elements provide optimal electronic or optical transduction of biological phenomena to yield highly effective plasmonic biosensors (Zheng *et al.*, 2012). Biomolecules, in particular, nucleic acids conjugated to magnetic nanoparticles, form nanobiohybrids that possess exceptional magnetic properties in addition to the inherent biological selectivity, improving disease diagnosis and therapy (Pershina *et al.*, 2014). A single-walled CNT complexed to a family of peptides known as bombolitin, forms a virtual chaperone sensor that transduces changes in conformation as near-infrared luminescence (Wong *et al.*, 2017). Simple, sensitive, and economic glucose biosensors based on the immobilization of the enzyme glucose oxidase on carbonaceous hybrid films comprising of three-dimensional graphene and multi-walled CNTs improved the electrochemical activity and stability of sensing (Bagheri *et al.*, 2017). Structures of lipid bilayer-coated silicon nanowires exhibit efficient barrier properties preventing the transport of redox-active species in solution, shielding the electrode surface, and hence, serve as a model for detection of membrane ion channels (Martinez *et al.*, 2009). Nanobiohybrids developed by coupling a metal-organic framework with live plants function as self-sustaining pre-concentrators, accumulating toxic metal and organic pollutants, resulting in the generation of luminescent signals that are digitized via a smartphone app for on-site environmental monitoring (Liang *et al.*, 2020). Nanostructured materials commonly reported for construction of nanobiohybrids can be grouped into the following categories and their significance in biosensing applications has been highlighted.

9.3.1 Metallic and metal oxide nanostructures

Metallic nanoparticles such as gold and silver have been extensively used in biosensor devices. These materials come in a wide range of shapes and sizes, with high surface energy and surface-to-volume ratio, making them suitable for easy surface functionalization. Owing to this nature, metallic gold and silver nanomaterials have been employed to electrically probe a number of biomolecules. Another important property of these nanostructures is the surface plasmon resonance effect. This is related to the coherent oscillations of the conductive surface electrons and has been correlated with the specific-binding of biomolecule concentrations. Colloidal gold nanospheres exhibit size-dependent color changes, varying from shades of red (diameter of ~5–50 nm) to purple (diameter of ~100 nm, a feature that has been exploited for use in colorimetric sensing. Gold nanoparticles are excellent current conductors enabling direct, efficient, and quick electron transfer between electrode materials and electroactive species, making them a material of choice in electrochemical sensing. Colorimetric nanosensors based on silver nanoparticles have been reported for the detection of highly toxic heavy metals such as lead, chromium, mercury, and many more.

Metal oxides are a class of economical nanostructures with ease of preparation and functionalization. These nanostructures have tunable morphologies and can be grown in different shapes and sizes. Semiconductive metal oxide structures like zinc oxide and titanium dioxide have been exploited for their photo-conductive properties in catalytic redox detection of a number of molecules. These nanoparticles have been shown to improve the sensitivity of sensors by transduction and amplification of biological signals. Dielectric nanomaterials such as silicon dioxide and manganese dioxide have been reported for use in immobilization of biomolecules (Banerjee *et al.*, 2021). Oxides of iron including magnetite and maghemite with low toxicity and unique magnetic properties have been shown to be easily functionalized with biomolecules. Also, these materials have found profuse use in environmental remediation due to the magnetic property which facilitates separation of pollutants and recovery of the nanostructure with ease.

9.3.2 Carbonaceous nanomaterials

The large surface area and mechanical strength along with excellent electrical and thermal conductivities make carbon nanomaterials a consistent choice in a spectrum of applications. The large surface area of carbon nanostructures enables loading of a variety of biological molecules on the carbon surface making it a viable system for biosensing. Graphene is a two-dimensional nanostructure with sp^2 hybridization consisting of horizontally stacked sheets of carbon. The nanostructure is chemically stable and on integration with electrodes has found significant use in improving the sensitivity of electrochemical sensors. In alternate cases, graphene has also been used to catalyze biochemical redox reactions, for example, detection of organophosphates by measuring the oxidation current resulting from the inhibition of the acetylcholinesterase enzyme immobilized on graphene sheets (Banerjee *et al.*, 2021).

CNTs are one-dimensional nanostructures consisting of rolled-up carbon sheets forming a cylindrical morphology. These nanotubes have been commonly used as probes for a range of biological molecules functioning either as analyte capture platforms or signal transducers. Due to its ability to cross biological membranes, CNT has also been used to probe biomolecules and mechanisms in-vivo. In electrochemical sensing, CNT enhances the transfer of electrons between the electrode surface and electroactive materials exhibiting good electrical conductivity and sensitivity, are chemically stable and also electrocatalytic making them ideal candidates for use. Functionalization of CNT with suitable fluorophores to enable optical detection, for example, peptide functionalized CNT for detection of nitroaromatics using near infra-red (NIR) photoluminescence has been described (Heller *et al.*, 2011).

9.3.3 Quantum dots

Quantum dots (QD) are zero-dimensional semi-conductive nanostructures with excellent optical transduction property. The nanostructures exhibit broad absorption bands with characteristically narrow fluorescence emission bands which is ideal for detection of multiple analytes at a single time. The emission wavelength of the QD can be readily adjusted by changing its size, shape, or composition, which results in the quantum confinement of charge carriers (Willner & Vikesland, 2018). QD can be readily functionalized and/or used to probe a plethora of biomolecules with high specificity. However, long-term use of semiconductor QD containing heavy metals (example, cadmium and selenium) may have adverse effects and hence, the use of carbon-dots has emerged as an eco-friendly alternative (Banerjee *et al.*, 2021).

9.3.4 Polymers

A large selection of polymeric materials, both synthetic and natural, has been widely used for developing sensors of disposable and wearable types. Polymers are transparent, flexible, and durable materials that can be formed into desired shapes and produced in bulk, at low cost. Polymers serve as matrices for immobilizing biological elements and can be used in combination with other nanomaterials to create biosensing devices. For example, highly sensitive biosensors consisting of the biopolymer chitosan with graphene oxide and hemoglobin for detection of nitromethane has been reported (Banerjee *et al.*, 2021). Disposable sensors based on the principle of immunochromatography have been reported for the quantification of human serum albumin (Jianrong *et al.*, 2004). The conductimetric sensor was generated using polyaniline modified colloidal gold which was used as the second antigen to obtain a sandwiched immune complex, generating a signal at the electrode.

The above discussed categories of nanoscale materials can either be used independently or in combination to develop novel nanobiohybrids with superior characteristic properties for use as sensing platforms in environmental pollutant detection and monitoring applications (Banerjee *et al.*, 2021; Dincer *et al.*, 2019; Jianrong *et al.*, 2004; Willner & Vikesland, 2018).

9.4 NANOBIOHYBRID TYPES: BASED ON RECOGNITION ELEMENTS

Nanobiohybrid-based electronic sensing constructs on a large scale use biological components as recognition elements to ensure improved selectivity and sensitivity. Advances in the field of research for optimization of nanobiohybrid sensors for environmental monitoring, both on-site and off-site (real-time analysis and online monitoring), has drawn the interest of many environmentalists. Here, we highlight the scope of some specific biological molecules commonly used as recognition elements in the design of nanobiohybrid materials for use as sensors in the monitoring of environmental water pollutants.

9.4.1 Proteins and peptides

Proteins and peptides form a class of biologically important macromolecular structures that have been long used as recognition elements in biosensors. The organic reactions involving proteinaceous structures, for example, enzyme-based sensing which has long been in use owing to the potential of enzymes to selectively detect the molecule of interest via enzyme-catalyzed reactions. The generated biological signals are transduced using suitable nanomaterials to produce electrochemical or colorimetric outputs that can be used for detection. The versatile nature of enzymes along with its ability to efficiently catalyze selected chemical processes has enabled it to emerge as a biorecognition element of choice for designing sensors. Enzymes such as laccase and tyrosinase have been reported for use in the detection of phenolic compounds (Rodríguez-Delgado *et al.*, 2015; Wee *et al.*, 2019). The direct interaction of toxic metal ions with metal-binding moieties such as the synechoccocal metallothionein and the mercury regulatory protein have been reported for use as affinity sensors to detect trace levels of metal concentrations by recording the changes in capacitance which varies in response to the quantity of ions contained in the sample (Corbisier *et al.*, 1999).

Protein-based biosensor strips for the detection of toxic aromatics in environmental wastewater has been reported by Ray *et al.* (2018). The nanobiohybrid consists of the MopR protein, a class of nitrogen regulating proteins from *Acinetobacter calcoaceticus* as the biological element immobilized on a solid support made of mesoporous silica nanoparticles. The property of the protein to undergo allosteric conformational changes in the presence of high concentrations of phenolics was employed for demonstrating the use of the sensor for on-site pollution monitoring (Ray *et al.*, 2018). Wong *et al.* (2017) reported the detection of nitroaromatics in ground water using infrared fluorescence sensors of functional single-walled CNTs complexed bombolitin II peptide nanobiohybrid structures embedded in live spinach plant leaves. The accumulation of nitroaromatics in the plant leaf and its interaction with the functional complex results in changes in conformation and this modulation of the peptide structure is translated to spectral modulations for detection (Wong *et al.*, 2017). A recent development in the use of proteins for biosensing applications is the technology of structure-guided protein engineering for developing highly selective recognition elements. The direct detection of hydrocarbons like benzene and its derivatives remained a challenge due to the lack of active functional groups in the hydrocarbon molecules; and the difficulties in isolating specific proteins with the inherent ability to identify the hydrocarbons. However, by understanding the three-dimensional structure of the binding domain(s) in combination with in silico and experimental investigations, Ray *et al.* (2018) designed highly specific and exclusive sensor devices to distinguish the structural variations between the group of pollutant molecules in environmental water samples.

9.4.2 Nucleic acids

Nucleic acid biosensors have functional nucleic acid molecules integrated with a signal transducing element to detect a range of molecules from complementary nucleic acid fragments or other biological molecules to metals and chemical substances with high specificity. Nucleic acids are oligonucleotide molecules that are an important class of biological recognition elements with versatile activities. The long chains of nucleotides are found universally in all living organisms and may be ribonucleic acid (RNA) or deoxyribonucleic acid (DNA). Among the two, for applications in material science and biotechnology, DNA is preferred over RNA due to its stability and relative resistance to hydrolysis. The functional nucleic acid molecules used in sensors may be of the following types:

- Naturally occurring genetic DNA sequences (genosensor).
- Short sequences of single-stranded synthetic oligonucleotides isolated from large libraries termed as aptamers.
- Nucleic acid sequences that possess catalytic activity termed as nucleic acid enzymes, such as ribozymes or DNAzymes.
- Aptazymes, which are a combination of aptamers and catalytic nucleic acids.

The mechanism of detection relies on the type of interaction of pollutant molecules with the nucleic acid sequences. For example, genosensing of *Escherichia coli* cells in water via hybridization of selective genetic sequences, combining the lac gene detection with the polymerase chain reaction was used in the amplification of DNA (Mo *et al.*, 2002). Reports on the use of DNA aptamers for chloroaromatics detection and RNA aptamers for detection of aromatic carcinogens based on aptamer–ligand recognition have been established (Tombelli *et al.*, 2007). Metal-ion-dependent enzymatic nucleic acids have been used for monitoring toxic ion concentrations in environmental samples (Palchetti & Mascini, 2008). Sensitive detection of toxic mercury (II) ions has been achieved by the method of surface-enhanced Raman spectroscopy using an oligonucleotide functionalized core–shell nanostructure having a gold core enveloped by a magnetic silica sphere. Moreover, by exploiting the specific base pairing property of DNA, the capture of the Hg(II) metal was made possible enabling its specific removal from the surrounding environmental sample (Liu *et al.*, 2014).

9.4.3 Carbohydrates

In nature, carbohydrates are of biological importance as molecules of energy or metabolic intermediates. Although available in surplus, the application of carbohydrates as recognition elements has been limited. For instance, synthetic preparation of sugars is extremely laborious when compared to existence of relative simpler amplification or cloning options for other biomolecules. Chemically these sugars may be simple or complex polyhydroxy molecules containing monosaccharide building units. This arrangement gives rise to diverse carbohydrate structures for example, it has been established that a single hexamer can generate more than 10^{12} conformations. Conjugates of carbohydrates known as glycoconjugates play key roles in mediating a number of physiological and pathological cellular functions. Unlike proteins or nucleic acids, carbohydrate receptors are relatively smaller in size and more stable. However, the molecular recognition signals of individual carbohydrates are generally weak and hence, these are usually present as clusters to increase the magnitude of interactions resulting from high ligand densities.

Lectins form a group of proteins that bind specifically to carbohydrates via hydrogen bonding, metal co-ordination, Van der Waals forces, and other hydrophobic interactions. These proteins have more than one binding site for the binding to the carbohydrate and therefore, increase the magnitude of interaction by 100-fold. They have been shown to be equivalent to antigen–antibody interactions with kD values approaching 1 nM. Investigations on interactions of gangliosides with cholera toxin have been reported with kD values of 1 nM to as low as 1 pM (Chevolot *et al.*, 2010; Zeng *et al.*, 2012).

9.4.4 Whole cells

For the real-time monitoring of toxins in aquatic environments, the use of living organisms, such as bacteria that naturally respond to the presence of certain specific molecules can enable a better assessment of the risk associated with exposure of humans to toxic pollutants. Whole cell biosensors are devises based on exploiting the natural ability of organisms which are responsive to the presence of pollutants in their living environment. For example, bacterial sensors of *Acinetobacter calcoaceticus* exploit the phenol–catabolic pathway of the species which gets triggered in the presence of phenol and activates a specific RNA polymerase that in turn induces the degradation of phenol. One advantage to using bacterial sensors is that the cell wall protects the sensor by restraining the direct contact with unpleasant harsh environments making it a robust system and the specific response gives an understanding about the bioavailability of the analyte (Roy *et al.*, 2021). The assessment of risk associated with a toxic contaminant is based on determining the concentration levels of the particular compound at the contaminated site. However, one should understand the importance of the term 'bioavailability' which can be defined as 'the fraction of a pollutant that can be spontaneously taken up by an organism from its surrounding environment at any given point of time' and is firmly associated with the toxicity of the particular compound of interest with relevance to the concerned organism. In a distinctive environment, the compounds' bioavailability is invariably nonequivalent to its total concentration in the same environment, but it is dependent on the bio-accessibility of the compound which can be defined as 'the amount of the compound that can potentially cross the cellular membrane if facilitated'. The bio-accessibility of compounds varies with organism and a bio-accessible fraction can be made bioavailable by engineering the inherent characteristics of the organism (Tecon & Van der Meer, 2008).

Bioavailable metal concentrations in environmental samples can be measured by using bacteria that produce a quantifiable signal when in proximity. Bacteria in soil can be engineered genetically to develop metal-responsive strains which can generate bioluminescent signals in the presence of selective metal ions. In order to effectively apply this principle in detection, it is important to understand the molecular basis and mechanism of regulation of engineered genes. Immobilization of the manipulated bacterium in solid matrices coupled with optical fibers has been used for sensor development (Corbisier *et al.*, 1999).

Organisms other than bacteria have also been reported for use in biosensing of environmental pollutants. Genetically modified variants of *Chlamydomonas reinhardtii* have been used as fluorescent probes for monitoring toxic chemicals in herbicide-contaminated environmental water samples. The detection is based on the activity of photosystem II, the multienzymatic chlorophyll–protein complex located in the thylakoid membrane that catalyzes the light-dependent photosynthetic primary charge separation and the electron transfer chain in cyanobacteria, algae, and higher plants (Scognamiglio *et al.*, 2009). Plant nanobiohybrids with photo-luminescent metal organic frameworks have been reported for the detection of toxic metal ions (cadmium, silver, iron, and copper) and organic pollutants (aniline). The sensors with fluorescence 'turn-on' and 'turn-off' signals were shown to have potential on-site monitoring capabilities (Liang *et al.*, 2020).

9.5 NANOBIOHYBRID SENSOR TYPES BASED ON TRANSDUCTION PATHWAYS

The suitability of a nanobiohybrid material for environmental sensing applications depends primarily on its ability to transduce the signals generated from a biological phenomenon and convert them into readable output. Depending on the type of signal transduced, a broad classification can be made as: electrochemical, optical, magnetic, gravimetric, and thermal.

9.5.1 Electrochemical nanobiohybrid sensors

The working principle of this biosensor class is that catalytic chemical reactions between the analyte and the bioelement bound to the electrode may produce or consume ions or electrons, thereby affecting the measurable electrochemical properties of the solution, such as an electric current or potential. The signals can be monitored by different means (Figure 9.1a): generating a measurable charge accumulation or potential (potentiometric), changes in impedance (impedimetric), conductive properties of the medium (conductometric), or changes in current (amperometric). Each method provides different signal types (Figure 9.1a) and have different advantages and disadvantages (Table 9.2). Electrochemical transducers have a lot of advantages, such as high sensitivity, their compatibility with modern miniaturization/microfabrication technologies, convenience to use, portability, minimal power requirements, low cost, simplicity of construction, and independence of sample turbidity, volume, and color (Curulli, 2021). The working electrode is mostly coated with a nanomaterial on which the bioreceptor is immobilized to provide better functionalization and enhanced perpetuation of electrical signals.

- Amperometric biosensors are more widely used due to their higher sensitivity and faster response than potentiometric sensors. However, they have lower selectivity and show susceptibility to interference from other electroactive species.
- Potentiometric biosensors use ion-selective electrodes (ISEs) surrounded by an ion-selective membrane, which helps in the selective movement of target ions and reduces interferences. The potential difference noted between the reference and working electrodes at nearly zero current is proportional to the analyte concentration.
- Conductometric biosensors measure the change in electrical conductivity of the sample solution between the different electrodes after a biochemical reaction. These biosensors require low driving voltage, inexpensive technology, are light-insensitive, and do not need a reference electrode.

Biosensors based on the ion-selective field-effect transistors (ISFET) are another class of electrochemical biosensors, suitable for direct ion detection. By varying the voltage at the gate, the flow of current from the source to the drain can be varied. The gate can be modified with a nanomaterial layer over which the bioreceptor is functionalized in order to enhance the sensitivity of the device to any electrochemical change. When the analyte solution is brought into contact with the gate electrode, a change in ions released due to interaction with the nanobiohybrid receptor can change the gate potential, which is proportional to the analyte concentration (Odobašić *et al.*, 2019).

Screen-printed electrodes (SPE) and paper-based electrodes are significant advancements of nanobiosensors as they can be mass-fabricated cost effectively and be utilized as single-use disposable kits that require low sample volume and eliminate the need for electrode cleaning and related biohazards. They have been widely employed as transducers of (bio)sensing devices. Some of the strategies by which nanomaterials can help in the electrochemical detection are: (1) direct detection methods, where the nanomaterials are directly detected using a redox process; (2) digestion methods, where the nanomaterials are dissolved to liberate ions, which can be measured directly by an ISE; (3) methods wherein an electrocatalytic reaction acts as an indirect indicator for nanomaterial estimation; (4) surface-enhanced processes, wherein the surface properties of the nanomaterials can in turn augment the signal; (5) nanoimpact methods, where individual nanomaterials can be

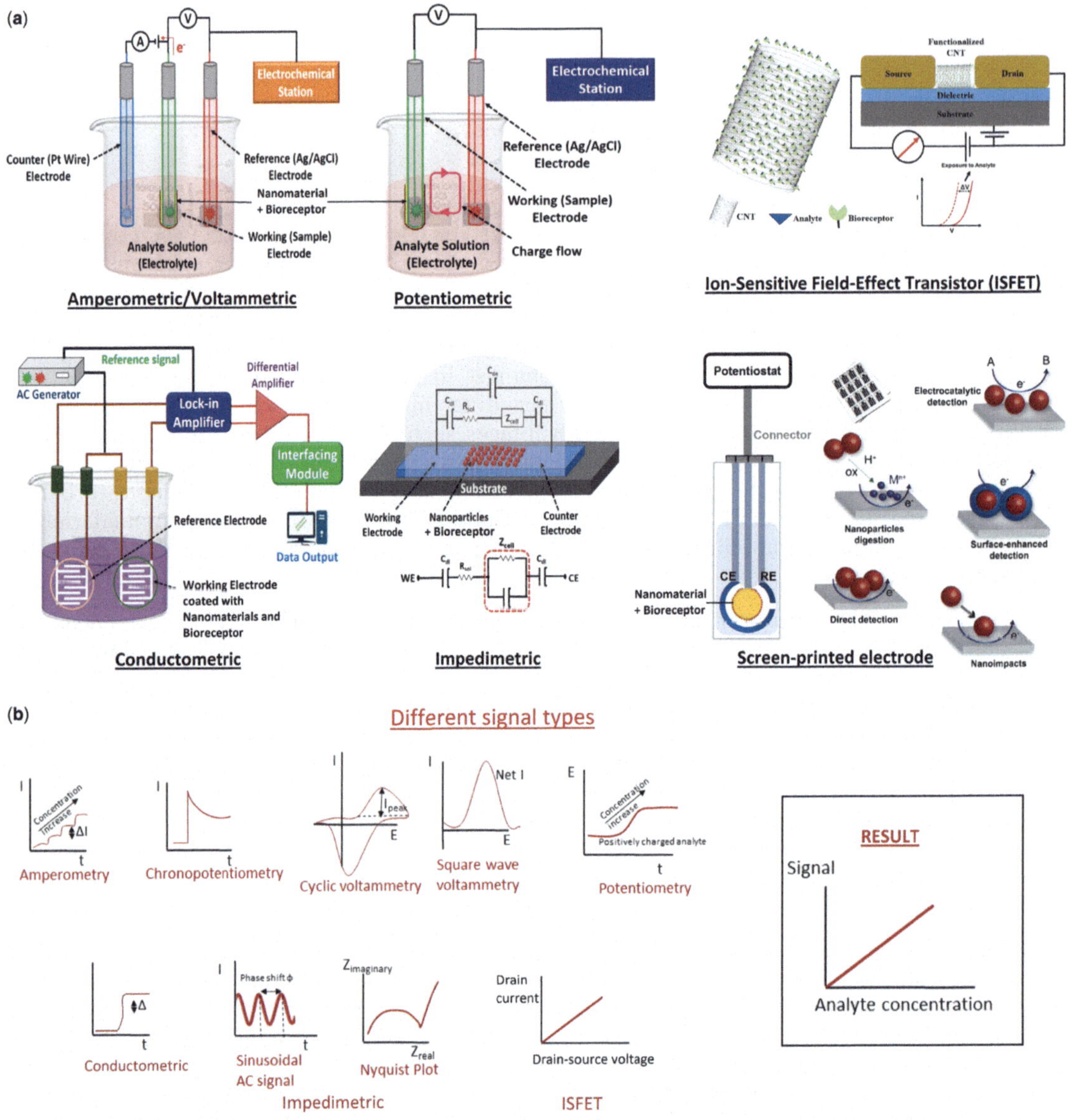

Figure 9.1 (**a**) Different types of electrochemical nanobiosensors and (**b**) the different signals obtained (Banerjee *et al.*, 2021; Curulli, 2021; Damiati & Schuster, 2020; Dincer *et al.*, 2019; Martín-Yerga, 2019; Shen *et al.*, 2014).

sensed by stochastic collisions on the surface of the electrode (Martín-Yerga, 2019). One example of the latter class of biosensors is a whole cell sandwich immunoassay for bacterial capture followed by amperometric sensing using a carbon SPE. The target analyte was recognized by capture onto antibody-coated Fe_3O_4@poly(dopamine) magnetic NPs and another detector antibody-labeled with horseradish peroxidase was used to create an immunocomplex, which was captured on the carbon SPE using a magnetic field. The assay was specific for *Legionella* and could sense as low as 10 CFU/

Table 9.2 Electrochemical biosensors with principles, advantages, and disadvantages.

Electrochemical Biosensors	Principles	Advantages	Disadvantages
Potentiometric	Electric potential	Decreased analysis time, good selectivity and sensitivity, and sample treatment not required.	Temperature, pH, and immunological cross-reaction variables all have an impact on sensitivity and lifespan.
Amperometric	Oxidation/reduction	Portability due to the portable system, high selectivity, sensitivity.	Regenerative between measurements.
Impedimetric	Change in impedance	High selectivity and sensitivity, simple operation, small device.	Complex construction, expensive labelling markers.
Conductometric	Change in conductance	Low cost, fast response.	Highly buffered solution may interfere.

Source: Modified from Kulkarni *et al.* (2022).

mL (after preconcentrating using a membrane) with a response time of <3 h in comparison to 10 days required for standard methods. This nanobiohybrid sensor can be used as a rapid screening technique for contaminated water sources (Martín *et al.*, 2015).

9.5.2 Optical nanobiohybrid sensors

This biosensor class utilizes a transducer to detect optical variations in the incoming light as a result of the analyte interaction with the bioreceptor, and the amplitude of these variations can be correlated with the analyte concentration. The major benefits of using optical biosensors include their noninvasive measurement, simplicity, insensitivity to electromagnetic interference, small instrumentation, and *in vivo* applicability due to their non-electrical nature. In these biosensors, a variation in absorption, transmission, refraction, reflection, phase, amplitude, frequency, and/or light polarization can be measured in response to physical/chemical changes noted in the bioreceptors. Several optical principles like surface plasmon resonance (SPR), evanescent wave (EW) fluorescence, optical waveguide interferometry, chemiluminescence, fluorescence, refractive index, and surfaced-enhanced Raman scattering can be used for the detection (Figure 9.2). They can be label-free or labeled detection based on whether labels with particular properties are used for detection of a specific target rather than direct measurement of the binding events. Several plasmonic nanomaterials like Au, Ag, and Pt nanoparticles are strong absorbers and scatterers of light and are extremely useful for the construction of optical nanobiohybrid sensors. They can be exploited for color-changing sensors, labeling, and surface-enhanced spectroscopy in order to monitor molecular binding interactions. By tuning the morphology and composition of these nanomaterials, the optical responses can be easily varied from the ultraviolet (UV) to near-infrared (NIR) regions of the electromagnetic spectrum. Nanomaterials can act as (Gao *et al.*, 2020)

(a) carrier for antibodies and enzymes and help to amplify the signals by concentrating the bioreceptors
(b) enzyme mimic to replace natural enzymes
(c) signal transducer

Absorption-based biosensors are simple and low-cost instruments that can differentiate and quantify analytes based on the specific wavelength transmitted by them. The light can be guided from the source to the sample and then to the detector using optical fibers in some cases. Based on the optical configuration, they can be classified as intrinsic or extrinsic based on whether the light is modulated

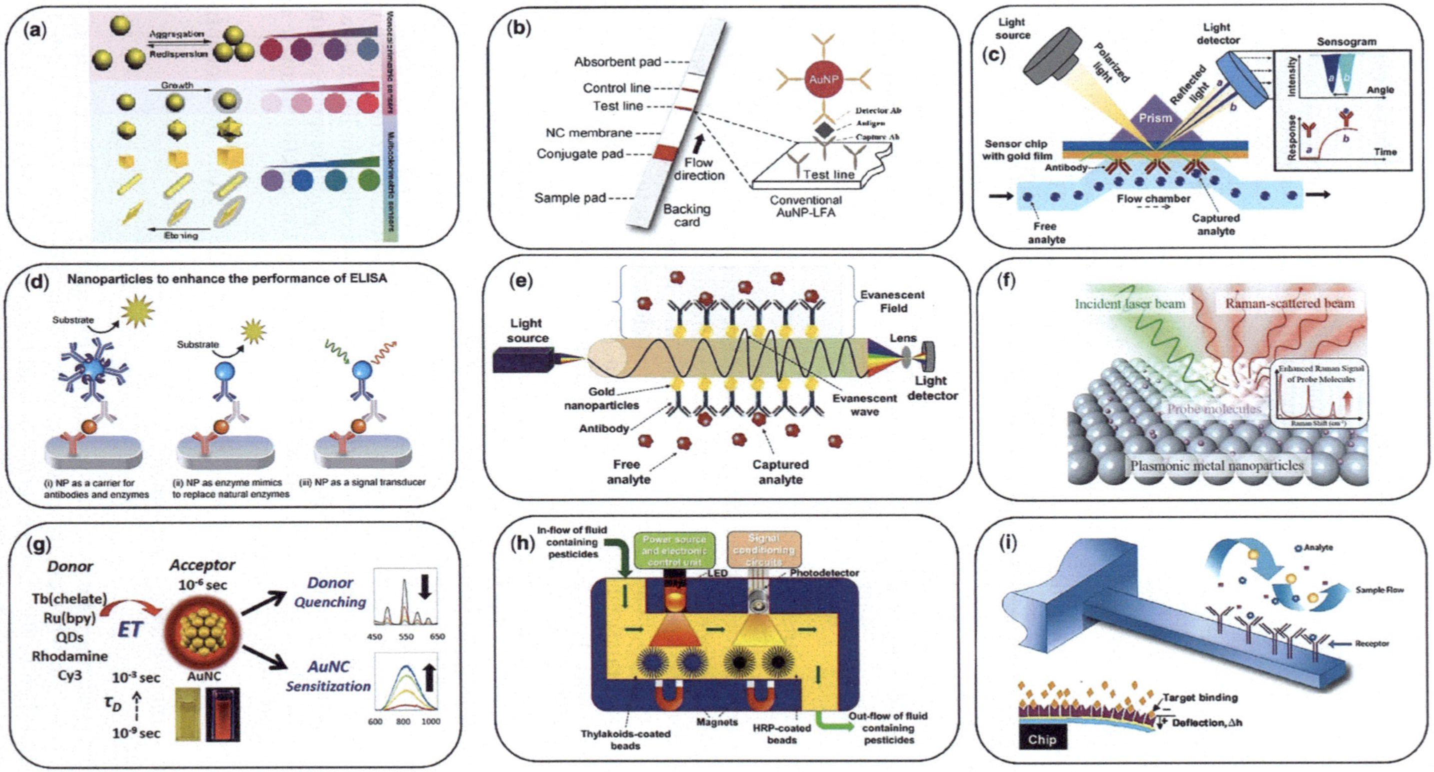

Figure 9.2 Different techniques used in optical nanobiosensors: (**a**) colorimetric (size- and shape-based color changes in nanomaterials), (**b**) nanoparticle-modified lateral flow strip, (**c**) surface plasmon resonance (SPR) nanobiosensor, (**d**) nano-enhanced ELISA, (**e**) nano-enhanced evanescent wave-based optical fiber, (**f**) nano-enhanced SERS, (**g**) FRET pairs using AuNC as quencher, (**h**) chemiluminescence nanobiosensor, (**i**) nanocantilever (Gao *et al.*, 2020; Jeon *et al.*, 2016; Mauriz, 2020; Oh *et al.*, 2016; Naresh & Lee, 2021; SoltanRezaee & Bodaghi, 2020).

in the propagating fiber itself or if the modulation takes place on another medium and the signal is recollected and transmitted. Tapered optical fibers can be frequently used for measuring different optical parameters using different techniques, one of which is the evanescent field sensing. When light passes to the sample from the optical fiber, evanescence waves are produced at the interface by total internal reflection (TIR), and there is an exponential decay of the signal intensity proportional to the distance from the interface. These evanescent waves can be used to understand the analyte binding and concentration (Touhami, 2014).

Another example of label-free sensing is the optical detection using a SPR biosensor. Here, an amplified light intensity illuminates the metal–dielectric interface at a specific angle and excites electrons in the metal to generate electromagnetic waves called surface plasmons. A reduced reflected light intensity relative to the incident light can be observed only at the resonance angle, wherein the incident light photons have similar momentum to that of the surface plasmons, which is necessary for creating the resonance. The propagation of the electromagnetic waves depends on the refractive index of the material or density of molecules close to the metallic surface that is functionalized with the bioreceptor. Specific molecular interactions or reactions caused by the analyte binding can cause refractive index changes that can be detected by the SPR as a function of resonance angle changes. By exploring the SPR properties for metallic nanomaterials like Ag and Au, a new phenomenon termed localized surface plasmon resonance (LSPR) could be witnessed (Naresh & Lee, 2021). Here, the main difference is that the plasmon resonance is governed by the total internal reflection occurring locally at the nanomaterial surface rather than at the metal surface in the case of SPR. This enables the colorimetric detection of different analytes based on the shift in wavelength or intensity of the LSPR peak due to changes in interparticle distance, size, shape, or refractive index of the bioreceptor-bound NPs after analyte interaction. The nanohybrids can also be immobilized on paper or a cellulose membrane to give paper-based assays and lateral flow strips that can help in the visual detection using naked eyes (Mauriz, 2020; Naresh & Lee, 2021). For example, several heavy metals can selectively etch the ends of gold nanorods and reduce their aspect ratio and finally convert them into spheres. This decrease in nanorod length could correspondingly result in a blue shift in the longitudinal SPR based on the concentration of Cr (VI) in water (Alex *et al.*, 2016).

Fluorescence-based biosensors can detect fluorescence intensity changes caused directly by the presence of fluorescent target analytes or indirectly by using fluorophore labels for indicating the analyte concentration, among which the latter techniques are employed more commonly due to the absence of inherent fluorescence for common analytes. Fluorescent molecules should be able to absorb UV/visible light of a specific frequency, get excited and emit light of lower frequency (visible) due to energy losses resulting from vibrational relaxation. The method has high sensitivity, selectivity, and fast response time. The commonly used fluorophores for this class of biosensors include dyes, QDs, and fluorescent proteins. The main strategies followed for detection are:

(1) Fluorescence quenching (turn-off) or enhancement (turn-on) due to the presence of target analyte.
(2) Fluorophores can be added as labels for enzyme-linked immunosorbent assays (ELISAs) for antigen quantification.
(3) Fluorescence resonance energy transfer (FRET) involves nonradiative energy transfer to an acceptor molecule at the ground state from an excited donor molecule via long-range multipole interactions. As FRET is a highly sensitive strategy to detect changes in small distance between biomolecules (angstroms to nanometers), it can be used in several aptamer-based detection techniques to form active nanodevices like molecular beacons, optical scissors, and duplex–tetraplex devices (Gaviria-Arroyave *et al.*, 2020). Gold NPs have been commonly used as fluorescence quenchers in several FRET biosensors for the detection of specific DNA sequences. In one study, gold NPs were attached to oligonucleotides labeled with a thiol group at one end and a fluorophore at the other end. This nanobiohybrid could readily self-assemble

to form a constrained arch-like structure on the NP surface, wherein the quenching happens due to the proximity of the fluorophore and quencher within the Förster distance. When the target analyte was added to this construct, a conformational change could restore the quenched fluorescence of the fluorophore due to an increase in separation distance between the fluorophore and quencher. This strategy can be used for the identification of bacteria and viruses in environmental samples (Oh *et al.*, 2016).

Luminescence-based optical sensors can be categorized as chemiluminescent and bioluminescent sensors (Naresh & Lee, 2021). In contrast to fluorescence, the triggered level of the target atoms/molecules can be acquired by an exothermic chemical reaction, and during the return of the excited species to the ground state, emission of light happens, which may or may not be accompanied by the liberation of heat. When these chemical reactions take place in biological entities, the phenomenon is termed as bioluminescence. The method is simple, sensitive and has a wide calibration limit. By using several nanomaterials for the chemiluminescence-based studies, the intrinsic sensitivity and applicability of these devices can be greatly enhanced (Mauriz, 2020).

Surface-enhanced Raman scattering (SERS) nanosensors are ultrasensitive nondestructive spectroscopic devices that can perform single molecular level identification of analytes (Naresh & Lee, 2021). Biorecognitive interactions can produce resonance enhancements in metal nanoclusters and are instrumental to study antigen–antibody, lectin–sugar, and protein–receptor interactions. Nanoclusters can be triggered to bind or dissociate based on the addition of analytes, and these nanomaterials are placed at a specific distance from a reflecting and preferably an electron-conducting substrate surface to obtain clear detectable signals (Jeon *et al.*, 2016).

Few other examples of optical nanosensors in combination with other transduction methods can also be noted. SoltanRezaee and Bodaghi (2020) have devised a mechanical cantilever-based nanosensor for the detection of analytes. Nanocantilever modified with the bioreceptor molecules can deflect proportional to the concentration of the analyte binding to the nanocantilever, and the deflection can be quantified optically using a laser (SoltanRezaee & Bodaghi, 2020). Liu *et al.* (2014) had constructed an opto-magnetic nanobiohybrid sensor for mercury detection using a sandwich assay. A core-shell nanomaterial made of magnetic silica spheres coated with gold was taken with Raman-labeled Au NPs and each bound to complementary DNA strands. These DNA strands can only fully hybridize in the presence of mercury, which reduces the inter-probe spacing and creates a plasmonic hotspot; the hybridized molecules can be magnetically separated and detected using SERS (Liu *et al.*, 2014).

9.5.3 Magnetic nanobiohybrid sensors

Magnetic nanoparticles (MNPs) can have interesting properties as they reach single domain and can be superparamagnetic (Fe_3O_4) and be made of various materials like greigite (Fe_3S_4), maghemite (g-Fe_2O_3), and various types of ferrites ($MeOFe_2O_3$, where Me = Ni, Co, Mg, Zn, Mn, etc.), all of which are useful in making powerful and versatile biosensors. A major advantage of this type of transducer is that they give low background signal (i.e., can be used for turbid samples) and can work regardless of the optical properties of the solution. Biomolecule-bound magnetic nanomaterials can help in separating and enriching the target molecule of interest using various methods. For magnetic cell sorting, magnetically tagged cells can be manipulated, isolated, and analyzed using magnetic field gradients. Thus, MNPs can facilitate the pre-concentration of analytes prior to use for detection by optical or electrochemical transducers (Wang *et al.*, 2022). Magnetic sensors have been created by employing various physical phenomena such as electromagnetic induction, Hall effect, giant magnetoresistance (GMR) (Mani *et al.*, 2011), nuclear magnetic resonance (NMR) (Luo & Alocilja, 2017), anisotropic magnetoresistance (AMR), tunnel magnetoresistance (TMR), and giant magnetoimpedance (GMI). [Some of these methods are shown in (Figure 9.3).] For high sensitivity applications, superconducting quantum interference devices (SQUID) are one of the primary choices (Khan *et al.*, 2021).

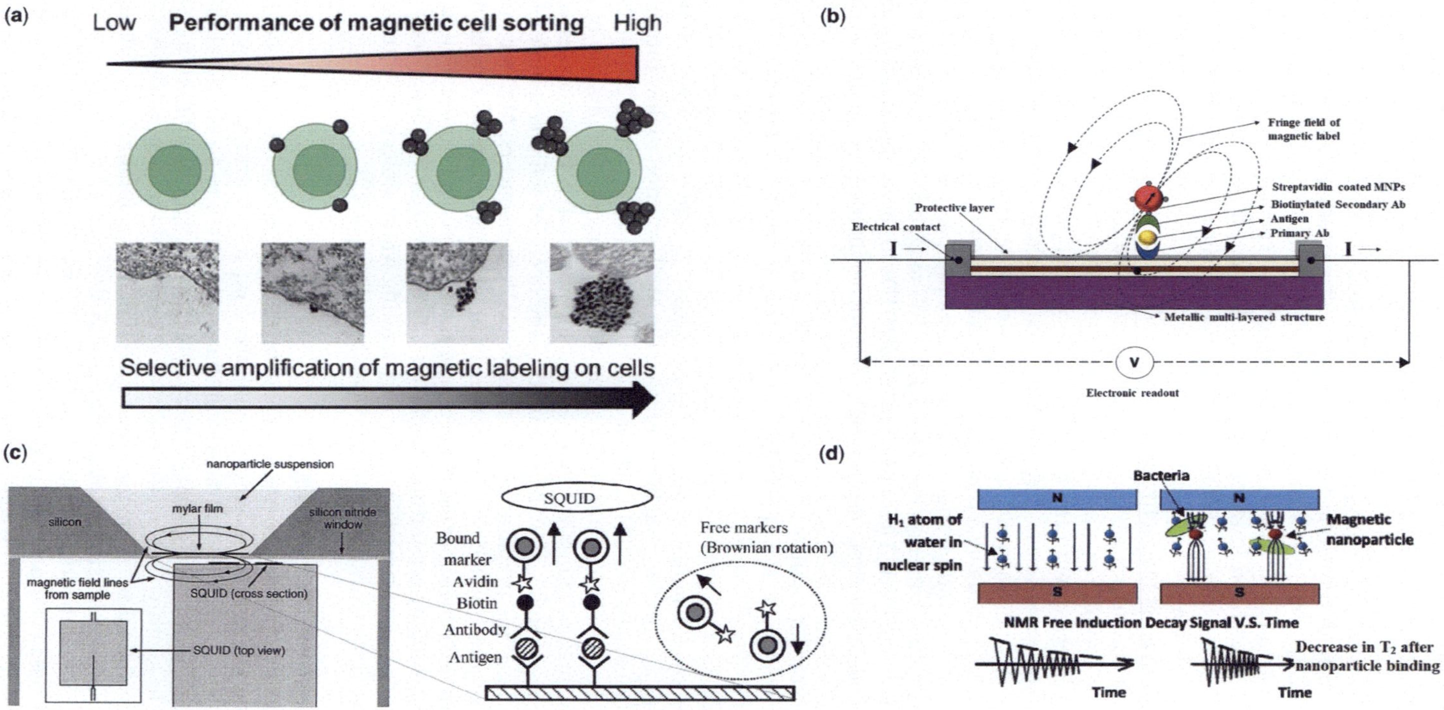

Figure 9.3 (**a**) Magnetic cell sorting (Wang *et al.*, 2022). Different technologies for target analyte detection such as (**b**) GMR (Mani *et al.*, 2011), (**c**) SQUID (Chemla *et al.*, 2000), and (**d**) NMR (Luo & Alocilja, 2017).

For magnetic immunoassays where target analytes are recognized using bioreceptor molecules that are magnetically tagged, their magnetic field can be directly quantified using a magnetometer to estimate the analyte concentration. In these techniques, when the magnetic tags used here are taken to nano dimensions, their magnetic properties are enhanced. In a special type of immunoassay, a 'microscope' based on a high-transition temperature direct current or DC SQUID in combination with superparamagnetic nanomaterials can be used in the rapid recognition of biomolecular target analytes. The device generally has one/two superconductors separated by thin insulating layers to create one/two parallel Josephson junctions, which can be manifested to measure extremely small magnetic fields in biological samples. When antibody-tagged superparamagnetic nanomaterials are added to samples in a well where 1-s pulses of magnetic field are employed parallel to the SQUID, the nanomaterials develop a net magnetization in the presence of the aligning field that relaxes upon turning off the field. Unbound nanomaterials with a diameter smaller than 100 nm can swiftly undergo Brownian relaxation and produce no measurable signal (so fast that the large background of unbound nanoparticles does not interfere with the quantification of nanoparticle binding events). The target-bound nanomaterials are captured as they undergo Néel relaxation with a measurable relaxation timescale of a few milliseconds-seconds, resulting in a slowly decaying magnetic flux, which can be measured by the SQUID. Thus, using a homogeneous assay, nanomaterial-bound target analytes can be distinguished without separation and removal of unbound particles. A SQUID-based bulk-type magnetometer has a detection limit down to around 1×10^{12} magnetic NPs/mL for particle sizes (10–20 nm) with 30 μL sample volumes. The sensitivity of SQUIDs to incredibly small localized magnetic signals can be enhanced by downsizing the SQUID dimensions to the nanometric scale (Chemla *et al.*, 2000).

Alternatively, GMR sensors can be used to detect 16-nm MNPs with a resolution of up to 23 particles. For these transducers, the application of external magnetic fields can alter the electrical resistance of the conductor. This phenomenon occurs strictly in thin nanoscale magnetic layers separated by nanoscale spacing between them. The sensitivity can be enhanced by trapping MNPs near the GMR sensor edge, where the strong fringe-fields are active and thus, improve the GMR response per MNP. This can be achieved by permitting the MNP solution to flow through microfluidic channels parallel to the edge of the GMR sensor under Poiseuille flow (Little, 2012).

NMR is based on the absorption and re-emission of electromagnetic radiation by certain nuclei when exposed to a magnetic field. NMR can be used for non-destructive analyses since the signal can penetrate turbid raw samples, thereby eliminating the need for tedious sample preparation and reducing analysis time. For the preparation of portable NMR transducers, MNPs can be used as the proximity sensors to augment molecular interactions. The binding of analytes to bioreceptor-bound MNPs can help to form MNP-conjugated clusters and the interference from unbound MNPs can be minimized well by magnetically separating the clusters. The formation of clusters causes a reduction in the bulk spin–spin relaxation time (T_2) of the surrounding water molecules, which can be used to assess the target analyte concentration (Hojjati-Najafabadi *et al.*, 2022).

9.5.4 Gravimetric nanobiohybrid sensors

This class of biosensors determines the analyte concentration based on the mass changes noted after the binding of the analyte to a bioreceptor immobilized on a nanotransducer, which results in a change in the vibration frequency of the transducer that is converted to an electrical signal (Touhami, 2014). A piezoelectric material (like quartz) can be subjected to light or allowed to vibrate using an electric field in order to identify the specific angle of emission of the electron waves. The frequency change corresponds to the mass of the absorbed material. The commonly used devices are quartz crystal microbalance (QCM) and surface acoustic wave (SAW).

Acoustic wave nanobiosensors provide good improvement in the analyte sensitivity. The mass-amplified QCM assay variant uses bioreceptor-functionalized sol particles that can attach indirectly to an electrode surface through complexation of the target that has been captured on an antibody-functionalized electrode surface. These are most commonly used for gaseous pollutants only. The significant mass changes after binding can considerably alter the vibrational frequency of the quartz crystal, thereby giving higher sensitivity of the analyte. The diameter of the sol is preferably taken in the nano range (5–100 nm) as it can greatly improve the signal. The bioassay can be performed in a competitive mode. The preferred diameter of sol particles is in the range of 5–100 nm, and the nanomaterials used are generally of higher densities (e.g., Au, CdS, Pt, TiO_2, and polymers) (Jianrong *et al.*, 2004).

Gravimetric biosensors are generally found to be inconvenient for usage in mixed aqueous samples due to the poor discrimination of specific compounds. Though piezoelectric biosensors have been widely used for sensing gaseous-phase pollutants, some studies are available for testing in water samples using QCM and SAW. Cross-linked polymer matrices and molecularly imprinted polymers (MIP) can exhibit good viscoelasticity and undergo swift uptake of the specific pollutant through swelling, which could result in a corresponding shift in mass and vibrating frequency of the polymer. The material can be tailored for detection of specific environmental pollutants such as toluene, benzene, xylene (BTX), and the discrimination of few PAHs in mixed samples. However, polymer-based gravimetric sensors may suffer from setbacks related to complex response mechanisms, insufficient frequency changes due to the various processing done during polymer coating, and requirement of good understanding prior to application in environmental samples (Zarejousheghani *et al.*, 2021). Sadek and his group have used various nanobiohybrids for the QCM-based detection of different environmental pollutants such as methylene blue, Pb (II), and Cd (II) (Figure 9.4a) (Al-Qasmi *et al.*, 2022). Hüseynli *et al.* (2019) have used QCM chips imprinted with co-complexed allyl mercaptane and Hg (II) ions for the detection of Hg (II) ions in wastewater (Hüseynli *et al.*, 2019).

The incorporation of nanomaterials in conventional piezoelectric transducers can yield several advantages since nanobiohybrid-based piezoelectric devices can:

(1) Endure greater deformations (this is essential for the assembly of robust, yet flexible and conformable devices)
(2) Have considerably larger piezoelectric coefficients
(3) Have larger force-to-displacement sensitivities
(4) Require lesser driving voltage and energy
(5) Be lighter and faster
(6) Have good transparency
(7) Use materials that are not piezoelectric in their bulk state, but achieve piezoelectric nature when they are used in the single-layer (or odd number of layers) state
(8) Allow the material to be engineered more effectively
(9) Be simpler to simulate at the atomic level and exploit the mechanisms that are active only when the dimensions become comparable or smaller to crucial characteristic lengths in the nanoscale regime (e.g. exciton Bohr radius, wavelength of light, phonon mean free path, and dimensions of magnetic domains)
(10) Take advantage of the unique fabrication techniques, especially wet-chemistry (i.e. simple, low-cost, and requires lower temperature, and the nanotransducer can be co-integrated with electronics and be fabricated as flexible PCBs), utilize natural materials (e.g. viruses or spider silk), and can be 3D printed.
(11) Be fabricated with lesser restrictions regarding the dimensions/shape/structure, possibilities to interrupt the anti-symmetry of the strain–piezopotential relation, mechanical input types, contact positions, and the option to intimately bring together various piezoelectric structures in a single device (e.g. two nanomaterials with opposite c-axes can be linked in series).
(12) Be designed as an array of transducers with exceptional spatial resolution.
(13) Facilitate the preparation of wireless nanotransducers for energy harvesting applications from environmental water samples (Falconi, 2019).

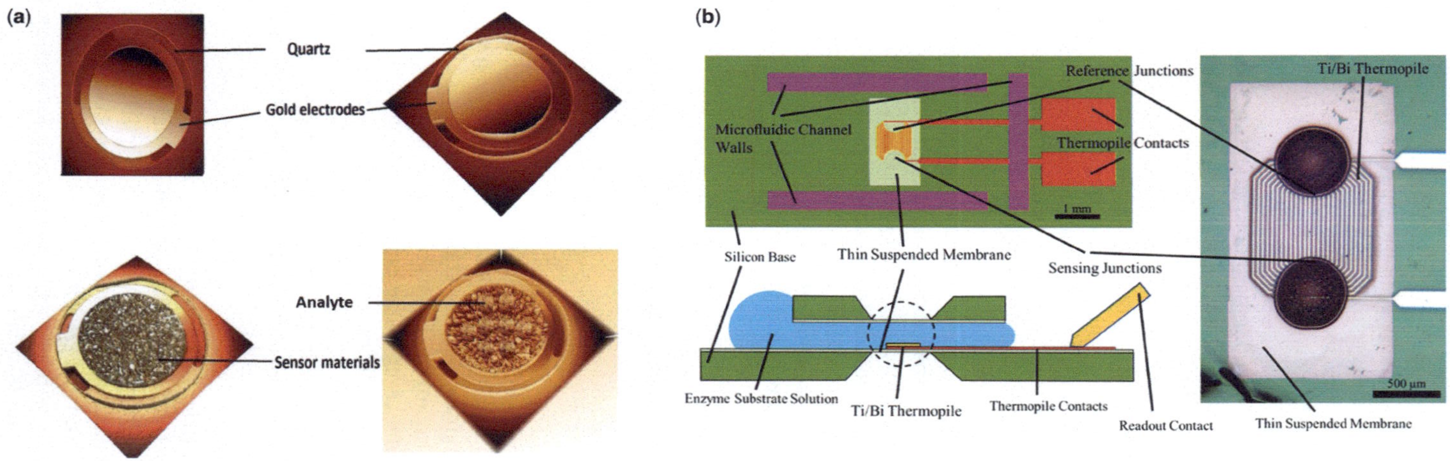

Figure 9.4 (**a**) Quartz crystal microbalance coated with nanobiohybrids for sensing Cd (Al-Qasmi *et al.*, 2022) (II) and (**b**) nanocalorimeter platform consisting of a microfluidic channel and a 27-junction Bi/Ti thermopile calorimeter sandwiched by thin polymer membranes positioned on a silicon base (Hüseynli *et al.*, 2019).

9.5.5 Calorimetric nanobiohybrid sensors

Thermal or calorimetric biosensors or enzyme thermistors utilize transducers that detect interactions between biological recognition molecules and an analyte that can result in a temperature change (absorption or evolution of heat). The total heat energy absorbed/released or the temperature change (ΔT) measured by calorimetric biosensors is directly related to the enthalpy (ΔH) and the total number of product molecules (n_p) resulting from the biochemical reaction and inversely related to the heat capacity (C_p) of the reaction, as shown by the equation $\Delta T = -(n_p \, \Delta H)/C_p$ (Naresh & Lee, 2021). Calorimetric biosensors can provide full thermodynamic profiling, such as the measurement of enthalpy, free energy, entropy, and the kinetic analysis of bioprocesses. Some advantages of calorimetric biosensors are that the signal is not disturbed by electrochemical and optical properties of the sample and that the instrument does not require frequent recalibration.

Enzymes are frequently used as the bioreceptors for thermal biosensors due to the exothermic nature of the reactions catalyzed by them. However, the application of thermal biosensors is limited as they lack specificity in quantifying temperature changes related to the biochemical reaction alone (non-specific heat changes can also be measured) and also may involve time-consuming experimental procedures. This can be overcome by the integration of nanomaterials to improve the sensitivity and response rate (Touhami, 2014). The configuration for flow injection analysis consists of a nanomaterial-immobilized biocatalyst reactor in combination with a differential temperature measurement system using a pair of thermal transducers like thermistors or thermopiles that are positioned across the enzyme column reactor. This differential temperature analyser reduces the effects of ambient temperature fluctuations and helps measuring the catalysis-specific thermal signal that is proportionate to the analyte concentration. The enthalpy shifts normally observed for enzymatic reactions are usually in the range from −10 to −200 kJ/mol, and this is sufficient to quantify the substrate at environmentally relevant levels for a range of pollutants.

Thermometric methods are advantageous for multiple-step catalysis as the sum of all enthalpies gives the assay's sensitivity (Naresh & Lee, 2021). Additionally, nanocalorimeter-based thermal ELISAs can quickly detect small temperature changes of nanojoules in enzyme-based immunoassays to deliver fast, quantifiable electronic readouts (Figure 9.4b) (Hüseynli *et al.*, 2019). The most commonly employed temperature sensors or transducers in these devices are thermistors or thermopiles. The thermistor is a sensitive thermal transducer, whose electrical resistance varies based on temperature changes. Thermopiles determine the temperature variation between two thermocouple junctions made using metals or semiconductors and are connected in series. Other recently used thermal transducers include liquid gas expansion, bimetallic strips, and pyroelectric systems, and microelectromechanical systems (MEMS). MEMS-based thermal transducers have multiple advantages like economic integration of miniaturized devices and their batch fabrication, simultaneous measurement of different samples in parallel, improved thermal isolation, low thermal mass, reduced sample volume, linear range, higher sensitivity, quick response time, and low power consumption (Naresh & Lee, 2021).

9.6 CONCLUSION

Sensors have become an integral part of many modern-day technologies and offer versatile applications in the fields of environmental and healthcare management. The exponential growth in the field of sensing is primarily attributed to the many advancements in the field of material sciences and nanotechnology. The variations in material properties at the nanoscale are important factors that enable sensitive detection of low levels of analytes with high precision and reliability in practical applications.

Owing to the depreciation in water quality over the years, continuous and meticulous detection of toxic pollutant levels in water has become an inevitable requirement for devising sustainable solutions. Biosensors based on novel nanobiohybrid materials with high sensitivity and selectivity have been widely explored for the use in this area. Nanobiohybrids, with augmentative characteristics of biological components and unique material properties in nanoscale have shown to be of value in design of sophisticated sensing systems. In this chapter, we have highlighted the various choices of biological and synthetic material components in the design of nanobiohybrids along with the various mechanisms of biological signal transduction for sensing applications. These sensing systems come in various shapes and sizes, with scope for miniaturization and portability or robust *in-situ* analysis of pollutants in water.

REFERENCES

Alex S., Chandrasekaran N. and Mukherjee A. (2016). State-of-the-art strategies for the colorimetric detection of heavy metals using gold nanorods based on aspect ratio reduction. *Analytical Methods*, **8**(10), 2131–2137, https://doi.org/10.1039/C5AY03428K

Al-Qasmi N., Al-Gethami W., Alhashmialameer D., Ismail S. and Sadek A. (2022). Evaluation of green-synthesized cuprospinel nanoparticles as a nanosensor for detection of low-concentration Cd(II) ion in the aqueous solutions by the quartz crystal microbalance method. *Materials*, **15**(18), 6240, https://doi.org/10.3390/ma15186240

Bagheri S., Termehyousefi A., Mansouri N., Amani Babadi A., Abd Karim M. and Adib Kadri N. (2017). Carbon-based nanobiohybrid thin film for amperometric glucose sensing. *ACS Biomaterials Science and Engineering*, **3**(9), 2059–2063, https://doi.org/10.1021/acsbiomaterials.7b00325

Banerjee A., Maity S. and Mastrangelo C. (2021). Nanotechnology for biosensors: a review. arXiv preprint arXiv:2101.02430.

Chemla Y., Grossman H., Poon Y., McDermott R., Stevens R., Alper M. and Clarke J. (2000). Ultrasensitive magnetic biosensor for homogeneous immunoassay. *Proceedings of the National Academy of Sciences*, **97**(26), 14268–14272, https://doi.org/10.1073/pnas.97.26.14268

Chevolot Y., Vidal S., Laurenceau E., Morvan F., Vasseur J.-J. and Souteyrand E. (2010). Carbohydrates as recognition receptors in biosensing applications. In: Recognition Receptors in Biosensors, M. Zourob (ed.), Springer, New York, NY, pp. 275–341.

Corbisier P., van der Lelie D., Borremans B., Provoost A., de Lorenzo V., Brown N. L., Lloyd J. R., Hobman J. L., Csöregi E., Johansson G. and Mattiasson B. (1999). Whole cell- and protein-based biosensors for the detection of bioavailable heavy metals in environmental samples. *Analytica Chimica Acta*, **387**, 235–244, https://doi.org/10.1016/S0003-2670(98)00725-9

Curulli A. (2021). Electrochemical biosensors in food safety: challenges and perspectives. *Molecules*, **26**(10), 2940, https://doi.org/10.3390/molecules26102940

Damiati S. and Schuster B. (2020). Electrochemical biosensors based on S-layer proteins. *Sensors*, **20**(6), 1721, https://doi.org/10.3390/s20061721

Dincer C., Bruch R., Costa-Rama E., Fernández-Abedul M., Merkoçi A., Manz A., Urban G. A. and Güder F. (2019). Disposable sensors in diagnostics, food, and environmental monitoring. *Advanced Materials*, **31**(30), 1806739, https://doi.org/10.1002/adma.201806739

Falconi C. (2019). Piezoelectric nanotransducers. *Nano Energy*, **59**, 730–744, https://doi.org/10.1016/j.nanoen.2019.03.027

Gao Y., Zhou Y. and Chandrawati R. (2020). Metal and metal oxide nanoparticles to enhance the performance of enzyme-linked immunosorbent assay (ELISA). *ACS Applied Nano Materials*, **3**(1), 1–21, https://doi.org/10.1021/acsanm.9b02003

Gaviria-Arroyave M., Cano J. and Peñuela G. (2020). Nanomaterial-based fluorescent biosensors for monitoring environmental pollutants: a critical review. *Talanta Open*, **2**, 100006, https://doi.org/10.1016/j.talo.2020.100006

Guo Z., Richardson J. J., Kong B. and Liang K. (2020). Nanobiohybrids: materials approaches for bioaugmentation. *Science Advances*, **6**(12), 330–348, https://www.science.org/doi/10.1126/sciadv.aaz0330

Heller D. A., Pratt G. W., Zhang J., Nair N., Hansborough A. J., Boghossian A. A., Reuel N. F., Barone P. W. and Strano M. S. (2011). Peptide secondary structure modulates single-walled carbon nanotube fluorescence as a chaperone sensor for nitroaromatics. *Proceedings of the National Academy of Sciences*, **108**(21), 8544–8549, https://doi.org/10.1073/pnas.1005512108

Hojjati-Najafabadi A., Mansoorianfar M., Liang T., Shahin K. and Karimi-Maleh H. (2022). A review on magnetic sensors for monitoring of hazardous pollutants in water resources. *Science of the Total Environment*, **824**, 153844, https://doi.org/10.1016/j.scitotenv.2022.153844

Huang X., Tan C., Yin Z. and Zhang H. (2014). 25th anniversary article: hybrid nanostructures based on two-dimensional nanomaterials. *Advanced Materials*, **26**(14), 2185–2204, https://doi.org/10.1002/adma.201304964

Hüseynli S., Çimen D., Bereli N. and Denizli A. (2019). Molecular imprinted based quartz crystal microbalance nanosensors for mercury detection. *Global Challenges*, **3**(3), 1800071, https://doi.org/10.1002/gch2.201800071

Jeon T., Kim D., Park S.-G., Kim S.-H. and Kim D.-H. (2016). Nanostructured plasmonic substrates for use as SERS sensors. *Nano Convergence*, **3**(1), 18, 10.1186/s40580-016-0078-6

Jianrong C., Yuqing M., Nongyue H., Xiaohua W. and Sijiao L. (2004). Nanotechnology and biosensors. *Biotechnology Advances*, **22**(7), 505–518, 10.1016/j.biotechadv.2004.03.004

Khan M. A., Jian S., Bodong L., Alexander P. and Jürgen K. (2021). Magnetic sensors – a review and recent technologies. *Engineering Research Express*, **3**, 022005, https://doi.org/10.1088/2631-8695/ac0838

Kulkarni M., Ayachit N. and Aminabhavi T. (2022). Biosensors and microfluidic biosensors: from fabrication to application. *Biosensors*, **12**(7), 543, https://doi.org/10.3390/bios12070543

Kuswandi B. (2019). Nanobiosensor approaches for pollutant monitoring. *Environmental Chemistry Letters*, **17**, 975–990, https://doi.org/10.1007/s10311-018-00853-x

Liang J., Zulkifli M., Choy S., Li Y., Gao M., Kong B., Yun J. and Liang K. (2020). Metal–organic framework–plant nanobiohybrids as living sensors for on-site environmental pollutant detection. *Environmental Science and Technology*, **54**(18), 11356–11364, https://doi.org/10.1021/acs.est.0c04688

Little C. A. (2012). GMR Fringe-Field Trapping and Detection of Magnetic Nanoparticles. Doctoral Dissertation, University of Colorado at Boulder.

Liu M., Wang Z., Zong S., Chen H., Zhu D., Wu L., Hu G. and Cui Y. (2014). SERS detection and removal of mercury(II)/silver(I) using oligonucleotide-functionalized core/shell magnetic silica sphere@Au nanoparticles. *ACS Applied Materials and Interfaces*, **6**(10), 7371–7379, https://doi.org/10.1021/am5006282

Luo Y. and Alocilja E. (2017). Portable nuclear magnetic resonance biosensor and assay for a highly sensitive and rapid detection of foodborne bacteria in complex matrices. *Journal of Biological Engineering*, **11**(1), 14, https://doi.org/10.1186/s13036-017-0053-8

Mani V., Chikkaveeraiah B. and Rusling J. (2011). Magnetic particles in ultrasensitive biomarker protein measurements for cancer detection and monitoring. *Expert Opinion on Medical Diagnostics*, **5**(5), 381–391, https://doi.org/10.1517/17530059.2011.607161

Martín M., Salazar P., Jiménez C., Lecuona M., Ramos M., Ode J., Alcoba J., Roche R., Villalonga R., Campuzano S. and Pingarrón J. M. (2015). Rapid *Legionella pneumophila* determination based on a disposable core-shell Fe_3O_4 @poly(dopamine) magnetic nanoparticles immunoplatform. *Analytica Chimica Acta*, **887**, 51–58, https://doi.org/10.1016/j.aca.2015.05.048

Martinez J., Misra N., Wang Y., Stroeve P., Grigoropoulos C. and Noy A. (2009). Highly efficient biocompatible single silicon nanowire electrodes with functional biological pore channels. *Nano Letters*, **9**(3), 1121–1126, https://doi.org/10.1021/nl8036504

Martín-Yerga D. (2019). Electrochemical detection and characterization of nanoparticles with printed devices. *Biosensors*, **9**(2), 47, https://doi.org/10.3390/bios9020047

Mauriz E. (2020). Clinical applications of visual plasmonic colorimetric sensing. *Sensors*, **20**(21), 6214, https://doi.org/10.3390/s20216214

Mo X. T., Zhou Y. P., Lei H. and Deng L. (2002). Microbalance-DNA probe method for the detection of specific bacteria in water. *Enzyme and Microbial Technology*, **30**(5), 583–589, https://doi.org/10.1016/S0141-0229(01)00484-7

Naresh V. and Lee N. (2021). A review on biosensors and recent development of nanostructured materials-enabled biosensors. *Sensors*, **21**(4), 1109, https://doi.org/10.3390/s21041109

Odobašić A., Šestan I. and Begić S. (2019). Biosensors for determination of heavy metals in waters. In: Biosensors for Environmental Monitoring, T. Rinken and K. Kivirand (eds.), BoD – Books on Demand, **2019**, London, United Kingdom, pp. 139–156

Oh E., Huston A., Shabaev A., Efros A., Currie M., Susumu K., Bussmann K., Goswami R., Fatemi F. K. and Medintz I. L. (2016). Energy transfer sensitization of luminescent gold nanoclusters: more than just the classical Förster mechanism. *Scientific Reports*, **6**(1), 35538, https://doi.org/10.1038/srep35538

Palchetti I. and Mascini M. (2008). Nucleic acid biosensors for environmental pollution monitoring. *The Analyst*, **133**(7), 846, https://doi.org/10.1039/b802920m

Pershina A., Sazonov A. and Filimonov V. (2014). Magnetic nanoparticles–DNA interactions: design and applications of nanobiohybrid systems. *Russian Chemical Reviews*, **83**(4), 299–322, https://doi.org/10.1070/RC2014v083n04ABEH004412

Ramezani M., Taghdisi S. M., Yazdian-Robati R., Oroojalian F., Abnous K. and Alibolandi M. (2020). Nanosensors for water safety. In: Nanosensors for Smart Cities, B. Han, V. Tomer, T. Anh Nguyen, A. Farmani and P. Kumar Singh (eds.), Elsevier, Amsterdam, Netherlands, pp. 285–301.

Ray S., Senapati T., Sahu S., Bandyopadhyaya R. and Anand R. (2018). Design of ultrasensitive protein biosensor strips for selective detection of aromatic contaminants in environmental wastewater. *Analytical Chemistry*, **90**(15), 8960–8968, https://doi.org/10.1021/acs.analchem.8b01130

Rodríguez-Delgado M., Alemán-Nava G., Rodríguez-Delgado J., Dieck-Assad G., Martínez-Chapa S., Barceló D. and Parra R. (2015). Laccase-based biosensors for detection of phenolic compounds. *TrAC Trends in Analytical Chemistry*, **74**, 21–45, https://doi.org/10.1016/j.trac.2015.05.008

Roy R., Ray S., Chowdhury A. and Anand R. (2021). Tunable multiplexed whole-cell biosensors as environmental diagnostics for ppb-level detection of aromatic pollutants. *ACS Sensors*, **6**(5), 1933–1939, https://doi.org/10.1021/acssensors.1c00329

Scognamiglio V., Raffi D., Lambreva M., Rea G., Tibuzzi A., Pezzotti G., Johanningmeier U. and Giardi M. T. (2009). *Chlamydomonas reinhardtii* genetic variants as probes for fluorescence sensing system in detection of pollutants. *Analytical and Bioanalytical Chemistry*, **394**(4), 1081–1087, https://doi.org/10.1007/s00216-009-2668-1

Shen M.-Y., Li B.-R. and Li Y.-K. (2014). Silicon nanowire field-effect-transistor based biosensors: from sensitive to ultra-sensitive. *Biosensors and Bioelectronics*, **60**, 101–111, https://doi.org/10.1016/j.bios.2014.03.057

SoltanRezaee M. and Bodaghi M. (2020). Simulation of an electrically actuated cantilever as a novel biosensor. *Scientific Reports*, **10**(1), 3385, https://doi.org/10.1038/s41598-020-60296-9

Tecon R. and Van der Meer J. (2008). Bacterial biosensors for measuring availability of environmental pollutants. *Sensors*, **8**(7), 4062–4080, https://doi.org/10.3390/s8074062

Tombelli S., Minunni M. and Mascini M. (2007). Aptamers-based assays for diagnostics, environmental and food analysis. *Biomolecular Engineering*, **24**(2), 191–200, 10.1016/j.bioeng.2007.03.003

Touhami A. (2014). Biosensors and nanobiosensors: design and applications.

Wang Z., Wang H., Lin S., Ahmed S., Angers S., Sargent E. and Kelley S. (2022). Nanoparticle amplification labeling for high-performance magnetic cell sorting. *Nano Letters*, **22**(12), 4774–4783, https://doi.org/10.1021/acs.nanolett.2c01018

Wee Y., Park S., Kwon Y., Ju Y., Yeon K.-M. and Kim J. (2019). Tyrosinase-immobilized CNT based biosensor for highly-sensitive detection of phenolic compounds. *Biosensors and Bioelectronics*, **132**, 279–285, https://doi.org/10.1016/j.bios.2019.03.008

Willner M. and Vikesland P. (2018). Nanomaterial enabled sensors for environmental contaminants. *Journal of Nanobiotechnology*, **16**(1), 95, https://doi.org/10.1186/s12951-018-0419-1

Wong M., Giraldo J., Kwak S.-Y., Koman V., Sinclair R., Lew T., Bisker G., Liu P. and Strano M. S. (2017). Nitroaromatic detection and infrared communication from wild-type plants using plant nanobionics. *Nature Materials*, **16**(2), 264–272, https://doi.org/10.1038/nmat4771

Zarejousheghani M., Rahimi P., Borsdorf H., Zimmermann S. and Joseph Y. (2021). Molecularly imprinted polymer-based sensors for priority pollutants. *Sensors*, **21**(7), 2406, https://doi.org/10.3390/s21072406

Zeng X., Andrade C., Oliveira M. and Sun X.-L. (2012). Carbohydrate–protein interactions and their biosensing applications. *Analytical and Bioanalytical Chemistry*, **402**(10), 3161–3176, https://doi.org/10.1007/s00216-011-5594-y

Zheng W., Chiamori H., Liu G., Lin L. and Chen F. (2012). Nanofabricated plasmonic nano-bio hybrid structures in biomedical detection. *Nanotechnology Reviews*, **1**(3), 213–233, https://doi.org/10.1515/ntrev-2011-0008

doi: 10.2166/9781789063592_0169

Chapter 10

Unlocking the potential of nanobiohybrids to combat environmental pollution

Amol D. Gholap[1], Navnath T. Hatvate[2], Sampada D. Sawant[1], Palak N. Khemka[1] and Lalitkumar K. Vora[3]*

[1]*Department of Pharmaceutics, St. John Institute of Pharmacy and Research, Palghar 401404, Maharashtra, India*
[2]*Institute of Chemical Technology Mumbai, Marathwada Campus, Jalna 431213, Maharashtra, India*
[3]*School of Pharmacy, Queen's University Belfast, 97 Lisburn Road, Belfast BT9 7BL, UK*
**Corresponding author: Mr. Amol Dilip Gholap, Department of Pharmaceutics, St. John Institute of Pharmacy and Research, Palghar 401404, Maharashtra, India; Email: amolgholap16@gmail.com*

ABSTRACT

The destructive effects of the world's population growth, which have led to the overuse of primary resources, are evident today. Recently, the world has hit an 8 billion population mark, which is accompanied by an increase in waste generation. Anthropogenic activities lead to environmental pollution that is a problem of great importance, and it is essential to develop economically sound, eco-friendly, and sustainable technologies to tackle this ever-increasing problem of the 21st century. Innovations are required since they can contribute to the creation of sophisticated, sustainable, and novel goods for ecological repair. Such properties can be seen in the current advancements of nanotechnologies and new performing nanomaterials. They have a substantial surface area, enzymatic activity, and responsiveness and are easily functionalized to chemically modify their characteristics. This chapter examines current advancements in bioremediation methods used in conjunction with various nanocomposites and nanobiohybrids to recover the environment from pollution. Special focus is placed on environmentally friendly and regenerative nanosolutions, as well as their secure properties.

Keywords: bioaugmentation, environment, nanobiohybrids, nanoparticles, pollutants

10.1 INTRODUCTION

10.1.1 Need for environmental bioremediation

Human activities in cities, agriculture, and industries generate plant waste and urban sewage are on the rise, creating dangerous and negative consequences on the earth's ecosystems as well as overall human health. These environmental contaminants and pollutants constantly threaten the earth's future. It is imperative that society finds innovative strategies to combat far more frequent contaminants and deploys new clean-up techniques.

There is exponential growth in new types of pollutants and their emerging resistance toward older remediation techniques (Morin-Crini *et al.*, 2022; Rathi *et al.*, 2021; Valdez-Carrillo *et al.*, 2020). Current techniques used to tackle pollution are not regarded as completely safe and do not protect the environment to the desired extent (Figure 10.1).

10.1.2 Nanotechnology for environmental remediation

Current clean-up efforts need to be improved by better techniques and to respond quickly when ecological crises such as oil spillage or forest fires occur. It is also essential to remove as many toxins

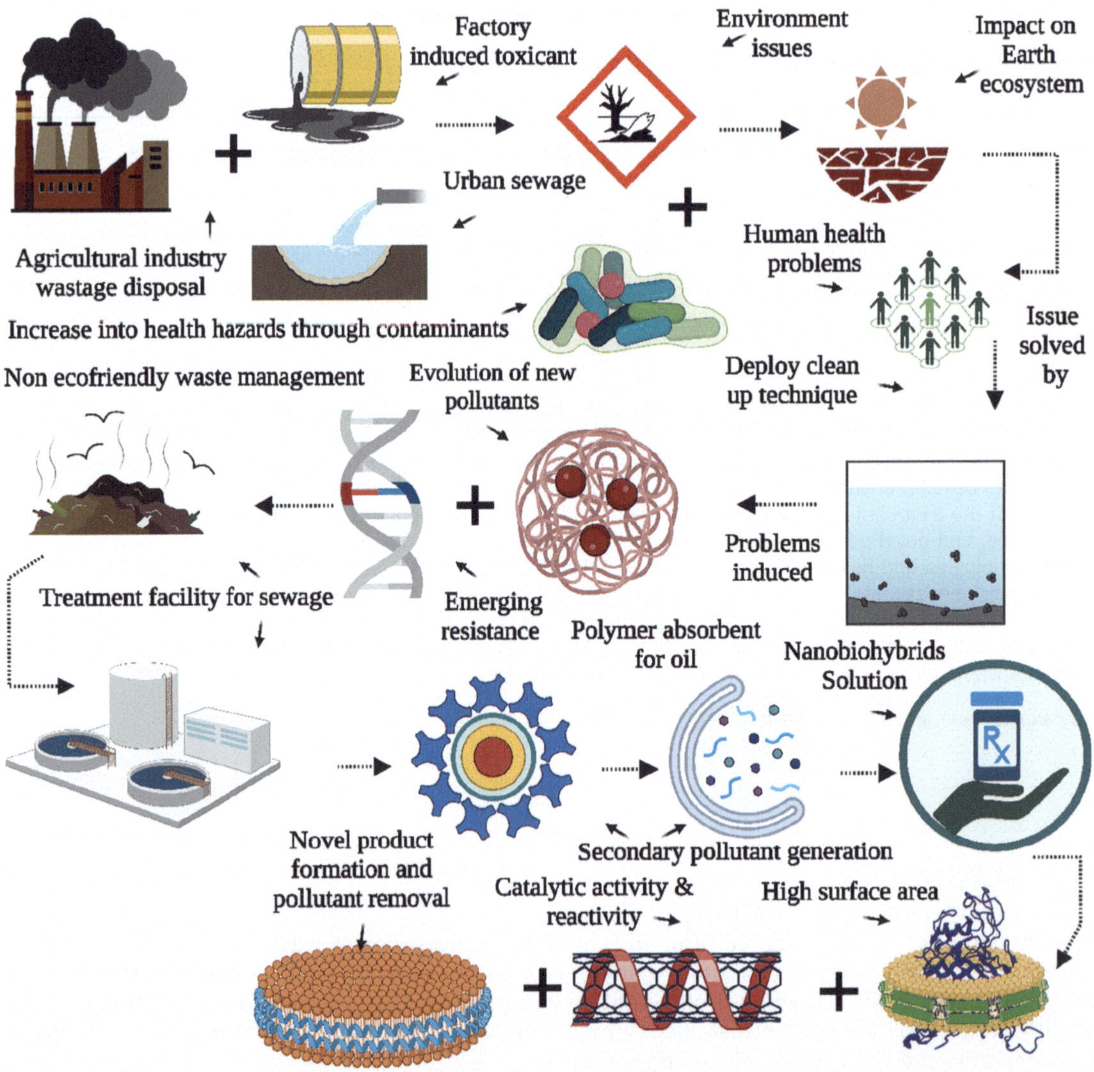

Figure 10.1 Need for nanobiohybrids for the removal of environmental pollutants produced from agricultural factories, waste materials, and urban sewage. Some modern clean-up techniques have been implemented to address these issues, but they cause secondary pollutant formation and hence limit their use. Nanobiohybrids are used to remove pollutants due to their advantages of catalytic activity, reactivity and high surface area.

as possible from the environment to protect the planet's future. Currently employed techniques for bioremediation and waste management are not always eco-friendly. For example, treatment facilities for sewage or polymer absorbents of oil contaminants generate so much secondary pollution that they do not actually compensate for the remediation action. Nanotechnologies and new nanomaterials have a wide range of uses in a variety of industries. Because of their qualities, they have a size effect, catalytic capacity, reactivity, and high surface area. In fact, nanomaterials are currently used to build sustainable, sophisticated, and novel products and processes for pollutant removal (AbouElmaaty *et al.*, 2021; Dontsova *et al.*, 2019; Ielo *et al.*, 2021b; Lemming *et al.*, 2012; Lu *et al.*, 2016; Santhosh *et al.*, 2016). Particular polymers as well as matrices are employed to create nanocomposites and mixed substances having specified biochemical and mechanical characteristics.

Within this context, the following suitable nanomaterials that may be employed as nanofillers can be categorized:

(i) Nanoparticles (NPs) of metal oxides, such as TiO_2, ZnO, Al_2O_3, Fe_2O_3, and SiO_2 produced through hydrothermal or sol–gel processes.
(ii) Metal nanomaterials, such as Pd, Pt, Ag, Al, Ru, Ni, Cu, and Au NPs characterized by different molecular, optical, and electric attributes.
(iii) Silica-based NPs, and zeolite with mesoporous shape.
(iv) Ceramic NPs.
(v) Bimetallic NPs.
(vi) Carbon NPs, that is, carbon nanotubes CNT, graphene sheets, and fullerene.
(vii) Bionanomaterials, polymeric nanomaterials, core–shell nanomaterials, and metal-organic frameworks.

Such composite sustainably grown NPs are created through diverse combinations involving both crystalline or amorphous stages, as well as through other chemical modifications (Descalzo *et al.*, 2006; Saleh, 2020). They can be classified into:

(1) Crossovers: molecular or atomic combinations from various elements with chemical reactions among them.
(2) Nanobiohybrids: involve molecular or atomic combinations of various elements with biological component (enzyme or microorganism) having (bio)chemical reactions among each other.
(3) Composite materials: combinations comprising elements comprised of a medium having micrometre scattering.
(4) Nanocomposites: submicrometric combinations (1–100 nm) with elements of similar nature. The nanocomposite method can include hybrid methods that use organic or inorganic constituent units and have nanometric dimensions.

10.1.3 Nanohybrids

According to the interaction of the two phases, nanohybrid materials can be classified into the following (García-Martínez & Collar, 2020):

Class I hybrids: Class I hybrids are characterized by the interaction between the two phases, which are very weak, for instance, van der Waal interactions, hydrogen connections, or low electrostatic forces.

Class II composites: Class II hybrids show a strong interaction between the phases in the order of first-order ionic bonds and covalent bonds. If the molecular contact among both inorganic and organic stages is weak, it can be categorized as class I behavior, which is a constant component that 'traps' a single scattered stage (a) or two consecutive interpenetrated stages (b).

If the molecular contact among both inorganic and organic stages is strong, distinct inorganic entities arise, for instance, groups chemically attached to a continuing organic layer or likewise (a') or two consecutive phases covalently coupled to Class II mixed substances. The categories comprising nanoblends as well as nanocomposites are founded on certain useful nanomaterials that operate as nanofillers when combined with both inorganic and organic polymeric composites. Soldier approaches are used to obtain these (Castellano *et al.*, 2019; Guido *et al.*, 2014; Plutino *et al.*, 2017; Trovato *et al.*, 2018, 2020).

The physical–mechanical and adsorbent capabilities of nanobiohydrids might be successfully employed for the management of sewage and groundwater that are polluted by organic molecules, heavy metals, petroleum hydrocarbons, or viruses along with other germs (Ucankus *et al.*, 2018).

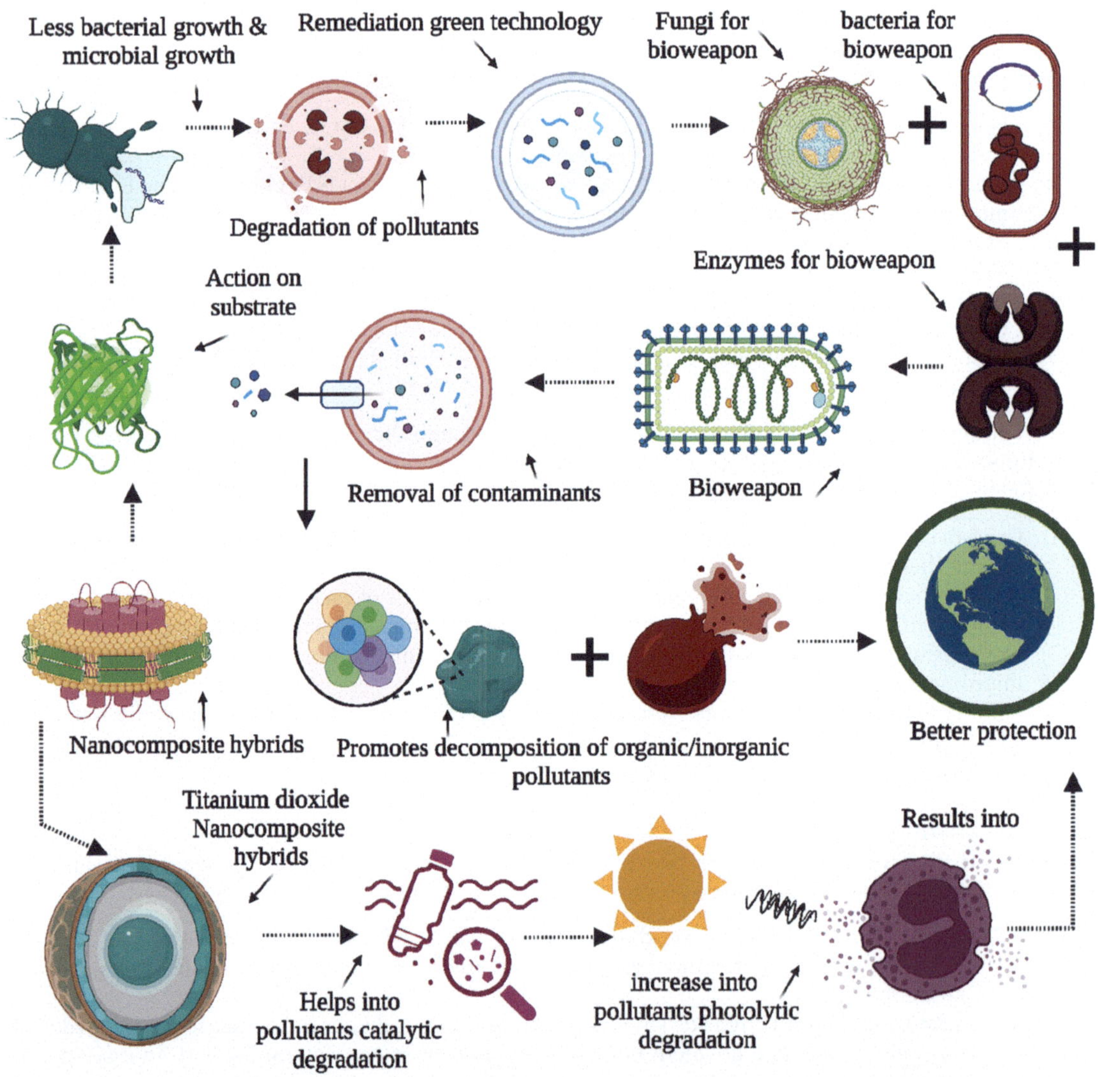

Figure 10.2 Impact of nanocomposite hybrids on bioremediation. The nanocomposite hybrids use substrate required for bacterial growth and enhance the degradation of pollutants by using enzymes, fungi, and bacteria to remove contaminants and promote the decomposition of organic and inorganic pollutants. The titanium dioxide nanocomposite induces the photocatalytic degradation of contaminants.

The properties of nanomaterials are versatile and tunable for different polymers, which has led to the innovation of new functional nanocomposites and nanohybrids for pollution tackling in regard to TiO_2 hybrid nanocomposites. This is an important property of these nanocomposites that helps the catalytic and photocatalytic degradation of contaminants in the environment (Szczepanik, 2017), but there is also the possibility of using them in environmental remediation.

These nanocomposites can also be used as a substrate that is ideal for microbial growth which promotes or increases the degradation of pollutants from the ecosystem (Vázquez-Núñez *et al.*, 2020). Remediation can also be regarded as a green technology. This method makes use of fungi, bacteria, and sometimes enzymes for the removal of contaminants from the environment, thus producing fewer side effects (Figure 10.2). Hence, this approach can be considered natural and sustainable, as it promotes the decomposition of inorganic and organic pollutants from the contaminated environment and follows an ecological path. There are generally two types of bioremediation that are undertaken. The first is *in situ* bioremediation, where the treatment of the environment is done at the same place as the contamination. The second is *ex situ* treatment, in which the polluted substance is being treated somewhere else (Cappello *et al.*, 2019; Catania *et al.*, 2020; Vázquez-Núñez *et al.*, 2020). Typical bioremediation methods are biosparging, bioaugmentation, bioventing, land agriculture, biostimulation, and recycling (Kumar *et al.*, 2018; Mahjoubi *et al.*, 2017).

In the scientific literature, numerous examples exist of how nanotechnologies and bioremediation methods might be combined (Scaffaro *et al.*, 2017), as reported in this chapter. Eco-friendly methods for environmental restoration can be used in conjunction with the characteristics of NPs (Corsi *et al.*, 2018). The creation of novel nanocomposites and hybrids with various functional features that can also be easily recycled and reused to prevent them from becoming a source of waste or secondary contamination represents new developments in this area. This chapter discusses the most recent advancements in bioremediation methods used in conjunction with various nanomaterials, nanocomposites, and nanobiohybrids for polluted areas. It focuses on nanohybrids and nanocomposites produced from metal NPs, metal oxide NPs, and silica-based nanomaterials. Humans focus on environmentally and regenerative nanosolutions to perform research and innovation activities to develop sustainable methods in the area of environmental services and biotransformation.

10.2 METAL NPS

Nanotechnology helps with the delivery of drugs which are poorly water soluable, and also provides the platform to bypass the liver, thus preventing the first pass metabolism (Gholap, 2012; Gholap & Musmade, 2011; Gholap & Rajurkar, 2012d; Gholap *et al.*, 2023a, 2023b, 2023c). It also helps with active targeting of several diseases like cancers with the help of elastic vesicles, nanocarriers, and metal NPs for effective disease treatment to patients (Gholap & Rajurkar, 2012a, 2012b, 2012c; Khedkar *et al.*, 2012a, 2012b).

10.2.1 Properties of metal NPs

Metal nanoparticles (MNPs) related to metals or noble metals such as Ru, Ag, Cu, Pt, Pd, Au, and Al differ from the bulk material due to their small size and high surface-to-volume ratio (Chandrakala *et al.*, 2022). MNPs can kill bacteria and have antibacterial capabilities that are applied by several processes, such as the formation of radical oxygen molecules, specific release of ATP and membrane interaction (Figure 10.3) (Slavin *et al.*, 2017). MNPs exhibit significant optical and electrical characteristics (suppression and absorption), plasmon resonance, Raman scattering and Rayleigh phenomenon (Doria *et al.*, 2012; Kelly *et al.*, 2003). They can also have enzymatic characteristics with activity that varies according to the chemical/physical eco-system morphology, oxidation state, content, and particle density (Cuenya, 2010).

These traits enable NPs to find application in various sectors, such as smart textiles, biosensing, nanocatalysis sensing (Elmaaty *et al.*, 2022), diagnostics, drug delivery therapeutics imaging,

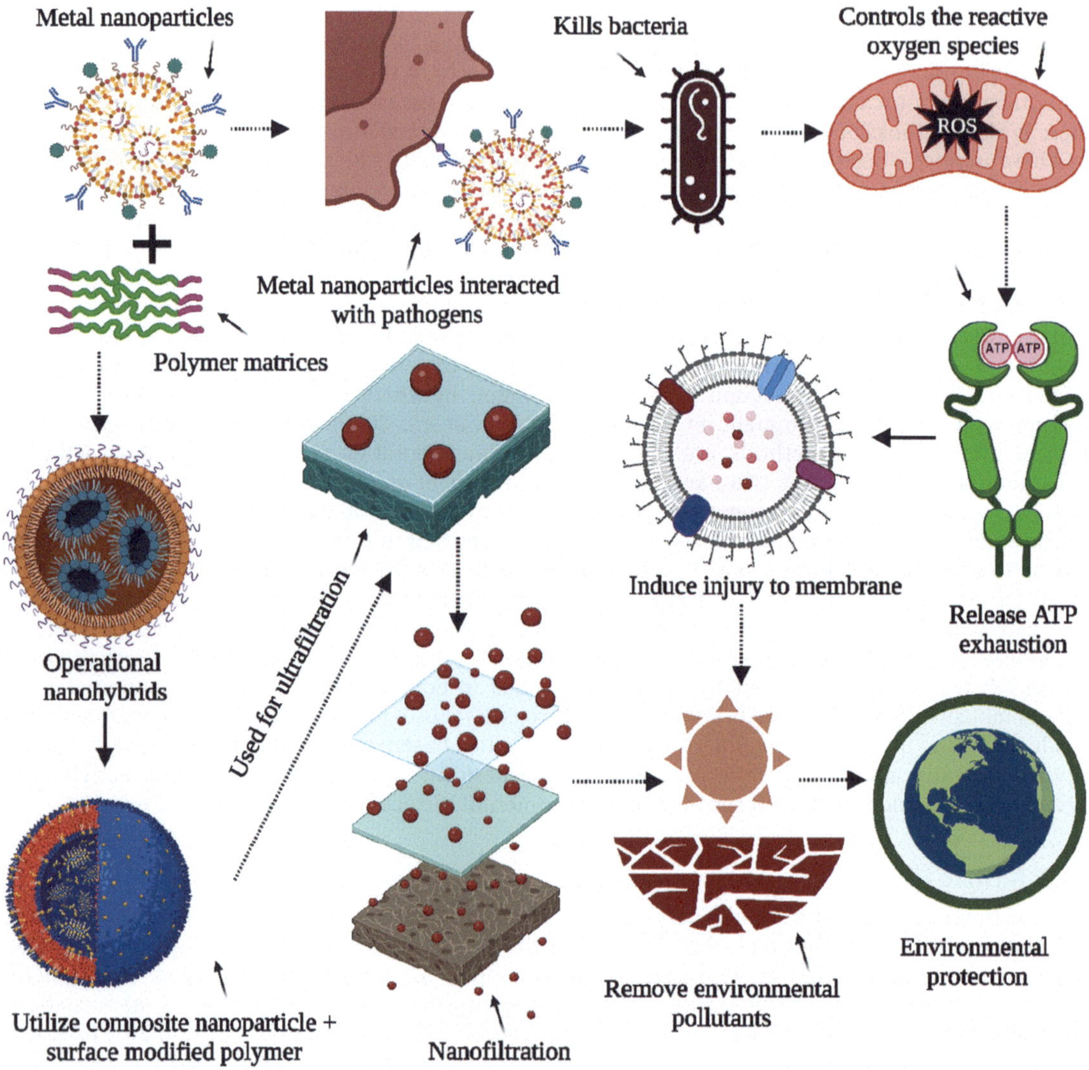

Figure 10.3 Potential of NP hybrids for environmental benefits. Metal NPs kill the bacteria and control the ROS through the release of ATP exhaustion and induce injury on membrane through molecular interactions. Metal NPs are mixed with polymer matrices to produce operational nanohybrids, which are later grafted into polymembranes to extend the capacity of ultrafiltration and nanofiltration to remove environmental pollutants.

biomedicine (imaging, molecular diagnostics, therapeutics and drug delivery) (De Luca *et al.*, 2018; Ielo *et al.*, 2021c, 2021d; Saturnino *et al.*, 2018), and the environmental remediation sector. To generate MNPs, top-down or bottom-up procedures can be used, including chemical (e.g., chemical vapor deposition, chemical reduction, coprecipitation, photochemical reduction, thermal breakdown and hydrolyzation), physical, and other methods (e.g., laser ablation, mechanical milling, ion sputtering, grinding, flame pyrolysis, and vapor deposition). Those techniques are under the top-down or bottom-up categories (Jamkhande *et al.*, 2019).

10.2.2 MNP composite applications in membrane technology

To create the final materials' physical characteristics, MNPs with distinct and useful characteristics can be combined with polymeric matrices to form operational nanohybrids or nanomaterials. This can result in the formation of different and composite NPs, including composite NPs, core/shell NPs, or surface-modified polymer composites, comprising separated NPs delicately scattered inside a polymer, and microsphere composite nanomaterials, which are larger nanocomposite spheres (Hanemann & Szabó, 2010).

Existing remediation techniques can be amalgamated with these functional materials using different approaches. The approaches can be grafted in the polymembrane for ultrafiltration and nanofiltration processes or imparting it to improve the antifouling characteristics (Ajmal *et al.*, 2016; Ali *et al.*, 2017; Çalışkan & Baran, 2022; Gnanaprakasam *et al.*, 2016; Hanemann & Szabó, 2010; Kang *et al.*, 2018; Li *et al.*, 2017; Mohammad *et al.*, 2015; Nechifor *et al.*, 2021a, 2021c; Pandey *et al.*, 2022; Sun *et al.*, 2022; Wali *et al.*, 2019; Yang *et al.*, 2015; Zhu *et al.*, 2014). One of the most severe issues with membrane technology is fouling. Fouling can compromise the filtration efficiency of the membrane and reduce its lifetime. It is typically induced by compounds present in sewage, such as bacterial proteins and extracellular substances (Ielo *et al.*, 2021a; Shahkaramipour *et al.*, 2017).

A poly(vinylidene fluoride) (PVDF) tissue was transplanted with silver NP-poly(carboxybetaine methacrylate (AgNP-PCBMA) nanocomposites, which demonstrated antibacterial as well as antiprotein fouling abilities, especially against *E. coli*, using a physiosorbed free-radical transplantation method. The water flow recovered proportion (FRR), as well as the bovine serum albumin (BSA) refusal proportion were raised by 40% and 60%, respectively, due to improved hydrophilicity performance. The combination of AgNPs and PCBMA's zwitterionic polymer brushes produced a synergistic effect that gave these properties (Li *et al.*, 2017).

10.2.3 MNP and degradation of environmental pollutants

Nanocomposite membranes have advantages such as the capacity to function at relativity lower temperatures to conduct both reduction and oxidation processes within a single model, to have a large contact reactive area and to have no pollutants in the workplace ecosystem. They also present a suitable technique for recovering response volume *via* ultrafiltration (Nechifor *et al.*, 2021b). Certain Au/Bi_2WO_6 nanocomposites, created using a hydrothermal technique paired with a fast reduction-deposition procedure using varied Au weight ratios may be employed as visible-light selectivity photocatalysts in moisture. They enable UV visible light-driven preferential reduction and oxidation of certain contaminants, particularly when exposed to visible light and in an aerobic environment. This hybrid nanostructure exhibits a strong ability to reduce Cr(VI) in oxidizing benzylic alcohol and water. The optimal catalysts for this oxidation/reduction process are 2.0 wt% and 1.0 wt% $Au/Bi_2 WO_6$, respectively (Yang *et al.*, 2015).

Olfactive photocatalysts in water can be used to oxidize harmful biological compounds such as methylene blue. Consider the employment of a heterostructure made of a plasmonic bimetal photocatalyst comprised of Pd-AgNPs as well as macroporous silica (macroPSi) as an example. Such heterostructure can enhance the efficiency of methylene blue degradation in water when subjected to ultraviolet light. The formation of this heterostructure involves a straightforward immersion process of macroporous silicon, which enables the deposition of monometallic and bimetallic NPs composed of Ag and Pd. In addition, it is possible to prepare monometallic photocatalysts, such as AgNPs/macroPSi and PdNPs/macroPSi. However, due to their enhanced particular surface area and plasmonic effect, bimetallic catalysts function better, with a greater efficiency (98.8%) as well as methylene blue biodegradability (0.033 min^{-1}) (Wali *et al.*, 2019).

MNPs are also used in a variety of other materials and composite structures (magnetic NPs) and utilize their catalytic properties, such as micelles, vesicles, and gels. Poly(methacrylic acid-co-acrylonitrile) microgels can be synthesized using inverse suspension polymerization. Subsequently, the nitrile groups in the microgels can be transformed into amidoxime groups, resulting in a more hydrophilic amidoximated microgel. Loading amidoximated microgels with an aqueous phase of metal salts, for example Cu(II) as well as Co(II) ions, and then treating them with sodium borohydride allows for *in situ* synthesis of MNPs depending upon Co and Cu. The resultant amid-p(Mac-co-AN)-M (M: Cu, Co) microgel composite shows outstanding enzymatic activity in the combined breakdown of nitrophenoles and anionic and cationic organic dyes (methylene blue, eosin Y, and methyl orange), which are often found in polluted water environments. These systems also demonstrate the capacity to be recycled for several enzymatic processes. The amid-p(Mac-co-AN)-Cu hybrids, in particular, can operate as protective catalyst devices for a maximum of four rounds, while the amid-p(Mac-co-AN)-Co composite exhibited no decrease in enzymatic performance for as many as seven loops. As a result, amid-p(Mac-co-AN)-Co composites seem to have greater robustness in comparable aqueous conditions than Cu compounds. The strong coordinating connection between Co nanostructures and amidoxime molecules is responsible for these experimental data (Ajmal *et al.*, 2016).

A pH-responsive multipurpose homopolymer vesicle comprising poly[2-hydroxy-3-(naphthalen-1-yl amino) propyl methacrylate] (PHNA) and supported by AuNPs can also aid in the reduction of 4-nitrophenol (Zhu *et al.*, 2014). The AuNPs and the PHNA vesicle supporter work together to create a synergistic effect in this system. Additionally, the π–π interaction between the naphthalene pendants in PHNA vesicles makes them effective as adsorbents for polycyclic aromatic hydrocarbons, reducing their concentration to less than 0.876 ppb within 1 h, making them useful for the remediation of polluted aquatic environments (Zhu *et al.*, 2014). In addition to its effectiveness as an adsorbent, the pH-responsive absorbent system (PHNA vesicle) decorated with AuNPs is also recyclable, as it can act as a nanoreactor for the reduction of 4-nitrophenol in water by adding $NaBH_4$ (Zhu *et al.*, 2014).

Nanotechnology is also being used to create nanocomposites that combine the antibacterial and enzymatic capabilities of MNPs only with the physical qualities of biopolymers such as cellulose. A material made up of AgNPs and cellulose that is very simple to produce is an example of this kind of nanomaterials. This material is composed by adsorbing cellulose from fruit waste with AgNPs, resulting in nanocomposites with antibacterial, oxidative, and photodegradation qualities. The nanocomposite made of AgNPs and cellulose is simple to produce, and it has several beneficial properties. For example, discs manufactured of this nanocomposite can decrease a *Staphylococcus* culture by more than 90% in just 150 min. This substance also has a moderate antioxidant property, minor 2,2-diphenyl--picrylhydrazyl (DPPH) antioxidant activity, and a medium photodegradation ability in the sunshine. After approximately 60 min, the photodegradation ability may destroy methylene blue by up to 63.16% (Ali *et al.*, 2017).

Outside remediation and pollutant decomposition, MNP compounds and hybridization have a variety of applications, such as the construction of specific detectors for compounds. A group of polymeric Ag/micelle nanomaterials (AgNPs) functions as a base for ground absorption spectroscopy (SERS) to extract and measure pesticides, such as phosmet, acetamiprid, and thiabendazole (Rando *et al.*, 2022). Polyurethane molecules have amphoteric properties, permitting them to absorb compounds while still maintaining the NP group. The procedure requires placing the polymeric AgNP/micelle substrate on the specimen surface, which might be spinach or apple, to remove and locate the desired compounds without the demand for any sample treatment (Xiang *et al.*, 2022).

10.3 METAL OXIDE NPS

10.3.1 Properties of metal oxide NPs

The field of environmental remediation dramatically benefits from the unique properties of metal oxide nanoparticles (MONPs), such as redox, photocatalytic, acid–base, filtration, and magnetic properties. When combined with polymeric supports, the properties of MONPs can enhance the mechanical strength, durability, and hydraulic properties of filtration systems. There are various types of MONP composites, including those consisting of polymer-supported NPs dispersed in polymeric matrices. These composites are helpful in removing heavy metals from contaminated water and have the potential to be reused. Furthermore, magnetically active polymeric particles are significant and may be employed again in related applications.

The literature has detailed several different synthesis methods for creating iron oxide NPs. They include hydrothermal/solvothermal synthesis, thermal breakdown, microemulsion, and the coprecipitation approach. MONPs, including zirconium oxide, cerium oxide, titanium dioxide, tin dioxide, ferric oxide, zinc oxide, and their nanocomposites with natural polymers are utilized to immobilize active enzymes that are utilized in bioremediation processes. Enzymes carry out degradation and detoxification of harmful and carcinogenic chemicals. MONPs are essential in facilitating the electron transfer process between the substrate and the active sites on the enzyme. The use of MONPs in biomonitoring procedures can also aid in speeding hydrocarbon oxidation, reducing their toxicity and fostering microbial growth. In fact, after using some MONPs for various remediation purposes, such as the removal of dangerous organic and inorganic compounds and oil/water separation, their magnetic properties, particularly those of ferrous oxide NPs, have been extensively used for the recovery of nanocomposite materials made up of different inorganic, organic, and polymeric matrices (Ali *et al.*, 2019).

10.3.2 Metal oxide NP composite and hybrid applications

10.3.2.1 Titanium-based composites

Several methods, including complexation, electrostatic attraction, and ion exchange, are used by MONP composites to absorb contaminants. Creating inner-sphere surface complexes is the basis for the adsorption of a new nanostructured ternary Fe, Ti, and Mn oxide. The Fe–Ti–Mn FTMO composite oxide is a prime example of adsorption based on the development of inner-sphere surface complexes. It was created by a one-step simultaneous oxidation and coprecipitation technique (Yin *et al.*, 2019). Due to the development of inner-sphere surface complexes at the water/oxide interface under both light and dark circumstances, the substance has a high absorption efficiency for As(V) and As(III), with a particular affinity for the latter.

Titanium dioxide is a crucial metal oxide that is well known for its exceptional photocatalytic abilities. As a result, it is frequently used to create nanocomposites and nanohybrid materials that are intended for applications in environmental remediation and the destruction of various contaminants. Using lignocellulosic biomass generated from olive pits (OP), a composite material made of magnetic iron oxide and titanium dioxide was created. The substance, known as TiO_2-OP@Fe_3O_4, was created *via* a hydrothermal magnetization technique at 180°C after a sol–gel procedure assisted by ultrasound. Methylene blue, rhodamine B, Congo red, and Cr(VI) were successfully removed from contaminated aquatic systems by TiO_2-OP@Fe_3O_4 using a combination of absorption and photocatalytic reactions triggered by visible light. The composite material demonstrated a 40-min Cr(VI) reduction rate of 100% at pH 3 and a concentration of 10 ppm.

As a matrix for both TiO_2 and Fe_3O_4, the lignocellulosic biomass is essential in keeping the two materials apart and from coming into contact. This matrix is important to prevent the dispersion of NPs in the treated water and to avoid an unfavorable heterojunction, which may lead to increased electron–hole charge recombination in pure TiO_2 and Fe_3O_4, ultimately causing lower photocatalytic performance. A natural alginate polymer can act as a matrix to hold the NPs in place. Fe_3O_4 magnetic NPs are produced and enclosed in alginate using a reverse coprecipitation technique to create alginate magnetic NP beads to remove blue textile colors from water (Djellabi *et al.*, 2019). Several kinds of magnetite NPs are used to create these composite beads, including a bead of ferrous sulfate, ferrous sulfate and biochar, and ferrous sulfate, biochar, and graphite. After 3 h of equilibrium, the ferrous sulfate, biochar, and graphite alginate beads displayed 82.4% removal of azo blue dye (25 ppm). At pH 8, 55.22% removal was observed for an initial dye concentration of 100 ppm. The utilization of these beads in nanobioremediation is conceivable (Lincy *et al.*, 2020).

10.3.2.2 Composites based on carbon materials

10.3.2.2.1 Graphene oxide-based nanocomposites

The GO–MO nanocomposite, which contains graphene oxide aggregates impregnated with Mn oxide NPs, is an illustration of such a composite. The inner-sphere complexation mechanism of the GO–MO composite resulted in a very high absorption rate (>99.9%) when it was evaluated using Cd(II) and Cu(II) as representative metals. Even when the column was operating, uptake was observed. After it was exhausted, the GO–MO composite can be regenerated by flushing it with a 10 BV acid–salt binary solution composed of 0.2 M HCl and 4 wt% $CaCl_2$. More than 97% of the preloaded Cu(II) and Cd can be desorbed by this procedure (II) (Wan *et al.*, 2018).

It is also possible to mix GO with magnetic Fe_3O_4 particles. A good illustration of a composite material that can be utilized to remove phosphate and nitrate ions from water is Fe_3O_4 NPs covered with a core–shell structure of graphene oxide and carboxymethylcellulose (Fe_3O_4@GOCMC). One-pot synthesis is a viable method for creating this hybrid. Using a 0.1 N NaOH solution allows for the regeneration of the nanocomposite material and the elimination of the adsorbed anions. The Fe_3O_4@GOCMC was dried after regeneration with 0.1 N NaOH solution and can be used again for up to four more adsorption–desorption cycles (Karthikeyan & Meenakshi, 2020).

10.3.2.2.2 Biochar-based composites

Biochar is a porous, environmentally friendly substance that can act as an absorbent and matrix for MONPs. It is a carbonaceous material produced by biomass (such as corncob waste) wet pyrolysis at low temperatures and oxygen levels. It is useful for a variety of applications, including soil conditioning, carbon sequestration, composting of solid waste, water and wastewater decontamination, catalysis, electrode materials, and electrode modification, due to its large surface area, porous structure, high carbon content, and cation-exchange capacity (Wang & Wang, 2019). Biochar can be utilized to create nanocomposites for the remediation of heavy metals and organic chemicals in the environment when mixed with NPs and stabilizers such as carboxymethyl cellulose. Fe NPs, for instance, can be utilized to create nanocomposites such as iron oxide/BC, iron sulfide/BC, and nano zero-valent iron (nZVI)/BC. Due to the redox, catalytic and magnetic properties of the Fe and Fe oxide NPs, these nanocomposites can improve the features of biochar by increasing its surface area, electron transfer efficiency, and abundance of functional groups.

Biochar/Fe oxide nanocomposites are suitable for encapsulating photosynthetic bacteria (PSB) to produce an environmentally benign and regenerative solution for contaminated wastewater due to their high absorption capacity and Fe NP catalytic, adsorption, and magnetic separation. Using a magnet bar, the PSB-loaded composites are simply retrieved. Wastewater nutrient removal and degradation abilities were enhanced with the use of PSB in combination with the Fe_3O_4/biochar composite. The composite, which supports *Rhodobacter capsulatus*, was able to remove the chemical oxygen requirement with an efficiency of 83.1%, NH_4^+ with 87.5%, and PO_4^{3-} with 92.1%. Even after five rounds of recycling, the composite remained effective. The presence of Fe_3O_4 NPs and Fe oxides stimulated the growth, metabolism, and enzyme activities of bacteria and PSB. These NPs effectively scavenge reducing equivalents because they function as extracellular electron acceptors (Djellabi *et al.*, 2019).

10.3.2.3 Magnetic nanocomposites

Nanocomposites made of polymers and NPs are also used in the cleanup and recovery of oil spills. One instance is the utilization of amphiphilic organic domains and a magnetic core that is present in magnetic shell cross-linked knedel-like NPs. These particles, which are useful for adsorbing hydrophobic guest molecules, are created by coassembling amphiphilic block copolymers of poly(acrylic acid)-block-polystyrene (PAA20-b-PS280) and oleic acid-stabilized magnetic iron oxide NPs. Thermal breakdown is used as a synthesis technique for these guest compounds. Amphiphilic block copolymers comprising PAA20-b-PS280 and hydrophobic MNPs are coassembled with solvents such as tetrahydrofuran, *N,N*-dimethylformamide, and water to produce in magnetic shell cross-linked knedel-like NPs nanocomposites. Amidation is used to cross-link the generated hybrid micelles. When these nanocomposites were examined for their ability to absorb crude oil in water, they revealed an oil sorption capacity 10 times greater than their initial dry weight. Following use, they can be regenerated by sonication and ethanol cleaning (Pavía-Sanders *et al.*, 2013).

To efficiently demulsify oil/water emulsions and remove/degrade oils, one approach of bioremediation uses a bionanocompound made of magnetite NPs coated with polyether amine (PEA), which may immobilize OmpA (a biosurfactant) and laccase (a remediation enzyme). Coprecipitation is used to create magnetite NPs, which are then activated by silanization with 3-aminopropyltriethoxysilane (APTES) and modified with glutaraldehyde. This process leads to the conjugation of the resultant magnetite NP-APGA with oxidized polyether amine (PEA). The biosurfactant OmpA and the remediation enzyme laccase were conjugated with magnetite NP-PEA to create bionanocompounds called magnetite NP-PEA-OmpA-Laccase. The biosurfactant OmpA and the remediation enzyme laccase were conjugated with magnetite NP-PEA to create bionanocompounds called magnetite NP-PEA-OmpA-Laccase (Rangel-Muñoz *et al.*, 2020).

10.4 CARBON-BASED NANOMATERIALS

10.4.1 Properties of carbon-based nanomaterials

Carbon-based nanomaterials have unique chemical and physical properties, which include exceptional electrical and heat conductivity, advanced optical properties, chemical stability, and high mechanical strength. These nanomaterials have various nanoscale dimensions and are made up of solid-state carbon allotropes with hybridized carbon atoms in sp^2 and/or sp^3. These dimensions include single-walled or multiwalled carbon nanotubes (SWCNTs and MWCNTs), graphene oxide, two-dimensional graphene and nanodiamonds, zero-dimensional fullerenes, and graphene quantum dots. There are many ways to make carbon-based nanomaterials, 'bottom-up' techniques including MWCNTs, chemical vapor deposition (CVD) on a substrate for SWCNTs and graphene. The 'top-down' techniques include ionic liquid-assisted electrochemical exfoliation for graphene and sonication liquid-phase exfoliation (Li *et al.*, 2019).

Nanomaterials based on carbon have a wide range of uses in many industries, including biomedical industries. They are particularly interesting because they have antibacterial qualities in addition to other special physical and chemical characteristics, cell and tissue imaging, particularly based on their two-photon and one-photon fluorescence properties and theranostics (Patel *et al.*, 2019), conversion and energy storage, catalysis (Bayatsarmadi *et al.*, 2017), synthesis for environmental remediation for the removal of organic molecules, oil/water separation material design and heavy metals (Patel *et al.*, 2019). Numerous studies have explored the potential of carbon-based nanomaterials, particularly their composites, for environmental remediation. These nanomaterials have been used to enhance the absorption capacity of water contaminants and to create regenerable materials. Materials with porous graphene structures, such as sponges, aerogels, and foams, have been developed for this purpose, as they exhibit superior physical, chemical (lipophilic/hydrophilic), and mechanical properties (Niu *et al.*, 2014).

The potential cellular adverse toxicity of specific nanomaterials based on carbon has been extensively studied, revealing size-dependent adverse effects and enhanced cytotoxic properties after acid treatments that modify their surface. This increased cytotoxicity is because of the prevalence of carbonyl, carboxyl, and/or hydroxyl groups on the interface of nanomaterials (Magrez *et al.*, 2006; Weijie *et al.*, 2020). Given the potential hazards associated with nanomaterials based on carbon, it is essential to create functional composite materials that are sustainable by design, while minimizing environmental dispersion. This will help ensure that these materials can be safely used without posing a risk to human health or the environment (Bhattacharya *et al.*, 2016).

10.4.2 Carbon-based nanomaterial composite and hybrid applications

There are many applications of composite nanomaterials based on carbon and hybrids in the field of bioremediation and remediation of different pollutants. The remediation of contaminated marine sites often involves separating oil/water mixtures, for which various approaches are used that utilize nanocomposites having chemically functionalized hydrophobic/hydrophilic characteristics that can be controlled. Nanomaterials based on carbon can show not only hydrophobic interactions but also hydrogen bonds, π–π interactions and electrostatic interactions to entrap pollutants present in the water. These characteristics are used to create adsorbent materials, which include nanosponges, to clean up toxins such as oil spills on water.

10.4.2.1 Composite materials

10.4.2.1.1 Carbon nanotubes

One such adsorbent substance is a hydrophobic floatable absorbent made of nanofibers and nanotubes of carbon, also known as CNTs, that were developed utilizing a chemical vapor deposition technique, that is, chemical vapor deposition on the surface of an enlarged vermiculite substrate. Studying the capacity of various oils to absorb moisture revealed an increase of approximately 600%. Nevertheless, unfavorable water absorption was decreased since the composite material's hydrophobic characteristics were greater than those of the original vermiculite [116]. Since they have been found to facilitate the *in situ* breakdown of pollutants in freshwater sediments contaminated with oil, carbon nanotubes (CNTs) offer potential for use in both *ex situ* and *in situ* bioremediation procedures (Abbasian *et al.*, 2016).

Because of their properties, including electrical conductivity, biocompatibility, lack of corrosiveness, and chemical stability, CNTs can be used in combination with conducting polymers such as polyaniline (PANI) to modify electrodes in microbial fuel cells (MFCs) for bioelectrochemical *ex situ* treatments. MFCs are a promising technology that can convert chemical energy into electrical energy in the presence of electroactive bacterial species while simultaneously removing toxic pollutants such as organic dyes, COD, and phenol from water. Nanomaterials such as carbon-based NPs and conductive polymers can be used as electrodes in MFCs, with the anodes collecting electrons produced by the biodegradation processes of pollutants and producing energy (Logan *et al.*, 2006).

CNT composites have been used in MFCs for the bioelectrochemical treatment of pollutants such as phenol. Multiwalled and bare nanotubes of carbon/polypyrrole, that is, MWCNT/Ppy-plated electrodes have been used in both the cathodic and anodic compartments under anaerobic and aerobic conditions. The most effective results have been achieved with biocathodic treatment under aerobic conditions, resulting in high COD and phenol removal efficiencies (Khan *et al.*, 2020). CNTs can also be used in combination with NPs of metal oxides for the bioremediation of azo dyes and Cr(VI). In this case, CNTs are used as a support material for anthraquinone-2-sulfonic acid or humic acids and Fe_3O_4 (AQS/Fe_3O_4/CNTs and HA/Fe_3O_4/CNTs). They can efficiently remove Cr(VI) and methyl orange from sewage water by grabbing electrons from the anaerobic metabolism of microbial bacteria. This procedure has high reusability and stability and performs best at a pH of 8.0. Additionally, the electron transfer capacity of CNTs is preserved even after the bioreduction of pollutants, and the immobilization of residual Fe and dissolved organic matter (DOM) in CNTs prevents their dispersion in the environment. This material can be reused multiple times due to its magnetic properties. Once the absorption of Cr(VI) and azo dyes is complete, a magnetic bar can be used to recover the material. It can be restored by drying after being washed with deionized water and ethanol (He *et al.*, 2020).

A hybrid composite of polyaniline and carbon nanotubes (PANi/CNT) was synthesized using a chemical oxidative polymerization method with the aim of generating biomass energy and treating sewage water. Specifically, the PANi/CNT composite was employed to create an anode electrode in MFCs. This system exhibits strong π–π interactions between PANi and CNTs, which promote the attachment of microbes to the electrode surface. The composite's charge-transfer capabilities enable it to produce bioelectricity while also removing 80% of the COD present in wastewater (Ali *et al.*, 2019).

10.4.2.1.2 Graphene oxide

CNT composites have been shown to be effective in developing remediation techniques for environmental pollutants such as organic dyes and heavy metals as well as hydrocarbons because of their photocatalytic, absorption, and electrocatalytic properties. However, there are also various examples of the use of other nanomaterials based on carbon, such as its graphene oxide derivative (GO) and graphene in this sector. GO is a hydrophilic layered carbon material prepared by oxidizing as well as exfoliating graphite along with oxidizing agents using Hummer's technique, including $NaNO_3$, H_2SO_4, and $KMnO_4$ (Wang *et al.*, 2019). It is characterized by charged oxygen-containing groups with very high density, such as alcohols, carboxylic groups, epoxides, and ketone carbonyls. Due to its surface having a negative charge, GO has a high adsorption capacity for various molecules through chemical forces and/or physical forces, such as π–π, electrostatic and hydrophobic interactions.

Two main approaches for obtaining nanocomposites and functionalizing GO are commonly used: using GO as a host material or incorporating GO into host materials. In the former approach, various methods are used to functionalize the surface of GO. In the latter approach, crosslinking between the polymeric matrix and GO is carried out using a nontoxic and simple synthesis to obtain a nanocomposite with a high absorption capacity and stable structural configuration that can be easily regenerated and recovered after use (Wang *et al.*, 2019).

Aerogels, which are porous natural materials, possess adsorption properties that make them useful in the remediation of water. To enable easy use and recovery in aquatic environments, aerogels with high hydrophobicity can be developed. An example is a novel composite made of graphene aerogel, Fe_3O_4, and polystyrene. The composite is produced by cost-effective and environmentally friendly solvothermal methods. Porous Fe_3O_4 NPs act as cross-linkers to plates of graphene oxide created by modifying Hummer's technique, whereas polystyrene allows the composite aerogel to develop a porous surface with increased hydrophobicity. The composite aerogel, which has an ultralow density, is capable of absorbing up to 40 times its own mass of crude oil after undergoing 10 cycles of water–oil separation. The aerogel is easy to recover because of its ability to float, and it also possesses magnetic properties thanks to the inclusion of Fe_3O_4 NPs, which allows for the exhausted aerogel to be collected using a magnet. Regenerating the aerogel is a simple process that involves squeezing (Zhou *et al.*, 2015).

GO and CNTs can be combined with bioremediation techniques to create anode electrodes that generate energy and simultaneously remove pollutants such as heavy metals. This remediation method involves designing and synthesizing new composite materials, including those made from natural resources. For example, GO can be produced from organic biomass, and its conductivity, chemical inertness, flexibility, mechanical robustness, biocompatibility, and high specific surface area make it an excellent candidate for producing MFCs with bioinspired anode materials. One such derivative of graphene, known as L-GO, is acquired from the biomass of organic oil palm through a soda pulping process for the extraction of powder lignin. The flakes of carbon are prepared by heating and then oxidizing organic oil palm biomass using Hummer's method. To improve the transportation of electrons in ZnO and MFCs, NPs are synthesized using lemon peel by utilizing a green solvothermal method and then combined with L-GO to produce a ZnO/L-GO composite. The ZnO/L-GO electrodes are submerged in a solution of polylactic acid (PLA) to mechanically strengthen the anode. The obtained anode is evaluated for its ability to generate energy and remediate Pb^{2+} ions from wastewater through an MFC. Bacterial species such as *Klebsiella* sp., *Lysinibacillus* sp. and *Leucobacter* sp. are observed on the anode surface (Yaqoob *et al.*, 2021).

10.4.2.1.3 Other graphene derivatives

A natural organic source such as sucrose, which is rich in carbon, can also be utilized for the synthesis of graphene derivatives. By treating sand-coated sucrose with acid and high temperature, a sand-based composite adsorbent with large porosity and surface area can be obtained. This composite material was tested for its ability to absorb Hg^{2+} from contaminated water using an economically and environmentally friendly approach. The composite material showed a high absorption capacity for Hg^{2+} (299.40 mg/g) resulting from the metal ion's interaction with the functional groups, as seen through FTIR measurements. The chelated bonds between the composite and Hg^{2+} can be less efficient at acidic pH, making it possible to re-obtain the material through a process of desorption, which allows its reuse (Bajpai *et al.*, 2017).

10.4.2.2 Hybrid composite materials

10.4.2.2.1 Metal organic frameworks

Green methods of solvothermal analysis can be utilized to fabricate advanced hybrid nanocomposite materials that are effective in remediating contaminated aquatic sites. For instance, a nanosized metal-organic framework (MOF) of nickel-benzene dicarboxylate (Ni-BDC) can be combined with carbon nanotubes (CNTs) or graphene oxide (GO) to form a complex hybrid material. Due to the existence of π–π – interactions, the Ni-BDC@GO nanocomposite in particular exhibits a strong methylene blue retention capacity (222 mg/g) and can be easily regenerated. The nanocomposites can be restored by using simple alcohol washing and centrifugation for repeated cycles of absorption (Ahsan *et al.*, 2020).

10.4.2.2.2 Metal oxide NPs

Metal oxide NPs such as TiO_2, ZnO, and Bi_2O_3 have been shown to degrade organic dyes such as methylene blue through photocatalytic action. Graphene, which can be easily obtained from GO through reduction processes (Rout *et al.*, 2022), has also been found to possess photocatalytic activity for removing contaminants when combined with metal oxide NPs. Adding graphene to composite nanofibers (TZB-Gr) generated by method-based sol–gel nozzle-less electrospinning increased the removal capability of organic dyes (Al-Dhahebi *et al.*, 2020). These composite nanofibers are formed of zinc oxide, titanium dioxide, graphene, and bismuth oxide. This is a result of the 2D graphene's interaction with nanofibers made of photoactive semiconductors. Under both UV-light irradiation and visible-light irradiation, the TZB-Gr composite nanofibers can activate organic dyes to produce ·OH and $\cdot O^{2-}$ radicals, which are highly effective oxidizing agents for degrading most of the pollutants having an organic nature (Kanjwal *et al.*, 2019). The TiO_2-graphene (P25-GR) composite, produced by a hydrothermal reaction of TiO_2 (P25) with GO having different graphene ratios, is an additional example of a nanocomposite that combines the properties of graphene with metal oxide NPs. Fluoranthene, phenanthrene, and benzo[a]pyrene were used as test compounds to determine this composite material's capacity to photodegrade and absorb polycyclic aromatic hydrocarbons (PAHs). Particularly at high PAH concentrations (2.0–4.0 g/mL) in alkaline circumstances, the P25-GR composite with 2.5% graphene content demonstrated charge transfer, the highest absorption and photocatalytic efficacy, eliminating approximately 80% of the PAHs after completion for 2 h (Bai *et al.*, 2017).

10.5 NANOBIOHYBRIDS ASSISTING NOVEL BIOLOGICAL FUNCTIONS

10.5.1 Extending biofunctions by nanobiohybrids

The change in biological functions by using human-generated objects has functioned from ancient times until now for improving life standards along with survival. These processes are supported by using different innovations across the globe, from clothes to weapons, to that of recent innovations, such as computers and smartphones. The continuous progress into nanotechnology-based components has given certain breakthroughs in the bioaugmentation process. Proper enhancement of biofunctions such as biostability along with biocatalytic efficiency are important and vital points for enhancement through advanced biomaterials. These biomaterials work at the nanolevel along with singular cellular and multicellular levels to provide advanced biofunctions. These materials deal with extrinsic functionalities, including the metabolic type of the pathways along with the electrical conductivity or optical properties of living systems.

10.5.2 Application of advanced nanobionics

10.5.2.1 Microbial cells

Nanobionic cells are an engineering type of single cell that interferes with that of a bioactive metallic organic framework (MOF) along with living cells. The former are comprised of an inner layer of beta galactosidase nanocoated along the outside of a zeolitic imidazole framework (ZIF)-8. This combination of outer layers was induced onto a yeast cell. This type of construction helps yeast to survive for approximately more than 5 days with the help of conversion of lactose to glucose by using

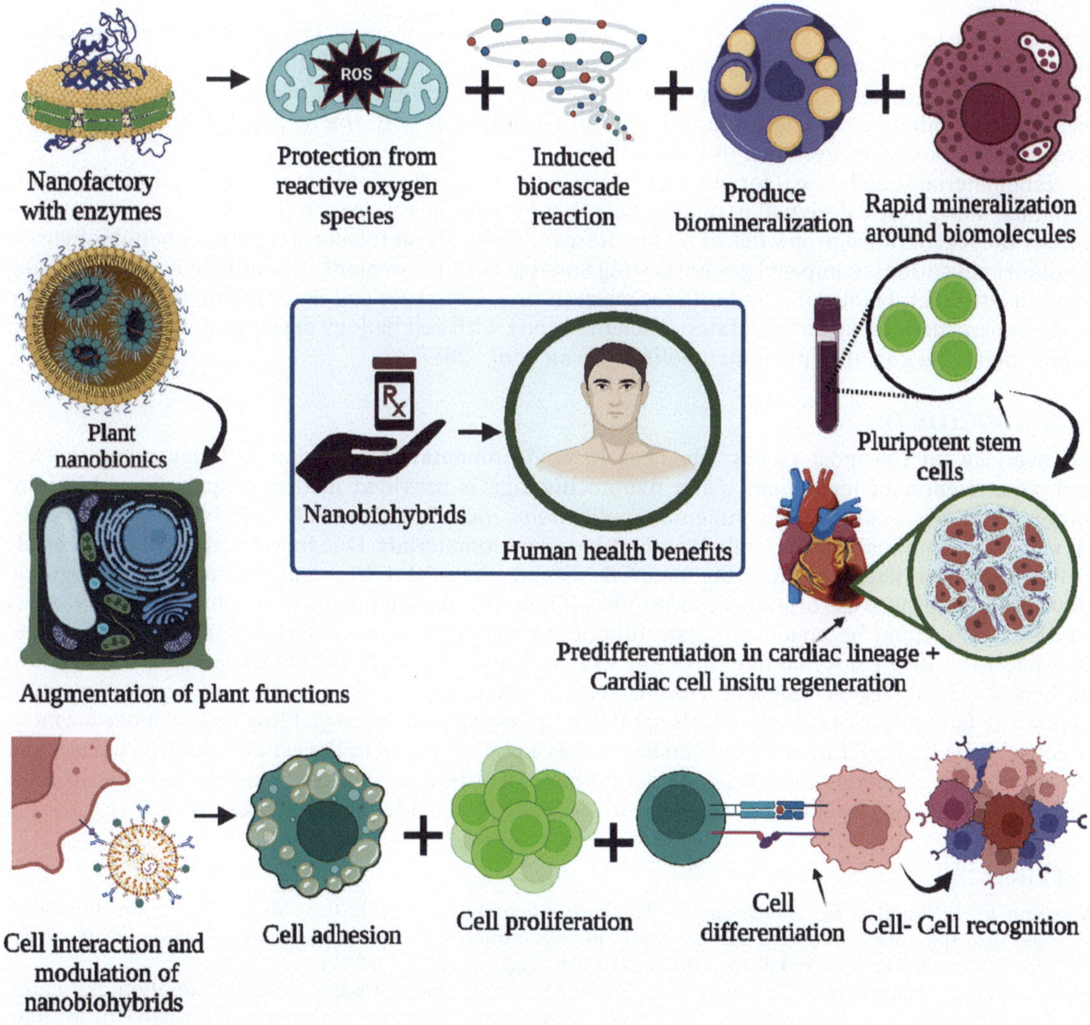

Figure 10.4 Nanobiohybrids assisted in novel biological functions, including stability of biomacromolecules, cell interaction, and modulation, alteration of cell fate, nanorobotics for enhancement of fracture strength, and nanohybrid supermaterials for augmentation of plant functions.

beta galactosidase. This enzyme helps in the conversion of lactose into glucose, which is required for cell metabolism of the yeast (Liang *et al.*, 2017).

10.5.2.2 Silk production

Silk is used in the textile industry due to its elasticity and extensibility. The effectiveness of silk is enhanced due to an increase in the mechanical properties of silk with the help of carbon nanomaterials, including single-walled carbon nanotubes (SWCNTs) as well as graphene. This type of material is directly fed to the silkworm, which results in hybrid silk (Wang *et al.*, 2016). Such hybrid silk has increased fracture strength along with the elongation required for excellent electrical conductance (Lepore *et al.*, 2017).

10.5.2.3 Plants

Plants are the ideal candidates for bioaugmentation along with bionics. The plant's vascular network, along with passive forces, is utilized for fluid transportation along with the dispersion of molecules or particles into these fluids (Gholap and Tupe, 2012). Several types of biosynthetic materials have been used to enhance the augmentation of plant biofunctions with the help of the unique biological properties of plants, as presented in Figure 10.4.

Nanomaterials can be used for biochemical detection as well as for light harvesting for the benefits of plants. Many newly designed nanomaterials that increase plant growth have been implemented to impact the regenerative properties of plants. Research focuses on the construction of nanobiohybrids for plant applications to impact light harvesting and regeneration of plants. Nanobiohybrids are helping to modulate several molecular and cellular mechanisms to improve cell-material interactions with the help of micro/nanoengineering. Material science along with cell biology develops new nanobiohybrid applications that can alter plant metabolism (Kwak *et al.*, 2017).

10.6 CONCLUSION

An overview of the most recent advances in environmental remediation utilizing adequate soil and water treatment techniques using nanotechnology is provided in this chapter. Four kinds of nanomaterial are used for environmental applications, including (1) metal oxide NPs, (2) metal NPs, (3) silica-based nanomaterials, and (4) carbon-based nanomaterials. Due to the various nanomaterials utilized as nanofillers, the composite materials display photocatalytic, catalytic, redox, and magnetic characteristics as well as sorption/desorption. Additionally, the chapter gives instances of how various nanomaterials could be mixed to create functional composite nanomaterials that have the unique characteristics of the original nanofillers as well as brand-new features. These nanocomposites have the benefit of being regenerable and reusable, with minimal decrement even after numerous adsorption cycles. The fabrication of nanomaterials applied to living cells has been yielding new avenues for more effective control of cell function and can be used as a powerful tool in the genetic engineering stream. However, several queries related to the actual biological functioning of nanomaterials due to their field emergence impact on biological functions still need to be adequately addressed.

REFERENCES

Abbasian F., Lockington R., Palanisami T., Megharaj M. and Naidu R. (2016). Multiwall carbon nanotubes increase the microbial community in crude oil contaminated fresh water sediments. *Science of the Total Environment*, **539**, 370–380, https://doi.org/10.1016/j.scitotenv.2015.09.031

AbouElmaaty T., Abdeldayem S., Ramadan S., Sayed-Ahmed K. and Plutino M. (2021). Coloration and multi-functionalization of polypropylene fabrics with selenium nanoparticles. *Polymers*, **13**(15), 2483, https://doi.org/10.3390/polym13152483

Ahsan M. A., Jabbari V., Imam M. A., Castro E., Kim H., Curry M. L., Valles-Rosales D. J. and Noveron J. C. (2020). Nanoscale nickel metal organic framework decorated over graphene oxide and carbon nanotubes for water remediation. *Science of the Total Environment*, **698**, 134214, https://doi.org/10.1016/j.scitotenv.2019.134214

Ajmal M., Demirci S., Siddiq M., Aktas N. and Sahiner N. (2016). Simultaneous catalytic degradation/reduction of multiple organic compounds by modifiable p(methacrylic acid-co-acrylonitrile)–M (M: Cu, Co) microgel catalyst composites. *New Journal of Chemistry*, **40**(2), 1485–1496, https://doi.org/10.1039/C5NJ02298C

Al-Dhahebi A. M., Gopinath S. C. B. and Saheed M. S. M. (2020). Graphene impregnated electrospun nanofiber sensing materials: a comprehensive overview on bridging laboratory set-up to industry. *Nano Convergence*, **7**(1), 27, https://doi.org/10.1186/s40580-020-00237-4

Ali A., Haq I. U., Akhtar J., Sher M., Ahmed N. and Zia M. (2017). Synthesis of Ag-NPs impregnated cellulose composite material: its possible role in wound healing and photocatalysis. *IET Nanobiotechnology*, **11**(4), 477–484, https://doi.org/10.1049/iet-nbt.2016.0086

Ali N., Zaman H., Bilal M., Shah A.-H. A., Nazir M. S. and Iqbal H. M. N. (2019). Environmental perspectives of interfacially active and magnetically recoverable composite materials – a review. *Science of the Total Environment*, **670**, 523–538, https://doi.org/10.1016/j.scitotenv.2019.03.209

Bai H., Zhou J., Zhang H. and Tang G. (2017). Enhanced adsorbability and photocatalytic activity of TiO_2 -graphene composite for polycyclic aromatic hydrocarbons removal in aqueous phase. *Colloids and Surfaces B: Biointerfaces*, **150**, 68–77, https://doi.org/10.1016/j.colsurfb.2016.11.017

Bajpai A. K., Dubey R. and Bajpai J. (2017). Synthesis, characterization, and adsorption properties of a graphene composite sand (GCS) and its application in remediation of Hg(II) ions. *Water, Air, & Soil Pollution*, **228**(9), 346, https://doi.org/10.1007/s11270-017-3511-5

Bayatsarmadi B., Zheng Y., Vasileff A. and Qiao S. (2017). Recent advances in atomic metal doping of carbon-based nanomaterials for energy conversion. *Small*, **13**(21), 1700191, https://doi.org/10.1002/smll.201700191

Bhattacharya K., Mukherjee S. P., Gallud A., Burkert S. C., Bistarelli S., Bellucci S., Bottini M., Star A. and Fadeel B. (2016). Biological interactions of carbon-based nanomaterials: from coronation to degradation. *Nanomedicine: Nanotechnology, Biology and Medicine*, **12**(2), 333–351, https://doi.org/10.1016/j.nano.2015.11.011

Çalışkan M. and Baran T. (2022). Palladium nanoparticles embedded over chitosan/γMnO_2 composite hybrid microspheres as heterogeneous nanocatalyst for effective reduction of nitroarenes and organic dyes in water. *Journal of Organometallic Chemistry*, **963**, 122284, https://doi.org/10.1016/j.jorganchem.2022.122284

Cappello S., Cruz Viggi C., Yakimov M., Rossetti S., Matturro B., Molina L., Segura A., Marqués S., Yuste L., Sevilla E., Rojo F., Sherry A., Mejeha O. K., Head I. M., Malmquist L., Christensen J. H., Kalogerakis N. and Aulenta F. (2019). Combining electrokinetic transport and bioremediation for enhanced removal of crude oil from contaminated marine sediments: results of a long-term, mesocosm-scale experiment. *Water Research*, **157**, 381–395, https://doi.org/10.1016/j.watres.2019.03.094

Castellano A., Colleoni C., Iacono G., Mezzi A., Plutino M. R., Malucelli G. and Rosace G. (2019). Synthesis and characterization of a phosphorous/nitrogen based sol-gel coating as a novel halogen- and formaldehyde-free flame retardant finishing for cotton fabric. *Polymer Degradation and Stability*, **162**, 148–159, https://doi.org/10.1016/j.polymdegradstab.2019.02.006

Catania V., Lopresti F., Cappello S., Scaffaro R. and Quatrini P. (2020). Innovative, ecofriendly biosorbent-biodegrading biofilms for bioremediation of oil-contaminated water. *New Biotechnology*, **58**, 25–31, https://doi.org/10.1016/j.nbt.2020.04.001

Chandrakala V., Aruna V. and Angajala G. (2022). Review on metal nanoparticles as nanocarriers: current challenges and perspectives in drug delivery systems. *Emergent Materials*, **5**(6), 1593–1615, https://doi.org/10.1007/s42247-021-00335-x

Corsi I., Winther-Nielsen M., Sethi R., Punta C., Della Torre C., Libralato G., Lofrano G., Sabatini L., Aiello M., Fiordi L., Cinuzzi F., Caneschi A., Pellegrini D. and Buttino I. (2018). Ecofriendly nanotechnologies and nanomaterials for environmental applications: key issue and consensus recommendations for sustainable and ecosafe nanoremediation. *Ecotoxicology and Environmental Safety*, **154**, 237–244, https://doi.org/10.1016/j.ecoenv.2018.02.037

Cuenya B. R. (2010). Synthesis and catalytic properties of metal nanoparticles: size, shape, support, composition, and oxidation state effects. *Thin Solid Films*, **518**(12), 3127–3150, https://doi.org/10.1016/j.tsf.2010.01.018

De Luca G., Bonaccorsi P., Trovato V., Mancuso A., Papalia T., Pistone A., Casaletto M. P., Mezzi A., Brunetti B., Minuti L., Temperini A., Barattucci A. and Plutino M. R. (2018). Tripodal tris-disulfides as capping agents for a controlled mixed functionalization of gold nanoparticles. *New Journal of Chemistry*, **42**(20), 16436–16440, https://doi.org/10.1039/C8NJ03086C

Descalzo A. B., Martínez-Máñez R., Sancenón F., Hoffmann K. and Rurack K. (2006). The supramolecular chemistry of organic–inorganic hybrid materials. *Angewandte Chemie International Edition*, **45**(36), 5924–5948, https://doi.org/10.1002/anie.200600734

Djellabi R., Yang B., Adeel Sharif H. M., Zhang J., Ali J. and Zhao X. (2019). Sustainable and easy recoverable magnetic TiO_2-lignocellulosic biomass@Fe_3O_4 for solar photocatalytic water remediation. *Journal of Cleaner Production*, **233**, 841–847, https://doi.org/10.1016/j.jclepro.2019.06.125

Dontsova T. A., Nahirniak S. V. and Astrelin I. M. (2019). Metaloxide nanomaterials and nanocomposites of ecological purpose. *Journal of Nanomaterials*, **2019**, 1–31, https://doi.org/10.1155/2019/5942194

Doria G., Conde J., Veigas B., Giestas L., Almeida C., Assunção M., Rosa J. and Baptista P. V. (2012). Noble metal nanoparticles for biosensing applications. *Sensors*, **12**(2), 1657–1687, https://doi.org/10.3390/s120201657

Elmaaty T. A., Raouf S., Sayed-Ahmed K. and Plutino M. R. (2022). Multifunctional dyeing of wool fabrics using selenium nanoparticles. *Polymers*, **14**(1), 191, https://doi.org/10.3390/polym14010191

García-Martínez J.-M. and Collar E. P. (2020). Organic–inorganic hybrid materials. *Polymers*, **13**(1), 86, https://doi.org/10.3390/polym13010086

Gholap A. D. (2012). Translational nanomedicine: Status assessment and opportunities. *International Journal of Research in Pharmaceutical and Biomedical Sciences*, **4**(3), 1454–1463.

Gholap A. D. and Musmade D. (2011). Advances in the field of nanooncology: an overview. *Inventi Impact: NDDS [ISSN 0976-3791]*, **4**, 1–3.

Gholap A. D. and Rajurkar V. (2012a). Lipid nanocapsules: a novel promising tool as drug delivery system. *Inventi Rapid: NDDS [ISSN 0976-3791]*, **4**, 1–4.

Gholap A. D. and Rajurkar V. (2012b). New threat of clinical toxicities evoked from nanocarrier systems. *Inventi Rapid: NDDS [ISSN 0976-3791]*, **4**, 1–4.

Gholap A. D. and Rajurkar V. (2012c). Potential applications of superparamagnetic nanoparticles technology. *Inventi Rapid: NDDS [ISSN 0976-3791]*, **4**, 1–3.

Gholap A. D. and Rajurkar V. G. (2012d). Recent technologies in nanomedicine for treatment of cancer. *Inventi Rapid: Pharm Tech [ISSN 0976-3783]*, **4**, 1–3.

Gholap A. D. and Tupe D. (2012). Potential applications of botanical extract. *Inventi Rapid: Ethnopharmacology [ISSN 0976-3805]*, **4**, 1–3.

Gholap A. D., Bhowmik D. D., Deshmukh A. Y. and Hatvate N. T. (2023a). Quintessential impact of dendrimer bioconjugates in targeted drug delivery, https://doi.org/10.1016/bs.coac.2023.02.009

Gholap A. D., Gupta J. S., Kamandar P. A., Banchhod G. V. and Hatvate N. T. (2023b). Antibody-drug conjugates for cancer therapy: An up-to-date review on the chemistry and pharmacology, https://doi.org/10.1016/bs.coac.2023.02.007

Gholap A. D., Said R. P., Pawar R. D., Ambore G. S. and Hatvate N. T. (2023c). Importance of carbohydrate-drug conjugates in vaccine development: a detailed review, https://doi.org/10.1016/bs.coac.2023.03.002

Gnanaprakasam P., Jeena S. E., Premnath D. and Selvaraju T. (2016). Simple and robust green synthesis of Au NPs on reduced graphene oxide for the simultaneous detection of toxic heavy metal ions and bioremediation using bacterium as the scavenger. *Electroanalysis*, **28**(8), 1885–1893, https://doi.org/10.1002/elan.201600002

Guido E., Colleoni C., De Clerck K., Plutino M. R. and Rosace G. (2014). Influence of catalyst in the synthesis of a cellulose-based sensor: kinetic study of 3-glycidoxypropyltrimethoxysilane epoxy ring opening by Lewis acid. *Sensors and Actuators B: Chemical*, **203**, 213–222, https://doi.org/10.1016/j.snb.2014.06.126

Hanemann T. and Szabó D. V. (2010). Polymer-nanoparticle composites: from synthesis to modern applications. *Materials*, **3**(6), 3468–3517, https://doi.org/10.3390/ma3063468

He C., Gu L., He H., Zhang Z., Wang X., Han F., Huang B. and Pan X. (2020). Dissolved organic matter modified magnetic carbon nanotubes enhance the bioremediation of azo dyes and Cr(VI). *Environmental Science: Water Research & Technology*, **6**(7), 1804–1815, https://doi.org/10.1039/C9EW00965E

Ielo I., Giacobello F., Castellano A., Sfameni S., Rando G. and Plutino M. R. (2021a). Development of antibacterial and antifouling innovative and eco-sustainable sol–gel based materials: from marine areas protection to healthcare applications. *Gels*, **8**(1), 26, https://doi.org/10.3390/gels8010026

Ielo I., Giacobello F., Sfameni S., Rando G., Galletta M., Trovato V., Rosace G. and Plutino M. R. (2021b). Nanostructured surface finishing and coatings: functional properties and applications. *Materials*, **14**(11), 2733, https://doi.org/10.3390/ma14112733

Ielo I., Iacopetta D., Saturnino C., Longo P., Galletta M., Drommi D., Rosace G., Sinicropi M. S. and Plutino M. R. (2021c). Gold derivatives development as prospective anticancer drugs for breast cancer treatment. *Applied Sciences*, **11**(5), 2089, https://doi.org/10.3390/app11052089

Ielo I., Rando G., Giacobello F., Sfameni S., Castellano A., Galletta M., Drommi D., Rosace G. and Plutino M. R. (2021d). Synthesis, chemical–physical characterization, and biomedical applications of functional gold nanoparticles: a review. *Molecules*, **26**(19), 5823, https://doi.org/10.3390/molecules26195823

Jamkhande P. G., Ghule N. W., Bamer A. H. and Kalaskar M. G. (2019). Metal nanoparticles synthesis: an overview on methods of preparation, advantages and disadvantages, and applications. *Journal of Drug Delivery Science and Technology*, **53**, 101174, https://doi.org/10.1016/j.jddst.2019.101174

Kang Y., Li L., Chen W., Zhang F., Du Y. and Wu T. (2018). Rapid in situ SERS analysis of pesticide residues on plant surfaces based on micelle extraction of targets and stabilization of Ag nanoparticle aggregates. *Food Analytical Methods*, **11**(11), 3161–3169, https://doi.org/10.1007/s12161-018-1290-2

Kanjwal M. A., Lo K. K. S. and Leung W. W.-F. (2019). Graphene composite nanofibers as a high-performance photocatalyst for environmental remediation. *Separation and Purification Technology*, **215**, 602–611, https://doi.org/10.1016/j.seppur.2019.01.044

Karthikeyan P. and Meenakshi S. (2020). In situ fabrication of magnetic particles decorated biopolymeric composite beads for the selective remediation of phosphate and nitrate from aqueous medium. *Journal of Environmental Chemical Engineering*, **8**(2), 103530, https://doi.org/10.1016/j.jece.2019.103530

Kelly K. L., Coronado E., Zhao L. L. and Schatz G. C. (2003). The optical properties of metal nanoparticles: the influence of size, shape, and dielectric environment. *The Journal of Physical Chemistry B*, **107**(3), 668–677, https://doi.org/10.1021/jp026731y

Khan N., Anwer A. H., Ahmad A., Sabir S., Sevda S. and Khan M. Z. (2020). Investigation of CNT/PPy-modified carbon paper electrodes under anaerobic and aerobic conditions for phenol bioremediation in microbial fuel cells. *ACS Omega*, **5**(1), 471–480, https://doi.org/10.1021/acsomega.9b02981

Khedkar A., Tupe D. and Gholap A. (2012a). Cubosomes: emerging drug delivery system. *Inventi Rapid: NDDS*, **4**, 1–4. Inventi:prndds/9987/21.

Khedkar A., Tupe D. and Gholap A. (2012b). Pharmacosomes: an emerging vesicular system. *Inventi Rapid: NDDS [ISSN 0976-3791]*, **4**, 1–3. Inventi:prndds/9986/21

Kumar V., Shahi S. K. and Singh S. (2018). Bioremediation: an eco-sustainable approach for restoration of contaminated sites. In: Microbial Bioprospecting for Sustainable Development, Springer, Singapore, pp. 115–136, https://doi.org/10.1007/978-981-13-0053-0_6

Kwak S.-Y., Giraldo J. P., Wong M. H., Koman V. B., Lew T. T. S., Ell J., Weidman M. C., Sinclair R. M., Landry M. P., Tisdale W. A. and Strano M. S. (2017). A nanobionic light-emitting plant. *Nano Letters*, **17**(12), 7951–7961, https://doi.org/10.1021/acs.nanolett.7b04369

Lemming G., Chambon J. C., Binning P. J. and Bjerg P. L. (2012). Is there an environmental benefit from remediation of a contaminated site? Combined assessments of the risk reduction and life cycle impact of remediation. *Journal of Environmental Management*, **112**, 392–403, https://doi.org/10.1016/j.jenvman.2012.08.002

Lepore E., Bosia F., Bonaccorso F., Bruna M., Taioli S., Garberoglio G., Ferrari A. C. and Pugno N. M. (2017). Spider silk reinforced by graphene or carbon nanotubes. *2D Materials*, **4**(3), 031013, https://doi.org/10.1088/2053-1583/aa7cd3

Li J., Zhang D., Ni X., Zheng H. and Zhang Q. (2017). Excellent hydrophilic and anti-bacterial fouling PVDF membrane based on Ag nanoparticle self-assembled PCBMA polymer brush. *Chinese Journal of Polymer Science*, **35**(7), 809–822, https://doi.org/10.1007/s10118-017-1944-3

Li Z., Wang L., Li Y., Feng Y. and Feng W. (2019). Carbon-based functional nanomaterials: preparation, properties and applications. *Composites Science and Technology*, **179**, 10–40, https://doi.org/10.1016/j.compscitech.2019.04.028

Liang K., Richardson J. J., Doonan C. J., Mulet X., Ju Y., Cui J., Caruso F. and Falcaro P. (2017). An enzyme-coated metal-organic framework shell for synthetically adaptive cell survival. *Angewandte Chemie International Edition*, **56**(29), 8510–8515, https://doi.org/10.1002/anie.201704120

Lincy A., Jegathambal P., Mkandawire M. and MacQuarrie S. L. (2020). Nano bioremediation of textile dye effluent using magnetite nanoparticles encapsulated alginate beads.

Logan B. E., Hamelers B., Rozendal R., Schröder U., Keller J., Freguia S., Aelterman P., Verstraete W. and Rabaey K. (2006). Microbial fuel cells: methodology and technology. *Environmental Science & Technology*, **40**(17), 5181–5192, https://doi.org/10.1021/es0605016

Lu H., Wang J., Stoller M., Wang T., Bao Y. and Hao H. (2016). An overview of nanomaterials for water and wastewater treatment. *Advances in Materials Science and Engineering*, **2016**, 1–10, https://doi.org/10.1155/2016/4964828

Magrez A., Kasas S., Salicio V., Pasquier N., Seo J. W., Celio M., Catsicas S., Schwaller B. and Forró L. (2006). Cellular toxicity of carbon-based nanomaterials. *Nano Letters*, **6**(6), 1121–1125, https://doi.org/10.1021/nl060162e

Mahjoubi M., Cappello S., Souissi Y., Jaouani A. and Cherif A. (2017). Microbial bioremediation of petroleum hydrocarbon-contaminated marine environments. In: Recent Insights in Petroleum Science and Engineering, M. Zoveidavianpoor (ed.), Rijeka, pp. 325–350. IntechOpen, https://doi.org/10.5772/intechopen.72207

Mohammad A. W., Teow Y. H., Ang W. L., Chung Y. T., Oatley-Radcliffe D. L. and Hilal N. (2015). Nanofiltration membranes review: recent advances and future prospects. *Desalination*, **356**, 226–254, https://doi.org/10.1016/j.desal.2014.10.043

Morin-Crini N., Lichtfouse E., Fourmentin M., Ribeiro A. R. L., Noutsopoulos C., Mapelli F., Fenyvesi É, Vieira M. G. A., Picos-Corrales L. A., Moreno-Piraján J. C., Giraldo L., Sohajda T., Huq M. M., Soltan J., Torri G., Magureanu M., Bradu C. and Crini G. (2022). Removal of emerging contaminants from wastewater using advanced treatments. A review. *Environmental Chemistry Letters*, **20**(2), 1333–1375, https://doi.org/10.1007/s10311-021-01379-5

Nechifor A. C., Cotorcea S., Bungău C., Albu P. C., Paşcu D., Oprea O., Grosu A. R., Pîrţac A. and Nechifor G. (2021a). Removing of the sulfur compounds by impregnated polypropylene fibers with silver nanoparticles-cellulose derivatives for air odor correction. *Membranes*, **11**(4), 256, https://doi.org/10.3390/membranes11040256

Nechifor A. C., Goran A., Grosu V.-A., Bungău C., Albu P. C., Grosu A. R., Oprea O., Păncescu F. M. and Nechifor G. (2021b). Improving the performance of composite hollow fiber membranes with magnetic field generated convection application on pH correction. *Membranes*, **11**(6), 445, https://doi.org/10.3390/membranes11060445

Nechifor G., Păncescu F. M., Grosu A. R., Albu P. C., Oprea O., Tanczos S.-K., Bungău C., Grosu V.-A., Pîrţac A. and Nechifor A. C. (2021c). Osmium nanoparticles-polypropylene hollow fiber membranes applied in redox processes. *Nanomaterials*, **11**(10), 2526, https://doi.org/10.3390/nano11102526

Niu Z., Liu L., Zhang L. and Chen X. (2014). Porous graphene materials for water remediation. *Small*, **10**(17), 3434–3441, https://doi.org/10.1002/smll.201400128

Pandey S., Son N. and Kang M. (2022). Synergistic sorption performance of Karaya gum crosslink poly(acrylamide-co-acrylonitrile) @ metal nanoparticle for organic pollutants. *International Journal of Biological Macromolecules*, **210**, 300–314, https://doi.org/10.1016/j.ijbiomac.2022.05.019

Patel K. D., Singh R. K. and Kim H.-W. (2019). Carbon based nanomaterials as an emerging platform for theranostics. *Materials Horizons*, **6**(3), 434–469, https://doi.org/10.1039/C8MH00966J

Pavía-Sanders A., Zhang S., Flores J. A., Sanders J. E., Raymond J. E. and Wooley K. L. (2013). Robust magnetic/polymer hybrid nanoparticles designed for crude oil entrapment and recovery in aqueous environments. *ACS Nano*, **7**(9), 7552–7561, https://doi.org/10.1021/nn401541e

Plutino M. R., Colleoni C., Donelli I., Freddi G., Guido E., Maschi O., Mezzi A. and Rosace G. (2017). Sol-gel 3-glycidoxypropyltriethoxysilane finishing on different fabrics: the role of precursor concentration and catalyst on the textile performances and cytotoxic activity. *Journal of Colloid and Interface Science*, **506**, 504–517, https://doi.org/10.1016/j.jcis.2017.07.048

Rando G., Sfameni S., Galletta M., Drommi D., Cappello S. and Plutino M. R. (2022). Functional nanohybrids and nanocomposites development for the removal of environmental pollutants and bioremediation. *Molecules*, **27**(15), 4856, https://doi.org/10.3390/molecules27154856

Rangel-Muñoz N., González-Barrios A. F., Pradilla D., Osma J. F. and Cruz J. C. (2020). Novel bionanocompounds: outer membrane protein A and laccase co-immobilized on magnetite nanoparticles for produced water treatment. *Nanomaterials*, **10**(11), 2278, https://doi.org/10.3390/nano10112278

Rathi B. S., Kumar P. S. and Show P.-L. (2021). A review on effective removal of emerging contaminants from aquatic systems: current trends and scope for further research. *Journal of Hazardous Materials*, **409**, 124413, https://doi.org/10.1016/j.jhazmat.2020.124413

Rout D. R., Chaurasia S. and Jena H. M. (2022). Enhanced photocatalytic degradation of malachite green using manganese oxide doped graphene oxide/zinc oxide (GO-ZnO/Mn_2O_3) ternary composite under sunlight irradiation. *Journal of Environmental Management*, **318**, 115449, https://doi.org/10.1016/j.jenvman.2022.115449

Saleh T. A. (2020). Nanomaterials: classification, properties, and environmental toxicities. *Environmental Technology & Innovation*, **20**, 101067, https://doi.org/10.1016/j.eti.2020.101067

Santhosh C., Velmurugan V., Jacob G., Jeong S. K., Grace A. N. and Bhatnagar A. (2016). Role of nanomaterials in water treatment applications: a review. *Chemical Engineering Journal*, **306**, 1116–1137, https://doi.org/10.1016/j.cej.2016.08.053

Saturnino C., Sinicropi M. S., Iacopettab D., Ceramella J., Caruso A., Muià N., Longo P., Rosace G., Galletta M., Ielo I. and Plutino M. R. (2018). N-Thiocarbazole-based gold nanoparticles: synthesis, characterization and anti-proliferative activity evaluation. *IOP Conference Series: Materials Science and Engineering*, **459**, 012023, https://doi.org/10.1088/1757-899X/459/1/012023

Scaffaro R., Lopresti F., Catania V., Santisi S., Cappello S., Botta L. and Quatrini P. (2017). Polycaprolactone-based scaffold for oil-selective sorption and improvement of bacteria activity for bioremediation of polluted water. *European Polymer Journal*, **91**, 260–273, https://doi.org/10.1016/j.eurpolymj.2017.04.015

Shahkaramipour N., Tran T., Ramanan S. and Lin H. (2017). Membranes with surface-enhanced antifouling properties for water purification. *Membranes*, **7**(1), 13, https://doi.org/10.3390/membranes7010013

Slavin Y. N., Asnis J., Häfeli U. O. and Bach H. (2017). Metal nanoparticles: understanding the mechanisms behind antibacterial activity. *Journal of Nanobiotechnology*, **15**(1), 65, https://doi.org/10.1186/s12951-017-0308-z

Sun Z., Zhang Y., Guo S., Shi J., Shi C., Qu K., Qi H., Huang Z., Murugadoss V., Huang M. and Guo Z. (2022). Confining FeNi nanoparticles in biomass-derived carbon for effectively photo-Fenton catalytic reaction for polluted water treatment. *Advanced Composites and Hybrid Materials*, **5**(2), 1566–1581, https://doi.org/10.1007/s42114-022-00477-4

Szczepanik B. (2017). Photocatalytic degradation of organic contaminants over clay-TiO_2 nanocomposites: a review. *Applied Clay Science*, **141**, 227–239, https://doi.org/10.1016/j.clay.2017.02.029

Trovato V., Colleoni C., Castellano A. and Plutino M. R. (2018). The key role of 3-glycidoxypropyltrimethoxysilane sol–gel precursor in the development of wearable sensors for health monitoring. *Journal of Sol-Gel Science and Technology*, **87**(1), 27–40, https://doi.org/10.1007/s10971-018-4695-x

Trovato V., Rosace G., Colleoni C., Sfameni S., Migani V. and Plutino M. R. (2020). Sol-gel based coatings for the protection of cultural heritage textiles. *IOP Conference Series: Materials Science and Engineering*, **777**(1), 012007, https://doi.org/10.1088/1757-899X/777/1/012007

Ucankus G., Ercan M., Uzunoglu D. and Culha M. (2018). Methods for preparation of nanocomposites in environmental remediation. In: New Polymer Nanocomposites for Environmental Remediation, C. M. Hussain and A. K. Mishra (eds), Elsevier, Netherlands, pp. 1–28, https://doi.org/10.1016/B978-0-12-811033-1.00001-9

Valdez-Carrillo M., Abrell L., Ramírez-Hernández J., Reyes-López J. A. and Carreón-Diazconti C. (2020). Pharmaceuticals as emerging contaminants in the aquatic environment of Latin America: a review. *Environmental Science and Pollution Research*, **27**(36), 44863–44891, https://doi.org/10.1007/s11356-020-10842-9

Vázquez-Núñez E., Molina-Guerrero C. E., Peña-Castro J. M., Fernández-Luqueño F. and de la Rosa-Álvarez M. G. (2020). Use of nanotechnology for the bioremediation of contaminants: a review. *Processes*, **8**(7), 826, https://doi.org/10.3390/pr8070826

Wali L. A., Alwan A. M., Dheyab A. B. and Hashim D. A. (2019). Excellent fabrication of Pd-Ag NPs/PSi photocatalyst based on bimetallic nanoparticles for improving methylene blue photocatalytic degradation. *Optik*, **179**, 708–717, https://doi.org/10.1016/j.ijleo.2018.11.011

Wan S., Ding W., Wang Y., Wu J., Gu Y. and He F. (2018). Manganese oxide nanoparticles impregnated graphene oxide aggregates for cadmium and copper remediation. *Chemical Engineering Journal*, **350**, 1135–1143, https://doi.org/10.1016/j.cej.2018.06.068

Wang J. and Wang S. (2019). Preparation, modification and environmental application of biochar: a review. *Journal of Cleaner Production*, **227**, 1002–1022, https://doi.org/10.1016/j.jclepro.2019.04.282

Wang C., Li X., Gao E., Jian M., Xia K., Wang Q., Xu Z., Ren T. and Zhang Y. (2016). Carbonized silk fabric for ultrastretchable, highly sensitive, and wearable strain sensors. *Advanced Materials*, **28**(31), 6640–6648, https://doi.org/10.1002/adma.201601572

Wang Y., Pan C., Chu W., Vipin A. and Sun L. (2019). Environmental remediation applications of carbon nanotubes and graphene oxide: adsorption and catalysis. *Nanomaterials*, **9**(3), 439, https://doi.org/10.3390/nano9030439

Weijie M., Chongnv W., Xuming P., Weixin J., Yuhang W. and Benhui S. (2020). TiO_2 nanoparticles and multi-walled carbon nanotubes monitoring and bioremediation potential using ciliates *Pseudocohnilembus persalinus*. *Ecotoxicology and Environmental Safety*, **187**, 109825, https://doi.org/10.1016/j.ecoenv.2019.109825

Xiang Y., Jiang L., Zhou Y., Luo Z., Zhi D., Yang J. and Lam S. S. (2022). Microplastics and environmental pollutants: key interaction and toxicology in aquatic and soil environments. *Journal of Hazardous Materials*, **422**, 126843, https://doi.org/10.1016/j.jhazmat.2021.126843

Yang J., Wang X., Chen Y., Dai J. and Sun S. (2015). Enhanced photocatalytic activities of visible-light driven green synthesis in water and environmental remediation on Au/Bi_2WO_6 hybrid nanostructures. *RSC Advances*, **5**(13), 9771–9782, https://doi.org/10.1039/C4RA15349A

Yaqoob A. A., Ibrahim M. N. M., Yaakop A. S., Umar K. and Ahmad A. (2021). Modified graphene oxide anode: a bioinspired waste material for bioremediation of Pb^{2+} with energy generation through microbial fuel cells. *Chemical Engineering Journal*, **417**, 128052, https://doi.org/10.1016/j.cej.2020.128052

Yin Y., Zhou T., Luo H., Geng J., Yu W. and Jiang Z. (2019). Adsorption of arsenic by activated charcoal coated zirconium-manganese nanocomposite: performance and mechanism. *Colloids and Surfaces A: Physicochemical and Engineering Aspects*, **575**, 318–328, https://doi.org/10.1016/j.colsurfa.2019.04.093

Zhou S., Jiang W., Wang T. and Lu Y. (2015). Highly hydrophobic, compressible, and magnetic polystyrene/Fe_3O_4/graphene aerogel composite for oil–water separation. *Industrial & Engineering Chemistry Research*, **54**(20), 5460–5467, https://doi.org/10.1021/acs.iecr.5b00296

Zhu Y., Fan L., Yang B. and Du J. (2014). Multifunctional homopolymer vesicles for facile immobilization of gold nanoparticles and effective water remediation. *ACS Nano*, **8**(5), 5022–5031, https://doi.org/10.1021/nn5010974

Part 4

Wastewater Treatment Using NBs

doi: 10.2166/9781789063592_0193

Chapter 11

Nanobiohybrids for pollution control and resource recovery

P. Kumari[1], *R. Kumari*[1], *R. Prabhakar*[1] *and S. R. Samadder*[1]

Department of Environmental Science and Engineering, Indian Institute of Technology (Indian School of Mines), Dhanbad 826004, Jharkhand, India
Email: samadder@iitism.ac.in

ABSTRACT

The rapid growth of human population, industrialization, and urbanization has led to an increase in pollution and resource depletion. This has created a need for highly efficient pollution abatement technologies and resource recovery methods. While there are already pre-existing techniques for both pollution abatement and resource recovery, they are often costly, involving the use of harmful chemicals, generating unsafe by-products, and are slow. On the other hand, nanobiohybrids (NBHs) have emerged as promising approaches with substantial benefits over traditional approaches and have gained global recognition in recent decades. The integration of functional nanomaterials with living systems has given rise to this fascinating area of study, which lies at the intersection of materials engineering and biological science. These NBHs have the ability to remove pollutants and recover resources from different sources. This chapter provides an overview of the potential application of NBHs in different types of pollution control and resource recovery techniques from various sources. Different NBHs have so far been extensively studied and successfully applied in the removal of pollutants from water and wastewater but their application in soil and air as well as in resource recovery is still incipient. Thus, the limitations and challenges of using NBHs have been identified and discussed in this chapter as well. The findings in this chapter provide new insights and future prospects for further research and development for the wider applicability of NBHs.

Keywords: nanobiohybrids, pollution control, resource recovery, pollutants, water and wastewater treatment

11.1 INTRODUCTION

The world is facing a major environmental crisis due to an increase in pollution that threatens human health, ecosystems, and the planet. It is caused by a variety of sources, including industrial activities, agricultural practices, and consumer behavior. The effects of pollution are far-reaching and can range from respiratory problems to climate change and loss of biodiversity. The need for effective and sustainable solutions for pollution control is urgent because it is burdensome to the environment.

On the contrary, the demand for resources has also grown all over the globe with the continued increase in the world population. These have led to increased resource consumption, waste, and pollution generation. Since the health and welfare of the society are closely linked to the availability of affordable and adequate resources, as well as a safe and clean environment; pollution abatement and resource recovery from the wastes form the core areas that need immediate attention. Resource recovery is the process of waste collection, extraction of valuable materials, treatment, waste reduction, and resource conservation. In recent years, resource recovery has become increasingly popular worldwide as governments, businesses, and individuals recognize the environmental and economic benefits of reducing waste.

Although there are numerous existing techniques for both pollution abatement and resource recovery, the quest to develop lost-cost techniques for pollution control and resource recovery is becoming increasingly important on a worldwide scale. In response to these challenges, researchers and scientists have been exploring efficient and sustainable technologies to tackle pollution and waste, which has led to the development of various innovative approaches. Among these, the development and design of nanobiohybrids (NBHs) has emerged as a promising option due to their unique properties and capabilities.

NBHs are a new class of materials that broadly encompasses both biologically derived components and synthetic components. The biological component includes biomolecules and biological systems such as proteins, DNA, tissues, living cells, or organisms (Richardson & Liang, 2018; Su *et al.*, 2016). The synthetic component consists of inorganic substances such as silver, gold, titanium, cadmium, iron oxide, carbon materials, or silica oxide (Lauth *et al.*, 2017), lipids and polymers organic materials (Niu *et al.*, 2017), or hybrid substances like metal-organic frameworks (MOFs) and metal-phenolic networks (MPNs) (Lykourinou *et al.*, 2011). The combination of biological systems and nanomaterials in NBHs enables them to perform complex functions that are not achievable by either component alone, such as enhanced catalytic activity, stability, and selectivity.

In recent years, the inclusion of inorganic nanoparticles (NPs) into organic materials has drawn considerable attention in a wide range of domains due to their distinctive properties such as biodegradability, large surface area, and magnetic or electrochemical properties (Mohammadalinejhad *et al.*, 2019). One of the key applications of NBHs is in the treatment of water and wastewater. Nanoparticles can be used to remove pollutants from water through processes such as adsorption, while biological systems can help break down organic pollutants into harmless substances. By combining these two approaches, NBHs can effectively remove a wide variety of pollutants from water (Nishu, 2023). Another area where NBHs have shown potential is in the recovery of valuable resources from waste streams, as they can capture, store, and convert waste into useful products. The application of NBHs in pollution control and resource recovery represents a promising approach for addressing some of the largest environmental challenges facing our world today. However, the selection of the biological and nanomaterial components affects several aspects of the final functionality of the NBH systems (Guo *et al.*, 2021).

The development of NBHs is still in its early stages. It requires a deep understanding of the properties of biological and nanoscale components, as well as the ability to manipulate these components in ways that create new, functional materials. Therefore, more study is required to completely understand the risks and potential benefits associated with synthesis technologies and to optimize their performance in real-world applications. To give a more comprehensive picture of NBHs, the combined benefits and drawbacks of NBHs are shown in Table 11.1 Both the benefits and drawbacks of NBHs imply that they hold potential for a wide range of applications. Nevertheless, careful consideration is required to ensure their usage will be safe and successful.

Herein, the objective of this chapter is to review the current state of knowledge and potential applications of NBHs in the fields of pollution abatement and resource recovery. In addition, different methods for the synthesis of NBHs are addressed, with an emphasis on their respective advantages and disadvantages. In the last section, the chapter highlights the challenges and opportunities for the application of NBHs in real-world environmental remediation scenarios. The purpose of this chapter is to contribute to a better understanding of NBHs for practical and industrial applications, notably in the areas of air pollution control, water and wastewater treatment, soil remediation, and recovery of resources. It will provide new valuable insights for researchers, policymakers, and industries interested in developing sustainable solutions for pollution control and resource recovery.

Table 11.1 Benefits and drawbacks of NBHs.

Benefits	Drawbacks
Enhanced stability NBHs have increased stability compared to their individual components, as the nanomaterials can protect the biological molecules from degradation or denaturation. Besides, the nanoscale components of NBHs exhibit more resistance to environmental factors.	**Complexity** The design and synthesis of NBHs involve the integration of different materials and processes. This can be complex, challenging, and require specialized expertise in both biology and nanotechnology.
Increased efficiency The combination of biological and nanomaterial components can result in enhanced efficiencies of the system, such as increased catalytic activity or improved sensing capabilities.	**Cost** The production of NBHs can be expensive, especially when using high-quality nanomaterials and biological molecules. It can also be costly due to the complexity of the process and the need for specialized equipment and materials.
Tailored functionality The characteristic of the NBHs can be tailored by controlling the shape, size, and surface chemistry of the nanomaterials, as well as the type and amount of biological molecules used.	**Safety concerns** There are concerns about the safety of nanomaterials, including their potential toxicity and environmental impact.
Biocompatibility NBHs are composed of both biological and nanoscale components. The use of biological components can enhance the biocompatibility of the hybrid material, reducing the risk of toxicity or immune reactions.	**Regulatory challenges** The regulation of NBHs can be challenging, as they fall into a regulatory grey area between biological and nanomaterials regulations.
Versatility NBHs can be designed for a wide variety of applications, from imaging and drug delivery to biosensors and bioremediation.	**Limited stability** Although NBHs can be more stable than their individual components, still the stability of NBHs can be a concern. The biological components may degrade over time, which can affect the overall performance of the material by limiting its shelf-life and practical use.
Multifunctionality NBHs can be designed to perform multiple functions, such as sensing, imaging, and drug delivery, in a single material.	**Potential toxicity** NBHs are composed of both biological and non-biological components, and the combination of these components can lead to the release of toxic substances. For example, some NBHs contain metals, such as gold or silver, which can be toxic to cells and tissues. Additionally, some NBHs contain polymers, which can also be toxic if they are not biodegradable.
	Limited biodegradability Another disadvantage of NBHs is their limited biodegradability. Many NBHs are composed of polymers, which are difficult to break down by the natural processes of the body. This can lead to the accumulation of NBHs in the body, which can cause adverse health effects.
	Ethical concerns The use of biological components in NBHs raises ethical concerns, such as the use of animal or human cells, and the potential for unintended consequences if the hybrid material interacts with living systems in unforeseen ways.

11.2 SYNTHESIS METHODS FOR NBHS

The methods of synthesis have a substantial influence on the morphology, size, and shape of NBHs, as well as their characteristics, functional properties, and overall performance. In-situ and ex-situ approaches are most often used for the synthesis of NBHs (Figure 11.1) (Mirtalebi *et al.*, 2019).

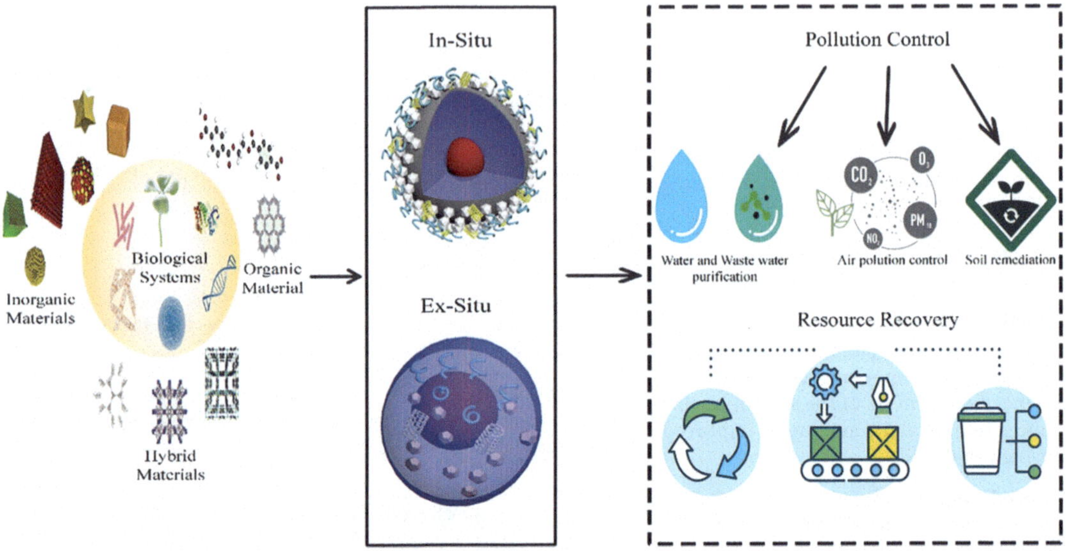

Figure 11.1 Integration of biological systems with functional nanoparticles using in-situ and ex-situ synthesis methods, as well as their applications in pollution control and resource recovery.

11.2.1 In-situ synthesis

The in-situ synthesis method offers a simple and efficient technique for the fabrication of NBHs with enhanced biological activity (Gao *et al.*, 2019; Shinde *et al.*, 2020). During the in-situ synthesis process, the metal ions are preloaded into the matrix first, where the ions are expected to disperse equally. The precursors are then exposed to the appropriate conditions in order to transform them into NBHs. For preparing NBHs via an in-situ approach, it is important to assess three different aspects, that is, the selection of non-toxic reducing agents, the solvent medium, and finally, the selection of environmentally friendly materials to stabilize the NBHs. One of the most commonly used methods for in-situ synthesis of NBHs is the chemical reduction method, which involves the reduction of metal salts in the presence of biological molecules such as proteins, enzymes, or DNA (Li *et al.*, 2013). Another in-situ approach is the use of biological molecules as reducing and stabilizing agents (Nadda & Kanwar, 2012).

The in-situ synthesis method of NBHs has its own set of merits and demerits. One of the significant advantages of this method is that it provides high stability to the NBHs due to the strong interactions between the biological matrix and the NPs. This ensures that the nanoparticles remain stable and do not agglomerate. This method is relatively simple and easy to perform and does not require complex equipment or lengthy procedures which make it a cost-effective and time-efficient technique for the preparation of NBHs. Additionally, the nanoparticles synthesized by this method can be tailored to exhibit specific properties such as biocompatibility, bioactivity, and biodegradability. These enhance the biological activity of the NBHs and make them suitable for a wide range of applications. However, the in-situ synthesis method offers limited control over the size, shape, and distribution of NPs, which may result in non-uniform NBHs with varying properties. The method may also be limited and incompatible with certain biological matrices and may pose a potential risk of toxicity to the biological matrix due to the use of chemicals and reagents. This may affect the overall stability and functionality of the NBHs.

11.2.2 Ex-situ synthesis

The ex-situ synthesis is a versatile method that is widely used for the preparation of NBHs (Guo *et al.*, 2014). In this method, the nanomaterials and biological components are synthesized separately at first, and they are then combined to form the NBHs (Patra & Baek, 2014). The first step in the ex-situ synthesis is the synthesis of the nanomaterials which can be achieved through a variety of methods such as chemical synthesis, physical deposition, or biological synthesis. Some commonly used nanomaterials include gold nanoparticles, silver nanoparticles, carbon nanotubes, and quantum dots. Once the nanomaterials are synthesized, the next step is to functionalize them with biological molecules. This can be achieved through the use of various coupling agents such as silanes, thiolated polymers, or phospholipids. These coupling agents can be used to attach a wide range of biological molecules such as proteins, enzymes, DNA, or cells to the surface of the nanomaterials. The resulting NBHs can be tailored for a variety of applications such as tissue engineering, drug delivery, biosensing, medicine, biotechnology, and environmental applications (Jaiswal *et al.*, 2023; Kumari & Samadder, 2022).

Ex-situ synthesis methods have several advantages. Firstly, it allows precise control of the size and shape of the NP, which is crucial for their biological applications. Secondly, this synthesis method is highly reproducible and scalable, which makes it suitable for industrial-scale production. However, ex-situ methods have also some drawbacks. One of the major concerns is the potential toxicity of the nanoparticles. The biomolecules used for binding the nanoparticles may also lose their activity during the synthesis process, affecting the overall functionality of the NBHs. Additionally, ex-situ synthesis methods may not be suitable for some biomolecules, which are highly sensitive to pH and temperature. The effectiveness of this approach depends on the specific application and the type of biomolecule used.

11.3 ROLE OF NBHS IN POLLUTION CONTROL

One of the most promising applications of NBHs for pollution control is their ability to absorb and degrade toxic substances into less harmful compounds, rather than simply removing them from the environment. This not only reduces the toxicity of the pollutants but also conserves valuable resources. This section primarily reviews the application of NBHs to remove and detect harmful pollutants present in water, air, and soil.

11.3.1 Water and wastewater purification

Water is an indispensable resource that is required to support life and plays a critical role in different industries, including the food, textile, pharmaceutical, and electronics industries. The usage of water has increased excessively in households, agriculture, and manufacturing by 80%, 70%, and 22%, respectively (Santhosh *et al.*, 2016). As a direct consequence, there is a significant increase in the volume of wastewater that contains potentially hazardous pollutants such as residues of chemical pesticides, microorganisms, toxic heavy metals, and synthetic dyes (Pradeep & Anshup, 2009). Lack of access to clean and safe water has been one of the most significant challenges confronting the world for the past decades.

To solve this issue, several traditional water and wastewater purification technologies have been implemented and used both on-site and off-site. Unfortunately, the rising concentration levels of these toxic pollutants and the strict regulatory standards have made it challenging for conventional wastewater treatment methods to be effective. Adsorption is one of the techniques for treating water that can concentrate impurities, yet it cannot entirely purify water. On the other hand, the chemical precipitation methods consume chemical agents and have a high potential for causing secondary pollution (Deng *et al.*, 2019). It is also feasible to remove the pollutants by using membrane filtration. However, the latter comes with a high up-front cost for regular operation and, more significantly, produces hazardous residues that are highly concentrated and challenging to eliminate (Saikia *et al.*, 2019). In addition, these conventional methods are expensive, time-consuming, and have low efficiency and affinity for new emerging water micropollutants. Therefore, it has become crucial to explore alternative solutions to tackle the pressing environmental challenges.

The paradigm could compel the scientific community to draw attention to the issue and come up with innovative techniques to ensure that everyone has access to clean and safe drinking water. Among the several existing water purification technologies, NBHs have shown tremendous potential for the purification of water and wastewater due to their extraordinary properties (Figure 11.1). Over the past few decades, various NBHs sourced from different origins have been explored for their effectiveness in treating wastewater. The distinctive size-dependent properties of nanomaterials, such as their large surface area, high stability, remarkable durability, and reusability, in conjunction with microbial cells, have made them a subject of great interest in the wastewater treatment field, particularly in the elimination of harmful pollutants. NBHs are an excellent illustration of multi-scale development and they address most of the problems that are present in existing water purification technologies. NBHs have been used as nanofiltration membranes in order to filter out various pollutants, including heavy metals and organic compounds. NBHs have also been used for monitoring and sensing the concentration of water pollutants in an efficient manner. Certain enzyme-NPs based NBH systems used in the purification of water and wastewater are summarized in Table 11.2.

Girelli *et al.* (2006) created an enzyme reactor by immobilizing tyrosinase on aminopropyl-controlled porous glass to remove phenols from aqueous solutions. The enzyme reactor was able to achieve a removal efficiency of 60% for phenol derivatives after around 7 h of operation (Girelli *et al.*, 2006). In another study, Wahid *et al.* (2013) created a NBH system made of graphene sheets and algal cells using a device called vortex fluidics. The initial nitrate content was completely removed in a span of only four days establishing the effectiveness of graphene-algae NBHs for nitrate removal. Xu *et al.* (2013) utilized an amidination process to achieve their goal of immobilizing the laccase enzyme on polyacrylonitrile (PAN) nanofibrous membranes. The remarkable removal efficiencies of 85% for 24,6-trichlorophenol from water was demonstrated by the enzyme-PAN NBH. In a pilot study, laccase

Table 11.2 NBH systems synthesized from different enzyme–metal nanoparticles applied in water and wastewater purifications.

NBHs	Enzymes Used	Enzyme Concentration	Reaction Conditions			Application	Reference
			Buffer	Temperature (°C)	Time (h)		
Enzyme–silver NPs	β-Galactosidase from *Aspergillus oryzae*	1 mg/mL	Sodium acetate, pH 4.5	37	36	Dye wastewater treatment	Ahmed *et al.* (2016)
Enzyme-MWCNT	Laccase from *Trametes versicolor*	10 mg/mL	Potassium phosphate buffer, pH 7.0	Room temp.	24	Phenolic compounds removal	Pang *et al.* (2015)
	Tyrosinase	1–5 mg/mL	Sodium phosphate buffer, pH 7.0	Room temp.	24	Phenolic compounds removal	Subrizi *et al.* (2014)
Enzyme-reduced graphene oxide	Horseradish peroxidase	2.5 mg/mL	Phosphate buffer, pH 7.0	Room temp.	–	Biodegradation of high phenol concentration	Besharati Vineh *et al.* (2018)
Enzyme-titanium oxide NPs	Peroxidase from bitter gourd	1 mL (1500 U)	Sodium citrate buffer, pH 7.4	Room temp.	24	Phenol and dye removal	Ahmad *et al.* (2013)

immobilized on silica nanoparticles was put into a membrane reactor to remove approximately 66% of phenolic endocrine-disrupting chemicals from wastewater flowing at 78 L/h over a period of 45 days. The cost of this treatment was comparable to that of the activated carbon sorption and ozonation process (Gasser *et al.*, 2014).

In another study, a 3,4-dioxygenase cleaving enzyme of *Pseudomonas putida* was successfully mounted on multi-walled carbon nanotubes (MWCNT) by Das *et al.* (2016) for the purpose of degrading the hazardous 3,4-dihydroxybenzoic acid (3,4-DHBA) pollutant prevalent in water. The enzyme-MWCNT NBH demonstrated a maximum 3,4-DHBA loading of 1060 μg/mg. Benavente *et al.* (2018) prepared iron-enzyme NBHs by coating iron NPs with an aqueous solution of an enzyme called lipase B from *Candida antarctica*. The developed heterogeneous iron-enzyme NBHs catalyst degraded the p-aminophenol present in water in the presence of hydrogen peroxide at room temperature. Under the irradiation of visible light, Wang *et al.* (2018) prepared a NBH mat using algae, silver, and titanium oxide which efficiently removed chromium-VI from wastewater. Similarly, silver-enzyme NBHs were fabricated for the complete degradation of industrially relevant dyes, including direct, reductive, and acid dyes (Ahmed *et al.*, 2016). The findings demonstrated that acid yellow 42, direct black 80, direct yellow 4, and direct red 23 were reduced to a degree that was exceeding 90%, whereas reduced red and reduced orange were decreased to a degree of about 70%. In another study, Uddandarao and Lens (2022) synthesized ZnS-Chemical NBHs and ZnS-Biological NBHs for the purpose of enhancing the elimination of methyl orange dye from wastewater. This was accomplished by chemical precipitation and biological sulfate reduction of *Aspergillus niger cell*-ZnS NBHs. ZnS-Chemical NBHs and ZnS-Biological NBHs degraded methyl orange at a removal efficiency of 97.5% and 98% within 90 min and 60 min, respectively.

On the surface of *Phanerochaete chrysosporium* cells, Qin *et al.* (2020) synthesized a ZnS nanoparticle layer which was used to remove lead and cadmium that were present in the water. When compared with *P. chrysosporium* cells, the results showed that the capacity of ZnS-cells NBH to adsorb Pb^{2+} and Cd^{2+} was increased by 140% and 160%, respectively. It was found that the removal of heavy metals by a hybrid of ZnS and *P. chrysosporium* cells was mostly reliant on ion exchange of ZnS nanofilms as well as complexation by surface functional groups of the cells. The removal of nitrates from wastewater using *Thiobacillus denitrificans*-CdS NBH was studied by Chen *et al.* (2019). *T. denitrificans*-CdS successfully and completely degraded nitrates into nitrous oxide while maintaining its purity of more than 96.4 ($\pm$0.4) % in gaseous products. This newly discovered denitrification process has shown promising implications for the removal of nitrogen from wastewater in an economically viable manner. These findings have the potential to pave the way to develop novel NP-enzyme hybrids that can be used in this area in the near future (Palomo, 2021).

In addition, NBHs materials have been used for the detection and biosensing of the presence of organic pollutants in water. These detectors and biosensors have been utilized to monitor the quality of the water supply and notify the appropriate authorities when levels of pollutants surpass acceptable levels (Ramesh *et al.*, 2022). For instance, tyrosinase enzyme conjugation on a carbon nanotube (CNT) was used by Pérez-Lopez and Merkoçi (2009) to develop CNT epoxy composite electrodes. In order to evaluate phenolic water contaminants like catechol, the authors compared the CNT-epoxy hybrid with a tyrosinase biosensor prepared from a graphite–epoxy composite. Higher signals were obtained from the CNT composite electrode, which reflects the superior electrocatalytic capabilities of CNTs to graphite sheets. With a 90% increase in signal, the CNT-based bio-composite responded better than graphite-based materials (Pérez-Lopez & Merkoçi, 2009). Different enzymes can thus be immobilized on various matrices for the purpose of biosensing and detecting phenolic compounds, rather than eliminating them from water and wastewater (Table 11.3).

11.3.2 Air pollution control

Air pollution is one of the most pressing environmental issues that the world is confronting in the modern era as a result of the emission of harmful gases and particles into the atmosphere. These

Table 11.3 Enzyme immobilization on different matrices for biosensing phenolic compounds in water and wastewater.

Enzymes	Matrix	Method	Optimal Reaction Condition		Application as Biosensor	Limit of Detection (micromole per litre)	Reference
			pH	Temp (°C)			
Tyrosinase (Tyr)	Polymerized polypyrrole-polyvinylsulphonate (PPy-PVS)	Entrapment procedure	8	30	Phenol	1.0	Arslan and Arslan (2011)
	poly (N-3-aminopropyl pyrrole-co-pyrrole) (PAPCP)	Carbodiimide coupling reaction	7.2	Room temp.	Phenol	1.2	Rajesh *et al.* (2004)
	Phosphate ion-doped polypyrrole film	Cyclic voltammetry	7	Room temp.	Phenol, catechol, 2-fluorophenol, 2-chlorophenol, 2-bromophenol, 2-iodophenol	0.84	Apetrei *et al.* (2011)
	Porous silicon (PS)	Oxidation	-	Room temp.	Catechol	5.0	Tembe *et al.* (2008)
	Carbon nanotube (CNT)	Electrochemical polymerization	7	-	Catechol	0.67	Ozoner *et al.* (2010)
	Polyaniline (PANI)	Direct electropolymerization (one step)	5.5	Room temp.	Catechol	0.01	Chen *et al.* (2013)
	Gold-NPs and dihexadecylphosphate (DHP)	Amperometry	6	Room temp.	Catechol	0.17	Vicentini *et al.* (2017)
	MWCNT, 1-butyl-3-methylimidazolium chloride, and DHP	Voltametric	7	Room temp.	Catechol	0.58	Vicentini *et al.* (2013)
	Magnetic nanoparticles and MWCNT	On–off magnetic switching assay	6.5	Room temp.	Catechol	7.6	Pérez-López and Merkoçi (2011)
	Titania	Sol–gel	7	80	Catechol, *p*-cresol, phenol, *p*-chlorophenol, and *p*-methylcatechol	0.01	Kochana *et al.* (2008)
Peroxidase, tyrosinase, acetylcholinesterase, and butyryl-cholinesteras	Acetylthiocholine chloride	Steady-state and flow-through measurements	7	Room temp.	Phenol, catechol, and pesticides	1.7	Solná *et al.* (2005)
Laccase	Zinc oxide	Sol–gel	6	25	Catechol	0.29	Qu *et al.* (2015)

emissions ultimately possess a detrimental effect on both human health and the ecosystem. Many countries have already implemented regulations to limit the total quantity of pollutants that can be discharged into the atmosphere in an effort to cut down the pollution level. In order to comply with these regulations, industries are often required to install air pollution control equipment, such as scrubbers and filters. Besides, various other norms such as emission standards for vehicles have also been implemented to control and improve the quality of air. Despite these numerous efforts, the quality of air particularly in developing countries does not seem to improve. Thus, it is necessary to look for alternative methods for air pollutant removal.

NBHs have been investigated for their potential utility in air purification in addition to their application in water purification (Figure 11.1). Liu *et al.* (2016) designed a cobalt-*Ralstonia eutropha* NBH system to fix carbon dioxide (CO_2) into liquid fuels and biomass. The designed NBH system demonstrated more than a 50% reduction in CO_2 emissions. In addition, when this NBH system was linked with a traditional solar system, it was discovered that the energy efficiency of the conventional system increased by more than 10%, resulting in an overall decrement in CO_2 emissions. In a similar manner, Wang *et al.* (2019) introduced a NBHs system consisting of cadmium nanoparticles on the surface of the photosynthetic *Rhodopseudomonas palustris* bacterium for the reduction of CO_2 emissions. The hybrid system provided an extensible and powerful platform for the reduction of carbon dioxide and the generation of valuable C_{2+} chemicals by the *R. palustris* NBH via the process of solar-to-chemical conversion. In another study, Ding *et al.* (2019) developed NBH by combining non-photosynthetic bacteria (*Azotobacter vinelandii* and *Cupriavidus necator*) with semiconductor NPs. These NBHs used air, carbon dioxide, and water as substrates in order to produce biofuels such as 2,3-butanediol, isopropanol, and hydrogen, as well as chemicals such as ammonia, formic acid, and ethylene. An inorganic–biological NBH system was reported by Sakimoto *et al.* (2016) through the process of bioprecipitation of cadmium NPs on the bacterium *Moorella thermoacetica*. The *M. thermoacetica*-Cd NBH exhibited a high capacity of producing acetic acid from carbon dioxide through the process of photosynthesis.

11.3.3 Soil remediation

Another advantage of using NBHs is their ability to remove pollutants from difficult-to-treat environments, such as contaminated soils. Researchers have focused on utilizing a number of NBH composites in the degradation and elimination of different types of pollutants from the soil (Figure 11.1). For example, Dai *et al.* (2011) focused on the degradation of polycyclic aromatic hydrocarbons (PAHs) from the soil. The NBH was fabricated with laccase-carrying electrospun fibrous membranes via emulsion electrospinning which showed 72.5%, 79.1%, 93.2%, and 95.1% removal efficiencies for the PAHs benzo[a]pyrene, benz[a]anthracene, fluoranthene, and phenanthrene. In another study, Kuang *et al.* (2013) studied the degradation of phenolic compounds in soil using a NBH composite of *Bacillus fusiformis* (BFN) and nanoscale zero-valent iron (nZVI) and Ni/Fe nanoparticles (Ni/Fe). Similarly, Singh *et al.* (2012) studied the degradation efficiency of the NBH system prepared using stabilized Pd/Fe^0 bimetallic nanoparticles (CMC-Pd/nFe^0) and a strain of *Sphingomonas* sp. *(NM05)* on the gamma isomer of 1,2,3,4,5,6-hexachlorocyclohexane (γ-HCH) present in the soil. The results revealed that the integration of these systems demonstrated a synergistic effect on the γ-HCH degradation approximately two times higher than the system containing only NM05 or CMC-Pd/nFe^0. In another study, Horváthová *et al.* (2018) highlighted that the combination of nZVI-initiated nano-bioremediation with the subsequent addition of bacterial strains *O. anthropic, S. maltophilia*, and *A. xylosoxidans* was found to be more efficient, resulting in an increase in polychlorinated biphenyls (PCB) degradation to 99%, 85%, and 75%, respectively.

Although there have been some encouraging outcomes, additional research that systematically explores various NBH systems is necessary prior to the utilization of the nanoparticle/bioprocess for co-contaminated sites. Unfortunately, very few studies have been done so far on the remediation of co-contaminated soils using NBHs due to the complexity of this environmental component. The simultaneous remediation of heavy metals and organic contaminants is difficult because of the distinct nature of the contaminants. Traditional remediation methods frequently cause soil damage, secondary pollution, and harm to flora and fauna. The NBHs, in contrast, have the potential to avoid secondary pollution and environmental risks associated with conventional remediation techniques. Thus, the use of NBH composites needs to be encouraged for their wider applicability in the removal of different persistent pollutants from the soil.

11.4 ROLE OF NBHS IN RESOURCE RECOVERY

NBHs have so far been extensively used in different types of pollution remediation. However, they can also be effectively utilized in a variety of ways for resource recovery. The application of NBHs in terms of resource recovery is thus far very limited despite possessing an immense potential for this, especially in the case of agricultural waste (Liang & Liang, 2023). Throughout the supply chain, agricultural waste streams contain valuable resources that can be recovered through the separation of nutrient-rich waste fractions and their subsequent use as fertilizer. This not only saves energy but also conserves valuable resources. By utilizing highly selective membranes, adsorbents, and paramagnetic materials, nanobiotechnology presents key opportunities for recovering nutrients not only within the agri-food value chain but also from various other sources such as wastewater, leachates, and runoff (Rahman *et al.*, 2023).

Another potential field of application of NBHs in terms of resource recovery includes the e-waste sector. These wastes can be considered as urban mines and secondary sources of numerous valuable metals and non-metals. Similarly, the nanoparticle–bacteria hybrids can be applied for resource recovery from food waste and sewage sludge. The valuable nutrients, such as nitrogen and phosphorus, from food waste and sewage sludge can be recovered using these NBHs, which can further help in reducing the costs associated with waste disposal and resource conservation (Chong *et al.*, 2019).

Recently, a photosensitized NBH system UUS-1/CdS has been created by Wang *et al.* (2023) to extract uranium from seawater using a whole-cell inorganic–biohybrid approach. The marine bacterial strain *Bacillus velezensis UUS-1* was modified with the deposition of CdS nanoparticles to form the photosensitized biohybrid system UUS-1/CdS which allowed a cascade adsorption–photoreduction strategy to enhance uranium extraction (Figure 11.2). Other potent applications of NBHs in various fields of resource recovery are still incipient and thus much attention is needed in this field. NBHs derived from different sources need to be explored more with the capability to selectively extract and recover specific metals from different waste streams, making them more efficient and effective than traditional methods of resource recovery.

11.5 LIMITATIONS AND CHALLENGES

Synthesis of NBHs is an emerging field that combines nanotechnology and biotechnology in order to develop innovative materials with unique properties. In spite of this, there are still a great number of constraints and challenges associated with the development and implementation of these materials. The synthesis of NBHs poses several challenges due to the complex nature of biological systems and the intricacies of nanomaterials. One of the major challenges is achieving a controlled and reproducible nanoparticle synthesis, which requires precise control over the size, shape, and composition of the nanomaterials. In addition, the interaction between the biological components and the nanomaterials must be carefully optimized in order to ensure that they are compatible with one another. Another issue involves scaling up the usage of NBHs for practical and commercial applications, assuring the safety of the materials, and preserving the activity of the biological components throughout the synthesis process. It is possible that the nanoparticles may deteriorate with time or the biological material may become inactive, which will result in a reduced effectiveness of NBHs.

NBHs are still relatively new materials; therefore, despite the fact that experiments conducted in the laboratory have demonstrated their potential, it is yet unknown how well they will perform when

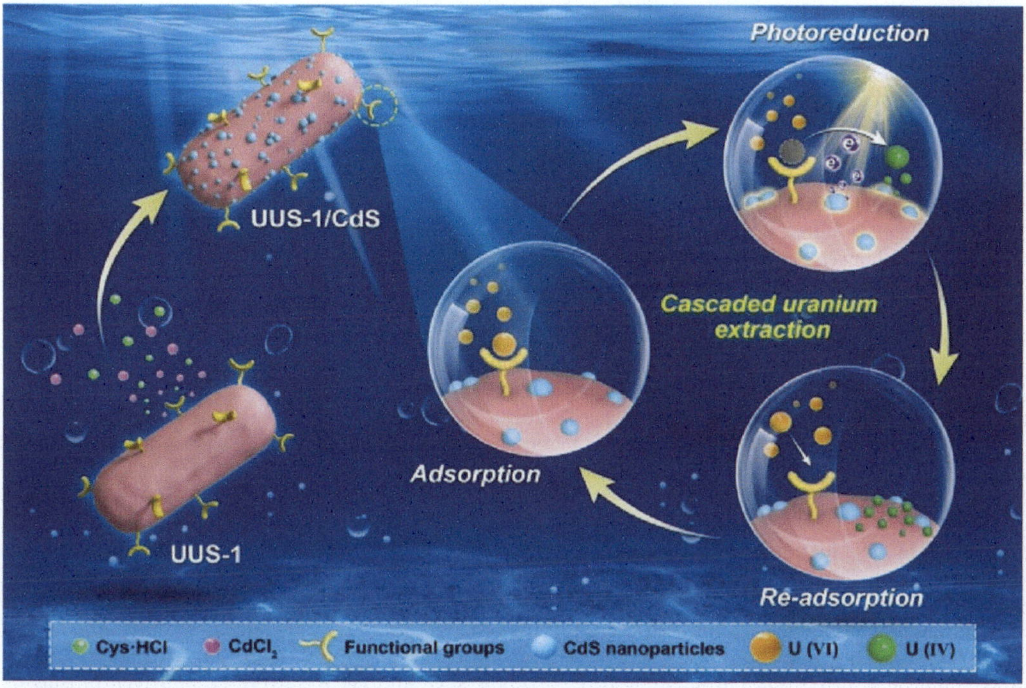

Figure 11.2 Cascaded enrichment mechanism of the photosensitized biohybrid system UUS-1/CdS for uranium extraction from seawater. Source: reprinted from Wang *et al.* (2023).

applied at larger scales. In addition, the current costs of nanomaterials and enzymes are rather high, which may pose substantial challenges to their widespread usage in many practical applications. The stability, robustness, and durability of the NBHs are three additional challenges that need to be addressed. To date, several studies have been conducted in order to demonstrate the cytotoxicity and ecotoxicity of nanomaterials, as well as their fate and mobility in the environment (Raval *et al.*, 2022). These studies have reported that certain NPs that are used in the NBH system may leach into the environment, which would be hazardous to human health and the environment. It is possible that incorporating NBHs into any application before the full extent of their potential hazards is understood could result in unforeseen negative consequences. Therefore, all the constraints and challenges must be overcome in order to realize the potential application of NBHs in a variety of different fields.

To overcome these barriers, there is a need for further collaboration between research institutions, industry, the government, and other stakeholders. This is necessary to fully understand their potential and limitations, as well as to ensure their use in the environment is both safe and effective. In spite of these limitations and challenges, the potential benefits of NBHs make them an intriguing area of research with significant promise for future applications.

ACKNOWLEDGMENTS

The authors would like to take this opportunity to extend their appreciation to the Department of Environmental Science and Engineering, Indian Institute of Technology (Indian School of Mines, Dhanbad, India) for all the support they provided during the writing of this chapter.

REFERENCES

Ahmad R., Mishra A. and Sardar M. (2013). Peroxidase-TiO_2 nanobioconjugates for the removal of phenols and dyes from aqueous solutions. *Advanced Science, Engineering and Medicine*, **5**, 1020–1025, https://doi.org/10.1166/asem.2013.1387

Ahmed S., Ahmad M., Swami B. L. and Ikram S. (2016). A review on plant extract mediated synthesis of silver nanoparticles for antimicrobial applications: a green expertise. *Journal of Advanced Research*, **7**, 17–28, https://doi.org/10.1016/j.jare.2015.02.007

Apetrei C., Rodríguez-Méndez M. L. and De Saja J. A. (2011). Amperometric tyrosinase based biosensor using an electropolymerized phosphate-doped polypyrrole film as an immobilization support. Application for detection of phenolic compounds. *Electrochimica Acta*, **56**, 8919–8925, https://doi.org/10.1016/j.electacta.2011.07.127

Arslan H. and Arslan F. (2011). Preparation of a polypyrrole-polyvinylsulphonate composite film biosensor for determination of phenol based on entrapment of polyphenol oxidase. *Artificial Cells, Blood Substitutes, and Biotechnology*, **39**, 341–345, https://doi.org/10.3109/10731199.2011.585616

Benavente R., Lopez-Tejedor D. and Palomo J. M. (2018). Synthesis of a superparamagnetic ultrathin $FeCO_3$ nanorods–enzyme bionanohybrid as a novel heterogeneous catalyst. *Chemical Communications*, **54**, 6256–6259, https://doi.org/10.1039/C8CC02851F

Besharati Vineh M., Saboury A. A., Poostchi A. A., Rashidi A. M. and Parivar K. (2018). Stability and activity improvement of horseradish peroxidase by covalent immobilization on functionalized reduced graphene oxide and biodegradation of high phenol concentration. *International Journal of Biological Macromolecules*, **106**, 1314–1322, https://doi.org/10.1016/j.ijbiomac.2017.08.133

Chen H., Li S., Wang S., Tan Y. and Kan J. (2013). A new catechol biosensor immobilized polyphenol oxidase by combining electropolymerization and cross-linking process. *International Journal of Polymeric Materials and Polymeric Biomaterials*, **62**(12), 620–626, https://doi.org/10.1080/00914037.2012.761629

Chen M., Zhou X. F., Yu Y. Q., Liu X., Zeng R. J. X., Zhou S. G. and He Z. (2019). Light-driven nitrous oxide production via autotrophic denitrification by self-photosensitized *Thiobacillus denitrificans*. *Environment International*, **127**, 353–360, https://doi.org/10.1016/j.envint.2019.03.045

Chong W. C., Mohammad A. W., Mahmoudi E., Chung Y. T., Kamarudin K. F. and Takriff M. S. (2019). Nanohybrid membrane in algal-membrane photoreactor: microalgae cultivation and wastewater polishing. *Chinese Journal of Chemical Engineering*, **27**(11), 2799–2806, https://doi.org/10.1016/j.cjche.2019.05.015

Dai Y., Yin L. and Niu J. (2011). Laccase-carrying electrospun fibrous membranes for adsorption and degradation of PAHs in shoal soils. *Environmental Science & Technology*, **45**, 10611–10618, https://doi.org/10.1021/es203286e

Das R., Hamid S. B. A. and Annuar M. S. M. (2016). Highly efficient and stable novel nanobiohybrid catalyst to avert 3,4-dihydroxybenzoic acid pollutant in water. *Scientific Reports*, **6**, 33572, https://doi.org/10.1038/srep33572

Deng S., Yan X., Zhu Q. and Liao C. (2019). The utilization of reclaimed water: possible risks arising from waterborne contaminants. *Environmental Pollution*, **254**, 113020, https://doi.org/10.1016/j.envpol.2019.113020

Ding Y., Bertram J. R., Eckert C., Bommareddy R. R., Patel R., Conradie A., Bryan S. and Nagpal P. (2019). Nanorg microbial factories: light-driven renewable biochemical synthesis using quantum dot-bacteria nanobiohybrids. *Journal of the American Chemical Society*, **141**, 10272–10282, https://doi.org/10.1021/jacs.9b02549

Gao M., Li J., Bao Z., Hu M., Nian R., Feng D., An D., Li X., Xian M. and Zhang H. (2019). A natural in situ fabrication method of functional bacterial cellulose using a microorganism. *Nature Communications*, **10**(1), 437, https://doi.org/10.1038/s41467-018-07879-3

Gasser C. A., Yu L., Svojitka J., Wintgens T., Ammann E. M., Shahgaldian P., Corvini P. F. and Hommes G. (2014). Advanced enzymatic elimination of phenolic contaminants in wastewater: a nano approach at field scale. *Applied Microbiology and Biotechnology*, **98**, 3305–3316, https://doi.org/10.1007/s00253-013-5414-8

Girelli A. M., Mattei E. and Messina A. (2006). Phenols removal by immobilized tyrosinase reactor in on-line high performance liquid chromatography. *Analytica Chimica Acta*, **580**, 271–277, https://doi.org/10.1016/j.aca.2006.07.088

Guo Q., Ghadiri R., Weigel T., Aumann A., Gurevich E., Esen C., Medenbach O., Cheng W., Chichkov B. and Ostendorf A. (2014). Comparison of in situ and ex situ methods for synthesis of two-photon polymerization polymer nanocomposites. *Polymers*, **6**, 2037–2050, https://doi.org/10.3390/polym6072037

Guo M., Huang K. and Xu W. (2021). Third generation whole-cell sensing systems: synthetic biology inside, nanomaterial outside. *Trends in Biotechnology*, **39**, 550–559, https://doi.org/10.1016/j.tibtech.2020.10.002

Horváthová H., Lászlová K. and Dercová K. (2018). Bioremediation of PCB-contaminated shallow river sediments: the efficacy of biodegradation using individual bacterial strains and their consortia. *Chemosphere*, **193**, 270–277, https://doi.org/10.1016/j.chemosphere.2017.11.012

Jaiswal J., Srivastav A. K., Rajput P. K., Yadav U. C. S. and Kumar U. (2023). Integrating synthesis, physicochemical characterization, and in silico studies of cordycepin-loaded bovine serum albumin nanoparticles. *Journal of Agricultural and Food Chemistry*, **71**(32), 12225–12236, https://doi.org/10.1021/acs.jafc.3c03608

Kochana J., Gala A., Parczewski A. and Adamski J. (2008). Titania sol-gel-derived tyrosinase-based amperometric biosensor for determination of phenolic compounds in water samples. Examination of interference effects. *Analytical and Bioanalytical Chemistry*, **391**, 1275–1281, https://doi.org/10.1007/s00216-007-1798-6

Kuang Y., Zhou Y., Chen Z., Mallavarapu M. and Naidu R. (2013). Impact of Fe and Ni/Fe nanoparticles on biodegradation of phenol by the strain *Bacillus fusiformis* (BFN) at various pH values. *Bioresource Technology*, **136**, 588–594, https://doi.org/10.1016/j.biortech.2013.03.018

Kumari P. and Samadder S. R. (2022). Valorization of carbonaceous waste into graphene materials and their potential application in water & wastewater treatment: a review. *Materials Today Chemistry*, **26**, 101192, https://doi.org/10.1016/j.mtchem.2022.101192

Lauth V., Maas M. and Rezwan K. (2017). An evaluation of colloidal and crystalline properties of $CaCO_3$ nanoparticles for biological applications. *Materials Science and Engineering: C*, **78**, 305–314, https://doi.org/10.1016/j.msec.2017.04.037

Li Q., Fan F., Wang Y., Feng W. and Ji P. (2013). Enzyme immobilization on carboxyl-functionalized graphene oxide for catalysis in organic solvent. *Industrial & Engineering Chemistry Research*, **52**, 6343–6348, https://doi.org/10.1021/ie400558u

Liang J. and Liang K. (2023). Nanobiohybrids: synthesis strategies and environmental applications from micropollutants sensing and removal to global warming mitigation. *Environmental Research*, **232**, 116317, https://doi.org/10.1016/j.envres.2023.116317

Liu C., Colón B. C., Ziesack M., Silver P. A. and Nocera D. G. (2016). Water splitting-biosynthetic system with CO_2 reduction efficiencies exceeding photosynthesis. *Science*, **352**, 1210–1213, https://doi.org/10.1126/science.aaf5039

Lykourinou V., Chen Y., Wang X.-S., MenG L., Hoang T., Ming L.-J., Musselman R. L. and Ma S. (2011). Immobilization of MP-11 into a mesoporous metal–organic framework, MP-11@mesoMOF: a new platform for enzymatic catalysis. *Journal of the American Chemical Society*, **133**, 10382–10385, https://doi.org/10.1021/ja2038003

Mirtalebi S. S., Almasi H. and Alizadeh Khaledabad M. (2019). Physical, morphological, antimicrobial and release properties of novel MgO-bacterial cellulose nanohybrids prepared by in-situ and ex-situ methods. *International Journal of Biological Macromolecules*, **128**, 848–857, https://doi.org/10.1016/j.ijbiomac.2019.02.007

Mohammadalinejhad S., Almasi H. and Esmaiili M. (2019). Simultaneous green synthesis and in-situ impregnation of silver nanoparticles into organic nanofibers by *Lythrum salicaria* extract: morphological, thermal, antimicrobial and release properties. *Materials Science and Engineering: C*, **105**, 110115, https://doi.org/10.1016/j.msec.2019.110115

Nadda A. and Kanwar S. (2012). Catalytic potential of a nitrocellulose membrane – immobilized lipase in aqueous and organic media. *Journal of Applied Polymer Science*, **124**(S1), E37–E44.

Nishu K. S. (2023). Smart and innovative nanotechnology applications for water purification. *Hybrid Advances*, **3**, 100044, https://doi.org/10.1016/j.hybadv.2023.100044

Niu J., Lunn D. J., Pusuluri A., Yoo J. I., O'malley M. A., Mitragotri S., Soh H. T. and Hawker C. J. (2017). Engineering live cell surfaces with functional polymers via cytocompatible controlled radical polymerization. *Nature Chemistry*, **9**, 537–545, https://doi.org/10.1038/nchem.2713

Ozoner S. K., Yalvac M. and Erhan E. (2010). Flow injection determination of catechol based on polypyrrole–carbon nanotube–tyrosinase biocomposite detector. *Current Applied Physics*, **10**, 323–328, https://doi.org/10.1016/j.cap.2009.06.017

Palomo J. M. (2021). Artificial enzymes with multiple active sites. *Current Opinion in Green and Sustainable Chemistry*, **29**, 100452, https://doi.org/10.1016/j.cogsc.2021.100452

Pang R., Li M. and Zhang C. (2015). Degradation of phenolic compounds by laccase immobilized on carbon nanomaterials: diffusional limitation investigation. *Talanta*, **131**, 38–45, https://doi.org/10.1016/j.talanta.2014.07.045

Patra J. K. and Baek K.-H. (2014). Green nanobiotechnology: factors affecting synthesis and characterization techniques. *Journal of Nanomaterials*, **2014**, 417305, https://doi.org/10.1155/2014/417305

Pérez-López B. and Merkoçi A. (2009). Improvement of the electrochemical detection of catechol by the use of a carbon nanotube based biosensor. *The Analyst*, **134**, 60–64, https://doi.org/10.1039/B808387H

Pérez-López B. and Merkoçi A. (2011). Magnetic nanoparticles modified with carbon nanotubes for electrocatalytic magneto switchable biosensing applications. *Advanced Functional Materials*, **21**, 255–260, https://doi.org/10.1002/adfm.201001306

Pradeep T. and Anshup (2009). Noble metal nanoparticles for water purification: a critical review. *Thin Solid Films*, **517**(24), 6441–6478, https://doi.org/10.1016/j.tsf.2009.03.195

Qin H., Hu T., Zhai Y., Lu N. and Aliyeva J. (2020). Sonochemical synthesis of ZnS nanolayers on the surface of microbial cells and their application in the removal of heavy metals. *Journal of Hazardous Materials*, **400**, 123161, https://doi.org/10.1016/j.jhazmat.2020.123161

Qu J., Lou T., Wang Y., Dong Y. and Xing H. (2015). Determination of catechol by a novel laccase biosensor based on zinc-oxide sol-gel. *Analytical Letters*, **48**, 1842–1853, https://doi.org/10.1080/00032719.2014.1003427

Rahman M. M., Ahmed L., Anika F., Riya A. A., Kali S. K., Rauf A. and Sharma R. (2023). Bioinorganic nanoparticles for the remediation of environmental pollution: critical appraisal and potential avenues. *Bioinorganic Chemistry and Applications*, **2023**, 1–26, https://doi.org/10.1155/2023/2409642

Rajesh Takashima W. and Kaneto K. (2004). Amperometric phenol biosensor based on covalent immobilization of tyrosinase onto an electrochemically prepared novel copolymer poly (N-3-aminopropyl pyrrole-co-pyrrole) film. *Sensors and Actuators B: Chemical*, **102**, 271–277, https://doi.org/10.1016/j.snb.2004.04.028

Ramesh M., Janani R., Deepa C. and Rajeshkumar L. (2022). Nanotechnology-enabled biosensors: a review of fundamentals, design principles, materials, and applications. *Biosensors (Basel)*, **13**, 40.

Raval B., Srivastav A. K., Gupta S. K., Kumar U., Mahapatra S. K., Gajjar P. N. and Banerjee I. (2022). Synthesis of exfoliated multilayer graphene and its putative interactions with SARS-CoV-2 virus investigated through computational studies. *Journal of Biomolecular Structure and Dynamics*, **40**(2), 712–721, https://doi.org/10.1080/07391102.2020.1817788

Richardson J. J. and Liang K. (2018). Nano-biohybrids: In vivo synthesis of metal–organic frameworks inside living plants. *Small*, **14**, 1702958, https://doi.org/10.1002/smll.201702958

Saikia J., Gogoi A. and Baruah S. (2019). Nanotechnology for water remediation. In: Environmental Nanotechnology: Volume 2, N. Dasgupta, S. Ranjan and E. Lichtfouse (eds.), Springer International Publishing, Cham, 195–211.

Sakimoto K. K., Zhang S. J. and Yang P. (2016). Cysteine–cystine photoregeneration for oxygenic photosynthesis of acetic acid from CO_2 by a tandem inorganic–biological hybrid system. *Nano Letters*, **16**, 5883–5887, https://doi.org/10.1021/acs.nanolett.6b02740

Santhosh C., Velmurugan V., Jacob G., Jeong S. K., Grace A. N. and Bhatnagar A. (2016). Role of nanomaterials in water treatment applications: a review. *Chemical Engineering Journal*, **306**, 1116–1137, https://doi.org/10.1016/j.cej.2016.08.053

Shinde P., Agraval H., Srivastav A. K., Yadav U. C. S. and Kumar U. (2020). Physico-chemical characterization of carvacrol loaded zein nanoparticles for enhanced anticancer activity and investigation of molecular interactions between them by molecular docking. *International Journal of Pharmaceutics*, **588**, 119795, https://doi.org/10.1016/j.ijpharm.2020.119795

Singh R., Misra V., Mudiam M. K. R., Chauhan L. K. S. and Singh R. P. (2012). Degradation of γ-HCH spiked soil using stabilized Pd/FeO bimetallic nanoparticles: pathways, kinetics and effect of reaction conditions. *Journal of Hazardous Materials*, **237–238**, 355–364, https://doi.org/10.1016/j.jhazmat.2012.08.064

Solná R., Sapelnikova S., Skládal P., Winther-Nielsen M., Carlsson C., Emnéus J. and Ruzgas T. (2005). Multienzyme electrochemical array sensor for determination of phenols and pesticides. *Talanta*, **65**, 349–357, https://doi.org/10.1016/j.talanta.2004.07.005

Su D., Liu X., Wang L., Ma C., Xie H., Zhang H., Meng X., Huang Y. and Huang X. (2016). Bio-inspired engineering proteinosomes with a cell-wall-like protective shell by self-assembly of a metal-chelated complex. *Chemical Communications*, **52**, 13803–13806, https://doi.org/10.1039/C6CC07655F

Subrizi F., Crucianelli M., Grossi V., Passacantando M., Pesci L. and Saladino R. (2014). Carbon nanotubes as activating tyrosinase supports for the selective synthesis of catechols. *ACS Catalysis*, **4**, 810–822, https://doi.org/10.1021/cs400856e

Tembe S., Chaudhari P., Bhoraskar S., D'souza S. and Karve M. (2008). Conductivity-based catechol sensor using tyrosinase immobilized in porous silicon. *Sensors Journal, IEEE*, **8**, 1593–1597, https://doi.org/10.1109/JSEN.2008.928925

Uddandarao P. and Lens P. N. L. (2022). Light driven *Aspergillus niger*-ZnS nanobiohybrids for degradation of methyl orange. *Chemosphere*, **298**, 134162, https://doi.org/10.1016/j.chemosphere.2022.134162

Vicentini F. C., Janegitz B. C., Brett C. M. A. and Fatibello-Filho O. (2013). Tyrosinase biosensor based on a glassy carbon electrode modified with multi-walled carbon nanotubes and 1-butyl-3-methylimidazolium chloride within a dihexadecylphosphate film. *Sensors and Actuators B: Chemical*, **188**, 1101–1108, https://doi.org/10.1016/j.snb.2013.07.109

Vicentini F., Santos H., Oliveira Neto E., Batagello C., Amundson J., Srougi M. and Mazzucchi E. (2017). Complete supine percutaneous nephrolithotomy from the surgeon's point of view with a gopro®. Ten steps for success. *Journal of Urology*, **197**, e1284–e1284, https://doi.org/10.1016/j.juro.2017.02.2997

Wahid M. H., Eroglu E., Chen X., Smith S. M. and Raston C. L. (2013). Entrapment of *Chlorella vulgaris* cells within graphene oxide layers. *RSC Advances*, **3**, 8180–8183, https://doi.org/10.1039/c3ra40605a

Wang L., Zhang C., Cheng R., Ali J., Wang Z., Mailhot G. and Pan G. (2018). *Microcystis aeruginosa* synergistically facilitate the photocatalytic degradation of tetracycline hydrochloride and Cr(VI) on PAN/TiO2/Ag nanofiber mats. *Catalysts*, **8**(12), 628.

Wang B., Jiang Z., Yu J. C., Wang J. and Wong P. K. (2019). Enhanced CO_2 reduction and valuable C^{2+} chemical production by a CdS-photosynthetic hybrid system. *Nanoscale*, **11**, 9296–9301, https://doi.org/10.1039/C9NR02896J

Wang M., Feng L., Luo G., Feng T., Zhao S., Wang H., Shi S., Liu T., Fu Q., Li J., Wang N. and Yuan Y. (2023). Ultrafast extraction of uranium from seawater using photosensitized biohybrid system with bioinspired cascaded strategy. *Journal of Hazardous Materials*, **445**, 130620, https://doi.org/10.1016/j.jhazmat.2022.130620

Xu R., Chi C., Li F. and Zhang B. (2013). Laccase–polyacrylonitrile nanofibrous membrane: highly immobilized, stable, reusable, and efficacious for 2,4,6–trichlorophenol removal. *ACS Applied Materials & Interfaces*, **5**, 12554–12560, https://doi.org/10.1021/am403849q

doi: 10.2166/9781789063592_0209

Chapter 12

Photoelectrocatalysis and nanohybrids in wastewater treatment

Sweta Sharma[1,2] *and Pooja Devi*[1,2*]

[1]*Academy of Scientific and Industrial Research, Ghaziabad, Uttar Pradesh, India*
[2]*Materials Science and Sensor Application, Central Scientific Instruments Organisation, Chandigarh, India*
Corresponding author: poojaiitr@csio.res.in

ABSTRACT

The rise in global industrialization, urbanization, and population growth has contributed to a significant increase in the production of large volumes of wastewater. The inadequate management of this valuable resource, rich in energy and minerals, has emerged as a worldwide concern. While conventional methods for treating wastewater are somewhat effective, they struggle to address persistent pollutants such as pesticides, textile dyes, pharmaceuticals, and personal care products. To tackle these challenges, there is a need for innovative solutions. Advanced oxidation processes (AOPs) present a promising alternative for breaking down these complex organic pollutants that are resistant to traditional chemical oxidants. Among these AOPs, photoelectrocatalysis (PEC) stands out because it not only effectively degrades pollutants, but it also offers the potential to harness energy from wastewater. This chapter discusses PEC, exploring its recent advancements, particularly focusing on the application of nanohybrids as photoelectrodes for enhancing wastewater treatment efficiency.

Keywords: wastewater, photo electrochemistry, nanohybrids, treatment

12.1 INTRODUCTION

In the absence of effective waste management, the rise in industrialization and population density has led to the release of harmful pollutants and hazardous wastes in natural bodies such as soil, water, and air. In particular, there is serious water pollution as a result of toxic waste being released from different industries such as paper, coal, pharma, and textile (Vithalani & Bhatt, 2022). As a result, the water quality has worsened over the past few decades. Though there have been conventional methods used for the treatment of wastewater before its discharge to water bodies, they are however limited by their lower mineralization efficiency, high operational cost, infrastructure and extensive time requirement (Samiksha & Kerkar, 2020). Alternatively, advanced oxidation processes (AOPs), such as photocatalysis (PC), photoelectrocatalysis (PEC) and Fenton processes, can help to break down the recalcitrant pollutants not degraded by conventional oxidation processes. The efficiency of these approaches is highly dependent on the catalyst/material characteristics utilized to carry out the degradation. Nanomaterials are one such promising class of materials highly used in AOPs due to their high surface-to-volume ratio, unique opto/electronic properties, and rich surface functionalities for the adsorption of specific pollutants (Kurian, 2021). Many times, certain metals and microorganisms, which cannot always be removed from water using conventional water treatment methods, are easily taken care of by nanomaterials (Fernandez-Ibanez *et al.*, 2021). In this direction, several categories of nanomaterials such as metal nanostructures, carbon materials, oxide nanomaterials and their composites have been utilized as adsorbents or catalysts for wastewater treatment (Garcia-Segura & Brillas, 2017).

Materials are generally combined together to have synergetic or improved properties and are classified in the category of nanohybrids or nanocomposites depending on the size of the components. Nanohybrids are a category of materials, which have two or more components in the nanoscale regime. As a result, they have improved mechanical qualities including tensile strength, elastic modulus, and fracture toughness; superior electrical conductivity; enhanced surface area and resultant properties for gas sorption, reactivity and catalytic activity; and many more. Thus, they have synergistic properties of the contributing components to achieve better performance. Their characteristics can be changed based on the composition, size, shape, and structure of the hybrid nanomaterial components (Joy *et al.*, 2018). A nanohybrid's structure and chemical makeup can be adjusted to provide particular optical qualities like increased absorption, fluorescence, or scattering. Nanocomposite, on the other hand, has nanomaterial added to a matrix such as a polymer, wherein the organic/inorganic hybrid tends to have better optical, thermal, and mechanical properties than their native matrix organic base. Due to the synergy of the qualities from each of the components combined into a single material, nanohybrids and nanocomposites have attracted interest in a variety of disciplines including wastewater treatment via PC and PEC (Zia & Riaz, 2021).

In particular, PEC in wastewater not only offers its treatment but also generates energy in the form of clean hydrogen in the process (Jakubow-Piotrowska *et al.*, 2022) Thus, solar energy can be captured in the form of chemical bonds, that is H_2, via the process called 'water-splitting' (WS) (Abe, 2011). Fujishima and Honda were the first to demonstrate WS using TiO_2 photoelectrodes to produce hydrogen and oxygen in a PEC cell (Abe, 2011). The PEC technique combines the benefits of PC and electrochemical methods, thus it has been recognized as a highly effective approach for wastewater treatment (Niu *et al.*, 2020). PEC reactions are essentially interfacial processes where contaminants are first diffusely deposited on the photoelectrode surface and then the catalytic oxidation reaction occurs immediately. The three main factors that affect PEC efficiency are light absorption efficiency (abs), interfacial charge injection (inj) and charge separation efficiency (sep) of the photo electrocatalyst. Therefore, creating an effective photoelectrode material with high n_{abs}, n_{sep}, and n_{inj} is of research interest in the catalysis community. Thus, this chapter has discussed the fundamentals of PEC with a focus on the wastewater environment. A section is dedicated to nanohybrid electrodes for PEC wastewater treatment, with further discussion on future directions.

12.2 PEC IN WASTEWATER TREATMENT

12.2.1 Principle of photoelectrocatalysis

A PEC system uses a photoelectrode or a photocatalyst, which is generally a semiconductor material that produces electrons and holes on exposure to light. These photoexcited electrons and holes further produce reactive oxygen species (ROS) and subsequently interact with water and organic contaminants to break them down into non-harmful compounds. PEC combines PC and electrocatalysis by transferring photoexcited electrons from the photocatalyst surface under external bias. Nanohybrids owing to the synergistic effects of the two nanomaterials can increase the effectiveness of PEC (McMichael *et al.*, 2021). For example, metal nanoparticles can function as an electron sink to increase the efficiency of semiconductor electrodes in a metal/semiconductor nanohybrid such as Pt/TiO_2 or Ag/TiO_2 by enabling better charge transport, separation and utilization in the PEC process.

12.2.2 PEC process

In general, the following processes occur at a photoelectrode surface in a PEC system for wastewater treatment:

(a) Photon absorption: Photocatalyst generates electron–hole pairs upon the absorption of incident light.
(b) Charge separation: The photocatalyst's internal electric field separates the photo-excited electrons and holes. While the holes remain in the valence band, the electrons are moved to the conduction band.
(c) Redox reaction: Water and organic contaminants react with the photoexcited electrons and holes to form ROS (O_2*, OH^-, OH*), which further mineralizes the pollutants to non-harmful compounds such as CO_2 and H_2O.
(d) Charges recombination: The photoexcited charge carrier can recombine if not well separated leading to poor degradation efficiency (Pahra *et al.*, 2022).

Thus, to control the above processes, several tactics, including the use of nanohybrids, optimization of the photocatalyst shape and structure, and adjustment of the reaction conditions can be used to increase the efficiency of PEC degradation. Nanohybrids can allow for achieving higher surface area for adsorption along with higher electron transfer and separation efficiency, which in turn can enhance PEC efficiency.

12.2.3 PEC systems

In general, in a PEC process, reduction happens at the cathode while oxidation happens at the anode. Accordingly, the photoelectrode can be designed to act as either a hydrogen evolution reaction (HER) catalyst or an oxygen evolution reaction (OER) catalyst (Singh *et al.*, 2023). When a photoelectrode is used as an HER catalyst as shown in Figure 12.1, the electrons at the electrode/electrolyte interface

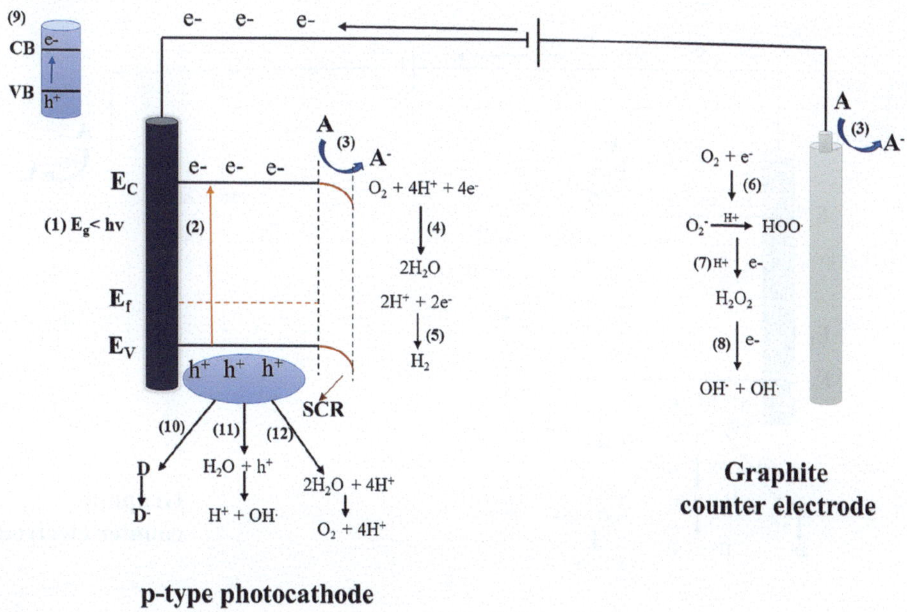

Figure 12.1 Photoelectrochemical mechanism of dye degradation by a p-type photocathode and counter electrode. For explanation of symbols, see text.

reduce water to generate H_2. While the holes participate in the oxidation process at the counter electrode, that is graphite. In general, a p-type semiconductor is used as a photoelectrode in a PEC–HER system.

12.2.3.1 p-Type semiconductors

In p-type semiconductors, the Fermi level (E_F) is near the valence band (VB), with the majority of charge carriers as holes. Under illumination, at the electrode–electrolyte junction, the electrons are transferred to the conduction band (CB) of the photocathode to majorly participate in the hydrogen generation. At the equilibrium, the negative charges are accumulated on the space charge region (SCR), also known as the depletion region, as a result of the formation of a downward band bending at the interface. The holes that are present in the VB of the photocathode form the ROS which can further degrade the organic pollutants. Pollutants oxidation and OER takes place on the counter electrode due to the presence of hydroxyl ion and superoxide radicals in the electrolyte. The above-discussed pathway of the PEC–HER reaction in wastewater is defined as: (1) photon absorption, (2) exciton formation, (3) electron transfer to an electron acceptor A A^-, (4) oxygen reduction to water, (5) proton reduction to hydrogen, (6) reduction of O_2 into superoxide (O_2^-), (7) H_2O_2 formation, (8) hydroxyl ion (OH)· formation, (9) dye sensitization, (10) donor e-transport D → D^-, (11) water oxidation in the form of OH· and (12) oxidation of water (McMichael *et al.*, 2021).

12.2.3.2 n-Type semiconductor

An n-type semiconductor is used in a PEC–OER system as shown in Figure 12.2. The Fermi level is near the CB in an n-type semiconductor. When the light is illuminated at the electrode–electrolyte junction, electrons are transported to the solution, positive charges begin to build up on the space charge region (SCR) at equilibrium. This region is called the depletion region and as a result the formation of an upward band bending at the interface. The degradation of the organic pollutants and OER takes place on the photoanode due to the presence of hydroxyl ions in the electrolyte, and the

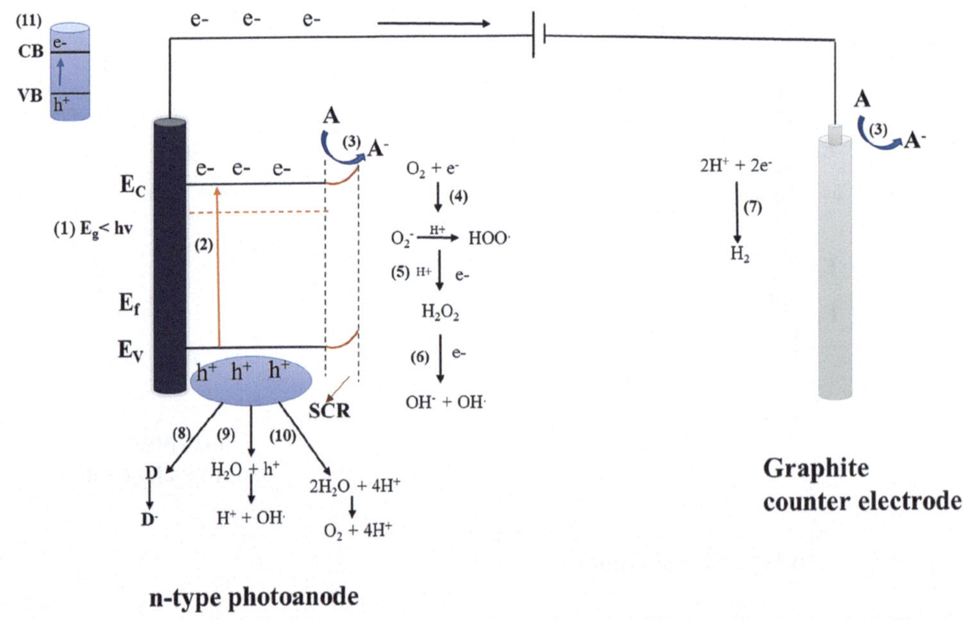

Figure 12.2 PEC-OER utilizing n-type photoanode in a wastewater system. For explanation of symbols, see text.

electrons are externally biased to participate in HER at the cathode. The pathway of the reaction is further defined in the form of steps as: (1) photon absorption, (2) exciton generation, (3) few electron transfer to an electron acceptor, (4) reduction of O_2 into superoxide, (5) creation of H_2O_2, (6) creation of OH, (7) proton reduction to hydrogen, (8) donor e-transport, (9) water oxidation in form of OH·, (10) OER and (11) dye sensitization (McMichael *et al.*, 2021).

12.3 NANOHYBRIDS IN PEC WASTEWATER TREATMENT

12.3.1 Synthesis strategies

Nanohybrids are made up of two or more different kinds of nanomaterials that work together to form a hybrid substance with distinct features. They are frequently utilized as photocatalysts in PEC water splitting to increase the reaction's effectiveness. Figure 12.3 lists various categories of materials used as nanohybrid in PEC wastewater splitting. The form, size, and surface area of the nanohybrids all impact the material choice and synthesis technique (Singh *et al.*, 2023). They can be synthesized by various routes such as hydrothermal, co-precipitation, and sol–gel, as discussed below:

(a) Co-precipitation technique: In this technique, a single solution is used to precipitate several nanoparticles at once. The resulting nanohybrids have a consistent structure and content. For example, co-precipitation has been used to create TiO_2–Au nanohybrids (Singh *et al.*, 2023).
(b) Deposition–precipitation method: The deposition–precipitation approach involves utilizing a reducing chemical to deposit one type of nanoparticle onto the surface of another type of nanoparticle. The final nanohybrids feature a core–shell structure, with the deposited nanoparticle serving as the shell. CdS–Au nanohybrids, for instance, have been created utilizing the deposition–precipitation technique (Tada *et al.*, 2018).
(c) Sol–gel: In this technique, metal alkoxides are hydrolyzed in a solution to create nanoparticles. To make the nanohybrids, the resultant nanoparticles are then mixed. The sol–gel technique enables fine control of the nanohybrids' shape and content. The sol–gel approach has been used frequently to create TiO_2–ZnO nanohybrids (Ali *et al.*, 2021).

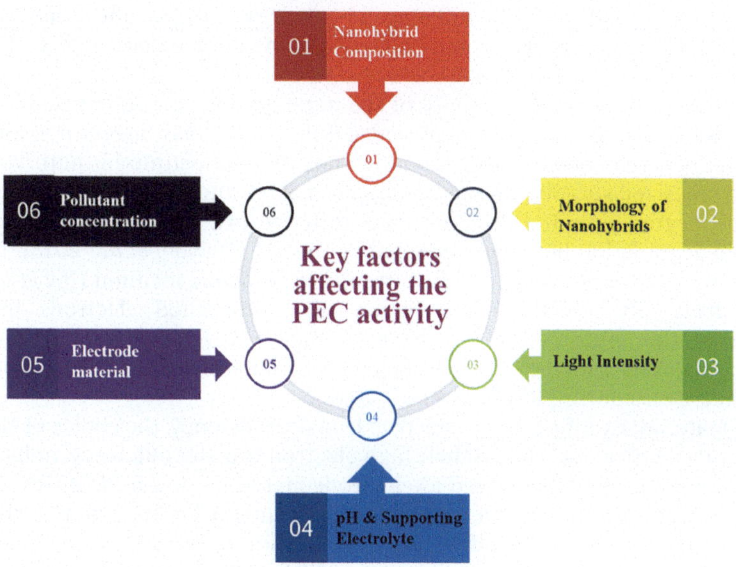

Figure 12.3 Factors affecting the wastewater treatment efficiency of PEC.

(d) Electrodeposition: In this technique, an electric field is used to electrodeposit one type of nanoparticle onto the surface of another type of nanoparticle. The final nanohybrids feature a core–shell structure, with the electrodeposited nanoparticle serving as the shell. This approach has been widely adopted in the synthesis of metal/oxide hybrids such as TiO_2–Au nanohybrids (Endrödi *et al.*, 2018).
(e) Hydrothermal: The hydrothermal approach involves creating nanoparticles in a solution at a high temperature and high pressure. To make the nanohybrids, the associated salt solutions are mixed and subjected to high-temperature conditions. Using the hydrothermal process, it is possible to create nanohybrids with special qualities including a large surface area and a crystalline structure. Hydrothermal synthesis is frequently used in the synthesis of oxides and their hetero structures such as TiO_2/NiO nanohybrids (Zhao *et al.*, 2020).

The overall choice of synthesis procedure relies on the desired features of the nanohybrids to design an efficient photoelectrode.

12.3.2 Parameter optimization

Several factors, such as the nanohybrid's composition, shape, pH, light intensity, electrode material, pollutant concentration, and supporting electrolyte, can affect how effective PEC employing nanohybrids is. The effectiveness of PEC as a technique for the degradation of organic contaminants in wastewater can be improved by optimizing these parameters (Fernandez-ibanez *et al.*, 2021).

The effectiveness of PEC can be further impacted by several operational factors discussed below:

(a) Nanohybrid composition: Properties of nanohybrids, such as the bandgap, surface area, and electron transfer efficiency, can be influenced by their composition. The efficiency of PEC can be impacted by the choice of materials and their ratios. Singla *et al.* (2023) reported a nanohybrid photoanode $BiVO_4/MoSe_2$ with a 3:1 composition. They found a 91.9% degradation efficiency along with 3.59 mmol/cm^2 hydrogen evolution rate (Singla *et al.*, 2023).
(b) Morphology of nanohybrids: Properties of nanohybrids, such as surface area and electron transfer efficiency, can be influenced by their morphology. Nanoparticles' light-scattering and absorption characteristics can be influenced by their size, shape, and composition. It has been well discovered that nanorods are more effective than nanoparticles in PEC (Fernandez-ibanez *et al.*, 2021).
(c) pH: The surface charge and reactivity of the nanohybrids can be impacted by the pH of the fluid. Depending on the materials employed, a different pH may be required for PEC employing nanohybrids. Tayyebi *et al.* (2018) reported that the monoclinic bismuth vanadate (c-BVO), which is synthesized via solvothermal calcination of tetrahedral $BiVO_4$. They have investigated the MB dye degradation and PEC results at different pH. They found the PEC photocurrent density and degradation both were affected by the pH (Tayyebi *et al.*, 2018). In another study, TiO_2–Au nanohybrids showed good PEC results in an acidic medium (Joy *et al.*, 2018).
(d) Light intensity: By affecting the number of photoexcited electrons produced by the photocatalyst, the light intensity can have an impact on PEC's efficiency. Depending on the materials employed and the light's wavelength, different materials will respond to light at different intensities during PEC (Joy *et al.*, 2018).
(e) Electrode material: By affecting the electron transfer efficiency, the choice of electrode material can impact PEC efficiency. Due to their high electron transfer efficiency, noble metals like gold and platinum are frequently utilized as electrode materials (Joy *et al.*, 2018).
(f) Pollutant concentration: The effectiveness of PEC might be impacted by the concentration of contaminants present in the effluent. The saturation of the active sites on the nanohybrids and higher pollutant concentrations can cause PEC's efficiency to decline (Fernandez-ibanez *et al.*, 2021).

(g) Supporting electrolyte: By affecting the solution's conductivity and the stability of the nanohybrids, the choice of supporting electrolyte can have an impact on PEC's effectiveness. NaCl and KCl are two often utilized supportive electrolytes (Joy *et al.*, 2018).

12.4 PEC–HER/OER FOR WASTE WATER TREATMENT WITH NANOHYBRIDS

12.4.1 Nanohybrid categorization

Nanohybrids used in PEC can be broadly categorized into different types of nanohybrids (Figure 12.4):

(a) Metal oxide–semiconductor nanohybrids, having a semiconductor metal oxide, such as zinc oxide (ZnO), or iron oxide (Fe_2O_3) combined with another semiconductor metal oxide, for example, titanium dioxide (TiO_2). The heterostructure helps in efficient charge transfer and separation (Kumar & Rao, 2016).
(b) Carbon-based nanohybrids, which consist of a semiconductor linked with a carbon-based substance, such as graphene or carbon nanotubes to facilitate the charge transport, increase semiconductor's electrical conductivity and stability (Dutta *et al.*, 2022).
(c) Metal–semiconductor nanohybrids, which include Au or Si. The metal aids in enhancing the semiconductor's capacity to separate charges and absorb light (Zhang *et al.*, 2018).
(d) Quantum dot-semiconductor nanohybrids, for example PbS or PbSe quantum dots interfaced with semiconductors like CdS or CdSe. The quantum dot aids in enhancing the semiconductor's capacity to absorb light (Mansur, 2010).
(e) Organic–inorganic nanohybrids, for example, a polymer, are combined with an inorganic substance, like a metal oxide or semiconductor, to form organic–inorganic nanohybrids. The stability and compatibility of the inorganic material are both enhanced by the organic material (Saveleva *et al.*, 2019).

12.4.2 Types of wastewater used

The nanohybrids listed above not only allow efficient degradation of wastewater pollutants, but also channel energy generation in a dual PEC system. These nanohybrids have been designed as photoanodes or photocathodes and are shown for the treatment of wastewater containing pharmaceuticals (diclofenac, sulfamethoxazole, norfloxacin, and tetracycline), dyes (reactive red, methyl orange, methyl blue, methylene blue, rhodamine B, and titan yellow), phenolics, and other organic contaminants in a PEC system for a variety of wastewater types, including municipal, industrial, and agriculture (Singh *et al.*, 2020).

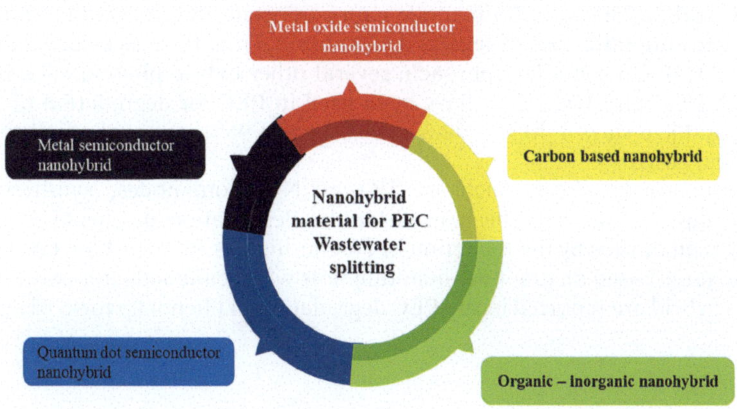

Figure 12.4 Nanohybrid material for PEC wastewater splitting.

TiO_2–Au nanohybrids have been employed in a PEC process for phenolic compound degradation of up to 90% (Qian *et al.*, 2020). Likewise, ZnO–Au and CdS–Pt nano hybrids are reported to degrade methylene blue and atrazine in textile effluent and agricultural wastewater, respectively. A Ni–TiO_2 photoanode is reported for the highest degree removal efficiency of acid red of 88% up to 97% under external bias 1.6 V (Ebrahim *et al.*, 2013). Bharath *et al.* (2021) reported a metal-semiconductor Au-WS_2 photoanode effectively put together to remove phenol and oxidize harmful As(III) into non-toxic As(V) under sunlight. The study reports Au-WS_2 photoanode with an oxidation efficiency of 99% for phenol and 95% for As(III) (Bharath *et al.*, 2021). To summarize, the metal/oxide hybrids take synergetic benefits of metal nanoparticles and semiconductor oxide properties, wherein the Au, Pt, and Ni act as an electron sink. Noble metal nanostructures such as Au, Ag, and Cu further help in enhanced light absorption due to their plasmonic properties. To further improve the degradation efficiency, researchers have also synthesized ternary hybrids. For example, Li *et al.* (2013) reported an n-ZnO/p-Cu_2O/n-Titanium nanotube array ternary heterojunction for tetracycline degradation to 85%. Two-dimensional materials have also been seen as a promising support matrix for the semiconductor oxides, nitrides, and sulfides to channel the charge separation in PEC. Reddy *et al.* (2023) hydrothermally synthesized a $ZnIn_2S_4$/g-C_3N_4 nanohybrid and demonstrated it for antibiotics degradation from wastewater up to 89.4% in 2 h. The system could also achieve a higher photocurrent density of 8.68 mA/cm^2 (Reddy *et al.*, 2023).

Another strategy of heterostructuring different oxide semiconductors has been successfully used in nanohybrids to achieve higher charge separation and thus degradation efficiency. Malathi *et al.* (2017) synthesized $BiFeWO_6$ doped bismuth oxyiodide (BiOI) nanocomposites using a facile one-step wet impregnation method. For an optimized 1% $BiFeWO_6$/BiOI nanohybrid, they could achieve RhB degradation of up to 92% in 90 min (Malathi *et al.*, 2017). Similarly, a hydrothermally synthesized $NiFe_2O_4$-$ZnWO_4$ nanohybrid exhibited a higher Cr(VI) reduction (93.5%) and removal of tetracycline (97.9%) and methylene blue (99.6%) over pristine $NiFe_2O_4$ and $ZnWO_4$ (Koutavarapu *et al.*, 2022). An organic–inorganic nanohybrid system based on poly (diphenylbutadiyne) (PDPB) and ZnO was shown to have 8.1 times enhanced photocurrent density compared to bare PDPB nano-fibers and also achieved a higher reduction of Cr ion (Ghosh *et al.*, 2018). Metal oxide semiconductor-based photo anodes, that is, F-doped $BiVO_4$@NiFe-LDH showed 86% degradation of tetracycline dye within 2 h along with achieving a high photocurrent density of 2.67 mA/cm^2 at 1.23 V versus RHE (Cong *et al.*, 2022). Furthermore, the change in the shape and morphology of oxides is found to enhance the PEC performance of nanohybrids. For instance, blue-TiO_2 nanotube (TNTs) arrays were found to achieve a photocurrent density of 5.7 mA/cm^2, which is four times higher than bare TNTs along with the removal of 4-chlorophenol (CP) 45% in 2 h and 38% TOC removal in 2 h (Koo *et al.*, 2017). In another work, a TiO_2–graphitic anode nanocomposite, synthesized by the sol–gel method, showed the complete mineralization of several drugs within 6 h of PEC, as evidenced by 90% reduced COD (Guo *et al.*, 2015). On a similar approach, several other hybrid photoanodes such as Au/TiO_2, Fe_2O_3/TiO_2, Cu_2O/TiO_2, and WO_3/TiO_2 have been used in PEC for degradation of MB, RhB, 2,4,6-trichlorophenol and have shown 40.3%, 92.3%, 99.9%, and 98%, removal efficiencies, respectively (Peleyeju & Arotiba, 2018).

Likewise, carbon-based hybrids such as TiO_2/gC_3N_4 photoanodes, synthesized by surface hybridization and dip coating, have been used for the degradation of phenol in wastewater. The removal efficiency is quantified by the reduction in TOC as high as 100% in 1.5 h (Wei *et al.*, 2017). The carbon heterostructures based on graphitic quantum dots with semiconductor oxide like g-C_3N_4QDs/TiO_2 photoanodes hybrid are reported in the PEC degradation of phenol up to 98.6% (Sun *et al.*, 2017).

12.5 CONCLUSIONS

Nanohybrids exhibit strong potential as viable contenders for enhancing the effectiveness of photoelectrochemical (PEC) wastewater treatment, owing to their heightened attributes and excellent capabilities in pollutant degradation. The synergistic interplay among diverse constituents within nanohybrids substantially bolsters the water treatment efficiency of PEC systems. However, there remain several challenges that necessitate resolution prior to their broad adoption. Chief among these concerns is the elevated cost associated with their synthesis and the intricate procedures involved, posing a barrier to their scalability. Additionally, there is a scarcity of comprehensive research addressing their long-term viability, stability, and durability when subjected to PEC conditions within a wastewater environment. Consequently, a more expansive array of studies is imperative to unravel their behavior and mechanism, enabling the formulation of enduring nanohybrids and the enhancement of their performance for widespread utilization in industrial applications.

ACKNOWLEDGEMENTS

The authors acknowledge the DST INSPIRE fellowship for doctoral study and the SERB for the funding support under WEA/2020/000022 grant for the work in this domain.

REFERENCES

Abe R. (2011). Recent progress on photocatalytic and photoelectrochemical water splitting under visible light irradiation. *Journal of Photochemistry & Photobiology, C: Photochemistry Reviews*, **11**(4), 179–209, https://doi.org/10.1016/j.jphotochemrev.2011.02.003

Ali M. M., Haque J., Kabir M. H., Kaiyum M. A. and Rahman M. S. (2021). Results in Materials Nano synthesis of ZnO – TiO_2 composites by sol-gel method and evaluation of their antibacterial, optical and photocatalytic activities. *11*.

Bharath G., Rambabu K., Hai A., Ponpandian N., Ejbye J., Dionysiou D., Abu M. and Banat F. (2021). Environmental dual-functional paired photoelectrocatalytic system for the photocathodic reduction of CO_2 to fuels and the anodic oxidation of furfural to value-added chemicals. *Applied Catalysis B: Environmental*, **298**, 120520, https://doi.org/10.1016/j.apcatb.2021.120520

Cong S., Yu J., Liu B., Teng W. and Tang Y. (2022). Composite photoanode for enhanced photoelectrocatalytic wastewater treatment and simultaneous hydrogen evolution. *New Journal of Chemistry*, **46**, 15227–15243, https://doi.org/10.1039/d2nj02210a

Dutta V., Verma R., Gopalkrishnan C., Yuan M., Batoo K. M., Jayavel R., Chauhan A., Lin K.-Y. A., Balasubramani R. and Ghotekar S. (2022). Bio-inspired synthesis of carbon-based nanomaterials and their potential environmental applications: a state-of-the-art review. *Inorganics*, **10**(10), 169, https://doi.org/10.3390/inorganics10100169

Ebrahim M., Pirkarami A., Soleimani M. and Bahmaei M. (2013). Photoelectrocatalytic degradation of acid dye using Ni-TiO_2 with the energy supplied by solar cell: mechanism and economical studies. *Journal of Environmental Management*, **121**, 210–219, https://doi.org/10.1016/j.jenvman.2013.01.041

Endrődi B., Kecsenovity E., Rajeshwar K. and Janáky C. (2018). One-step electrodeposition of nanocrystalline TiO_2 films with enhanced photoelectrochemical performance and charge storage. *ACS Applied Energy Materials*, **1**(2), 851–858, https://doi.org/10.1021/acsaem.7b00289

Fernandez-ibanez P., Mcmichael S., Cabanillas A. R., Alkharabsheh S., Moranchel A. T. and Byrne J. A. (2021). ScienceDirect new trends on photoelectrocatalysis (PEC): nanomaterials, wastewater treatment and hydrogen generation. *Current Opinion in Chemical Engineering*, **34**, 100725, https://doi.org/10.1016/j.coche.2021.100725

Garcia-Segura S. and Brillas E. (2017). Applied photoelectrocatalysis on the degradation of organic pollutants in wastewaters. *Journal of Photochemistry & Photobiology, C: Photochemistry Reviews*, **31**, 1–35, https://doi.org/10.1016/j.jphotochemrev.2017.01.005

Ghosh S., Remita H. and Basu R. N. (2018). Visible-light-induced reduction of Cr(VI) by PDPB-ZnO nanohybrids and its photo-electrochemical response. *Applied Catalysis B: Environmental*, **239**, 362–372, https://doi.org/10.1016/j.apcatb.2018.08.034

Guo X., Li D., Wan J. and Yu X. (2015). Electrochimica acta preparation and electrochemical property of TiO_2/nano-graphite composite anode for electro-catalytic degradation of ceftriaxone sodium. *Electrochimica Acta*, **180**, 957–964, https://doi.org/10.1016/j.electacta.2015.09.055

Jakubow-Piotrowska K., Witkowski B. and Augustynski J. (2022). Photoelectrocatalytic hydrogen generation coupled with reforming of glucose into valuable chemicals using a nanostructured WO_3 photoanode. *Communications Chemistry*, **5**, 125, https://doi.org/10.1038/s42004-022-00745-w

Joy J., Mathew J. and George S. C. (2018). Nanomaterials for photoelectrochemical water splitting. *International Journal of Hydrogen Energy*, **43**(10), 4804–4817, https://doi.org/10.1016/j.ijhydene.2018.01.099

Koo M. S., Cho K., Yoon J. and Choi W. (2017). Photoelectrochemical degradation of organic compounds coupled with molecular hydrogen generation using electrochromic TiO_2 nanotube arrays. *Environmental Science and Technology*, **51**(11), 6590–6598, https://doi.org/10.1021/acs.est.7b00774

Koutavarapu R., Reddy C. V., Syed K., Reddy K. R., Saleh T. A., Lee D. Y., Shim J. and Aminabhavi T. M. (2022). Novel Z-scheme binary zinc tungsten oxide/nickel ferrite nanohybrids for photocatalytic reduction of chromium (Cr (VI)), photoelectrochemical water splitting and degradation of toxic organic pollutants. *Journal of Hazardous Materials*, **423**, 127044, https://doi.org/10.1016/j.jhazmat.2021.127044

Kumar S. G. and Rao K. S. R. K. (2016). Comparison of modification strategies towards enhanced charge carrier separation and photocatalytic degradation activity of metal oxide semiconductors (TiO_2, WO_3 and ZnO). *Applied Surface Science*, **391**, 124–148, https://doi.org/10.1016/j.apsusc.2016.07.081

Kurian M. (2021). Advanced oxidation processes and nanomaterials – a review. *Cleaner Engineering and Technology*, **2**, 100090, https://doi.org/10.1016/j.clet.2021.100090

Li J., Lv S., Liu Y., Bai J., Zhou B. and Hu X. (2013). Photoeletrocatalytic activity of an n-ZnO/p-Cu_2O/n-TNA ternary heterojunction electrode for tetracycline degradation. *Journal of Hazardous Materials*, **262**, 482–488, https://doi.org/10.1016/j.jhazmat.2013.09.002

Malathi A., Arunachalam P., Grace A. N. and Madhavan J. (2017). A robust visible-light driven $BiFeWO_6$/BiOI nanohybrid with efficient photocatalytic and photoelectrochemical performance. *Applied Surface Science*, **412**, 85–95, https://doi.org/10.1016/j.apsusc.2017.03.199

Mansur H. S. (2010). Quantum dots and nanocomposites. *WIREs Nanomedicine and Nanobiotechnology*, **2**(2), 113–129, https://doi.org/10.1002/wnan.78

McMichael S., Fernández-Ibáñez P. and Byrne J. A. (2021). A review of photoelectrocatalytic reactors for water and wastewater treatment. *Water*, **13**(9), 1198, https://doi.org/10.3390/w13091198

Niu Q., Gu X., Li L., Zhang Y. and Zhao G. (2020). 3D CQDs-{001}TiO_2/Ti photoelectrode with dominant {001} facets: efficient visible-light-driven photoelectrocatalytic oxidation of organic pollutants and mechanism insight. *Applied Catalysis B: Environmental*, **261**, 118229, https://doi.org/10.1016/j.apcatb.2019.118229

Pahra S., Sharma S. and Devi P. (2022). Fundamental understanding and figure of merits for electrocatalytic and photoelectrocatalytic H_2 production. *Green Energy Harvesting: Materials for Hydrogen Generation and Carbon Dioxide Reduction*, 46–74, https://doi.org/10.1002/9781119776086.CH3

Peleyeju M. G. and Arotiba O. A. (2018). Recent trend in visible-light photoelectrocatalytic systems for degradation of organic contaminants in water/wastewater. *Environmental Science: Water Research and Technology*, **4**(10), 1389–1411, https://doi.org/10.1039/c8ew00276b

Qian H., Hou Q., Duan E., Niu J., Nie Y. and Bai C. (2020). Honeycombed Au@C-TiO_2-X catalysts for enhanced photocatalytic mineralization of acid red 3R under visible light. *Journal of Hazardous Materials*, **391**, 122246, https://doi.org/10.1016/j.jhazmat.2020.122246

Reddy C. V., Kakarla R. R., Cheolho B., Shim J. and Aminabhavi T. M. (2023). Heterostructured 2D/2D $ZnIn_2S_4$/g-C_3N_4 nanohybrids for photocatalytic degradation of antibiotic sulfamethoxazole and photoelectrochemical properties. *Environmental Research*, **225**, 115585, https://doi.org/10.1016/j.envres.2023.115585

Samiksha M. and Kerkar S. (2020). Application of eco-enzyme for domestic waste water treatment. *International Journal for Research in Engineering Application & Management (IJREAM)*, **5**, 114–116, https://doi.org/10.35291/2454-9150.2020.0075

Saveleva M. S., Eftekhari K., Abalymov A. and Douglas T. E. L. (2019). Hierarchy of hybrid materials – The place of inorganics *-in-* organics in it, their composition and applications. *Frontiers in Chemistry*, **7**, 1–21, https://doi.org/10.3389/fchem.2019.00179

Singh P., Singh S., Jaggi N., Kim K. H. and Devi P. (2020). Recent advances in bacteriorhodopsin-based energy harvesters and sensing devices. *Nano Energy*, **79**, 105482, https://doi.org/10.1016/j.nanoen.2020.105482

Singh P., Sharma S. and Devi P. (2023). 2D nanomaterial photoelectrodes for photoelectrochemical degradation of pollutants and hydrogen generation BT. In: Two-Dimensional Materials for Environmental Applications, N. Kumar, R. Gusain and S. Sinha Ray (eds.), Springer International Publishing, Germany, pp. 299–325, https://doi.org/10.1007/978-3-031-28756-5_10

Singla S., Singh P., Basu S. and Devi P. (2023). $BiVO_4$/$MoSe_2$ photocatalyst for the photocatalytic abatement of tetracycline and photoelectrocatalytic water splitting. *Materials Chemistry and Physics*, **295**, 127111, https://doi.org/10.1016/j.matchemphys.2022.127111

Sun B., Lu N., Su Y., Yu H., Meng X. and Gao Z. (2017). Decoration of TiO_2 nanotube arrays by graphitic-C_3N_4 quantum dots with improved photoelectrocatalytic performance. *Applied Surface Science*, **394**, 479–487, https://doi.org/10.1016/j.apsusc.2016.10.121

Tada H., Naya S. and Fujishima M. (2018). Water splitting by plasmonic photocatalysts with a gold nanoparticle/cadmium sulfide heteroepitaxial junction: a mini review. *Electrochemistry Communications*, **97**, 22–26, https://doi.org/10.1016/j.elecom.2018.10.005

Tayyebi A., Soltani T. and Lee B. (2018). Effect of pH on photocatalytic and photoelectrochemical (PEC) properties of monoclinic bismuth vanadate. *Journal of Colloid and Interface Science*, **534**, 37–46, https://doi.org/10.1016/j.jcis.2018.08.095

Vithalani P. C. and Bhatt N. S. (2022). Photoelectrochemical water treatment of sewage. In: Industrial Wastewater Treatment, Vol. **106**, T. Karchiyappan, R. R. Karri and M. H. Dehghani (eds), Water Science and Technology Library, Germany, pp. 399–420, https://doi.org/10.1007/978-3-030-98202-7_16

Wei Z., Liang F., Liu Y., Luo W., Wang J., Yao W. and Zhu Y. (2017). Photoelectrocatalytic degradation of phenol-containing wastewater by TiO_2/g-C_3N_4 hybrid heterostructure thin film. *Applied Catalysis B, Environmental*, **201**, 600–606, https://doi.org/10.1016/j.apcatb.2016.09.003

Zhang N., Han C., Fu X. and Xu Y. J. (2018). Function-oriented engineering of metal-based nanohybrids for photoredox catalysis: exerting plasmonic effect and beyond. *Chem*, **4**(8), 1832–1861, https://doi.org/10.1016/j.chempr.2018.05.005

Zhao H., Li C., Liu L., Palma B., Hu Z., Renneckar S., Larter S., Li Y., Kibria G., Hu J. and Su B. (2020). n-p Heterojunction of TiO_2-NiO core-shell structure for efficient hydrogen generation and lignin photoreforming. *Journal of Colloid and Interface Science*, **585**, 694–704, https://doi.org/10.1016/j.jcis.2020.10.049

Zia J. and Riaz U. (2021). Photocatalytic degradation of water pollutants using conducting polymer-based nanohybrids: a review on recent trends and future prospects. *Journal of Molecular Liquids*, **340**, 117162, https://doi.org/10.1016/j.molliq.2021.117162

Index

A

Actinomycetes, 105
Adsorbent, 3, 15, 88, 91–92, 96, 117–122, 172, 176, 180–181, 202, 210
Advanced oxidation process (AOP), 22, 209
Algae, 22, 33, 36–37, 43, 47, 100, 102–104, 119, 154, 198–199

B

Bacteria, 3, 5, 11, 13, 22, 28–31, 41, 43–44, 95, 100, 102–106, 114, 119, 129–130, 139, 154, 156, 160, 172–176, 178–181, 201–202
Battery, 62–73, 77–80
Bimetal, 27, 30, 105–106, 165, 171, 175, 201
Bioaugmentation, 135–139, 173, 182, 184
Biogenic metal, 99
Biological recognition, 148, 150, 153, 165
Biomaterials, 96, 130, 135, 140–141, 182
Biomolecules, 22, 26, 35, 104, 148–151, 153, 159, 194, 197
Bioremediation, 4–5, 105, 129, 169, 171–173, 177, 179–181, 195, 201
Biosensors, 148–150, 152–160, 162–163, 165, 195, 199

C

Cadmium, 11, 66–67, 77, 103, 115, 119, 122, 148, 151, 154, 194, 199, 201
Capacitance, 73–74, 76, 152
Capacitor, 61–62, 73–81
Carbon nanotubes (CNT), 14, 14–15, 74, 91, 117, 121, 121, 123, 137, 149–151, 171, 171, 179–180, 180–182, 184, 197, 199, 199, 215
Catalyst, 10, 13–17, 39, 41–42, 44, 88, 92, 94–96, 118, 121–123, 130, 165, 175–176, 199, 209–211, 213–214

D

Degradation, 3–5, 14–16, 22, 39–40, 42, 44, 65, 80, 88, 102, 104, 115, 120, 122–123, 132, 154, 172–173, 175–178, 180, 195, 198–199, 201, 210–212, 214–217
Dye, 5, 11, 15–16, 22, 41, 95, 114, 116, 118–120, 159, 176–177, 180–182, 197–199, 211–216

E

Emerging pollutants, 88, 102
Energy, 1, 3–6, 10–11, 13, 33, 38–39, 41, 46–47, 61–81, 91, 99–100, 104, 114, 120, 123, 129–130, 132–133, 139–140, 148–150, 153, 159, 163, 165, 179–181, 201–202, 210, 215
Environmental monitoring, 148, 150, 152
Environmental pollutants, 88, 149, 154, 163, 170, 174–175, 181
Enzyme, 3, 5, 10, 13–16, 23, 28–29, 31, 33, 44, 46, 94–96, 101, 103–104, 106, 133–137, 139, 150–153, 157, 159, 165, 171–173, 177–179, 184, 196–200, 203

F

Fungi, 11, 13, 22, 28, 32, 34–35, 90, 102–105, 119, 172–173

G

Gold, 13, 23, 25, 28–29, 31, 35–37, 40, 44, 103, 105, 150–153, 159–160, 194–195, 197, 200, 214
Graphene, 10, 14, 65, 74, 90–91, 93, 117, 119, 121–123, 131, 137, 139, 149–152, 171, 178–179, 181–182, 184, 198, 215
Graphene oxide, 10, 90, 93, 117, 119, 152, 178–179, 181–182, 198

I

Iron, 3–5, 13, 21–22, 25–27, 33–34, 36, 40–41, 43, 46, 69, 71, 81, 88, 90, 93–94, 96, 101–107, 114–123, 129–130, 132, 135, 140–141, 145, 147–155, 160, 163, 165, 169–171, 173–182, 184, 193–199, 201–203, 210, 215, 217

L

Lead, 11, 15, 31, 38, 44, 46, 62–63, 65–70, 74, 77, 103–105, 114–116, 120, 122, 133, 135, 140, 148, 151, 177, 179, 195, 199, 211
Lithium, 64–66, 69–70, 77–81

M

Magnetic nanoparticles, 3, 28, 85, 117–119, 150, 160, 200
Manganese, 69, 104, 117, 151
Microbes, 4, 44, 47, 101–102, 104–107, 129, 131–132, 180

N

Nanobiohybrids, 3–4, 10, 15, 17, 44–46, 87–89, 96, 147–150, 152, 154, 163–166, 169–171, 173, 182–184, 193–194
  Carbon based nanobiohybrids, 1–94
  Inorganic nanobiohybrids, 105, 129–141
Nanofiltration, 4, 11, 174–175, 198
Nanomaterials, 3, 10, 13, 15–16, 22, 33, 38, 44, 47, 88, 91–92, 96, 99–101, 103–104, 106–107, 114, 117–118, 121–123, 129–130, 148–152, 155, 157–160, 162–163, 165, 171–173, 175–176, 179–181, 184, 194–195, 197–198, 202–203, 210, 213
Nanoparticle, 3, 10, 13, 21–26, 28–34, 36, 38, 40–42, 44, 46–47, 85, 87, 90, 96, 99, 117–120, 122, 149–152, 157–158, 160, 162, 171, 173, 176, 194, 196–202, 210, 213–214, 216
  Nanoparticle synthesis, 29, 32, 202

P

Palladium, 36, 103, 105
Parameter optimization, 214
Photobioreactors, 44
Photocatalysis, 4–6, 14–15, 21–22, 33, 38–39, 123, 131–132, 134, 209
Photoelectrocatalysis, 7, 209–210
Plants, 3, 11, 22–24, 28, 35, 47, 62, 90, 104, 116, 133, 148, 150, 154, 184
Pollutants, 4–6, 10–11, 13–15, 21–22, 38–40, 42, 87–88, 91–92, 96, 102, 104, 106, 114–117, 119–121, 123, 129, 140, 147–152, 154, 163, 165–166, 169–170, 172–175, 180–182, 194, 197–199, 201–202, 209–212, 215
Purification, 1, 3–4, 9–17, 88, 96, 114–115, 197–198, 201

R

Radionuclides, 114, 116, 119
Renewable energy, 61–62, 73, 77, 80–81
Resource recovery, 193–194, 196, 202
Reversed osmosis, 11

S

Semiconductor, 5, 25–27, 30–31, 33, 38, 132–134, 136, 139, 149, 151, 165, 182, 201, 210, 212, 215–216
Sensors, 148–160, 162–163, 165, 195, 199
Silver, 13, 25–31, 34–38, 41, 90, 93, 103, 105, 150–151, 154, 175, 194–195, 197–199
Sol-gel, 40, 42–88, 90, 93, 100, 117, 171, 177, 182, 200, 213, 216
Supercapacitor, 61–62, 73–81
Sustainability, 3, 10, 116
Synthesis, 1, 3, 5, 15, 22–26, 28–30, 32–34, 36, 38, 42, 44, 47, 88–90, 93, 100–107, 121, 123, 130, 136, 139, 149, 176–179, 181, 194–197, 201–202, 213–214, 217

T

Toxicity, 23, 46–47, 88, 100, 106, 123, 134, 148, 151, 154, 177, 179, 195–197, 203

V

Virus, 11, 13, 148, 160, 163, 172

W

Wastewater, 3–5, 11, 13–14, 21, 43, 87–88, 91–92, 95–96, 102, 104–107, 113–119, 121–122, 129–130, 133, 139–140, 148, 152, 163, 178, 180–181, 191, 194, 197–200, 202, 209–217
Wastewater treatment, 3–5, 43, 87–88, 91–92, 96, 102, 104–107, 113–114, 116–118, 129–130, 133, 139–140, 191, 194, 197–198, 209–211, 213, 216
Water pollution, 11, 15, 87, 96, 114–116, 209
Water treatment, 3–5, 10–11, 43, 47, 87–88, 91–92, 96, 102, 104–107, 113–114, 116–119, 121–123, 129–130, 139–140, 148, 184, 191, 194, 197–198, 209–211, 213, 215–217

Y

Yeast, 28, 32–34, 102, 104, 135, 183–184

IWA Publishing's authorised EU representative for General Product Safety Regulations is Diane D'Arras, 15 rue Duret, 75116 Paris, France, e-mail: safety@iwap.co.uk.

Printed and bound by CPI Group (UK) Ltd, Croydon, CR0 4YY
12/05/2026
02108869-0009